WELDING
skills and practices

Joseph W. Giachino, William Weeks
Western Michigan University
Kalamazoo, Michigan

American

CHICAGO, ILL. 60637

050279

preface

This edition of WELDING SKILLS AND PRACTICES provides current and authoritative methods and techniques written for students; apprentices, technical students, or the teachers, instructors in metals trades, manufacturing process supervisors, or practitioners of an allied technology or trade. This comprehensive textbook takes a simple, basic manipulative approach that fits the secondary, post secondary, or community college field.

The 5th Edition has been updated to include new processes and new materials, and to better fit existing curriculum demands. It updates all of the 4th Edition, with some rearrangement, plus four new Chapters:

1) Welding Safety
2) Welding Plastics
3) Metric Dimensioning
4) Welding Certification

The Joint Design Chapter is improved. Oxy-Mapp, Oxy-Hydrogen and Air-Oxygen have been added to the Chapter on Oxy-Acetylene Processes. Safety has been reinforced throughout the book. The two Chapters on Pipe Welding have been combined and simplified.

While Metal-Arc Welding remains the introductory welding process, the Oxy-Acetylene sequence of six Chapters has been moved up to the middle of the book. Those teachers who prefer to start with Oxy-Acetylene will find it convenient to do the four introductory Chapters, then Chapters 21 through 28. Others will start with Shielded Metal-Arc Welding Chapters 5 to 16 after the introduction.

The entire book of WELDING SKILLS AND PRACTICES has been up-dated, put into a new and revised format with easy-to-read type face and many new illustrations. The resulting book is compact and easy to handle.

The Publishers

contents

SPECIAL WELDING PROCESSES

SUPPLEMENTARY WELDING DATA

introduction to welding

CHAPTER 1 an essential skill

Welding is essential to the expansion and productivity of our industries. Welding has become one of the principal means of fabricating and repairing metal products. It is almost impossible to name an industry, large or small, that does not employ some type of welding. Industry has found that welding is an efficient, dependable, and economical means of joining metal in practically all metal fabricating operations and in most construction, Fig. 1-1.

Fig. 1-1. Fabrication is simplified by automatic welding techniques. (Douglas Aircraft Corp.)

1

Fig. 1-2. Many parts of airplanes are joined by some welding process. (Douglas Aircraft Corp.)

WHERE WELDING IS USED

In tooling-up for a new model automobile, a manufacturer may spend upward of a million dollars on welding equipment. Many buildings, bridges, and ships are fabricated by welding. Where construction noise must be kept at a minimum, such as in the building of hospital additions, the value of welding as the chief means of joining steel sections is particularly significant.

Without welding, the aircraft industries would never be able to meet the enormous demands for planes, rockets, and missiles, Fig. 1-2. Rapid progress in the exploration of outer space, Fig. 1-3, has been made possible by new methods and knowledge of welding metallurgy.

Probably the most sizeable contribution welding has made to society is the manufacture of special products for household use. Welding processes are employed in the construction of such items as television sets, refrigerators, kitchen cabinets, dishwashers, and other similar products.

As a means of fabrication, welding has proved fast, dependable, and flexible. It lowers production costs by simplifying design and eliminates costly patterns and machining operations.

Welding is used extensively for the manufacture and repair of farm equipment, mining and oil machinery, machine tools, jigs and fixtures, and in the construction of boilers, furnaces, and railway cars. With improved techniques for adding new metal to worn parts, welding has also resulted in economy for highly competitive industries, Figs. 1-4 through 1-10.

Fig. 1-3. Welding made possible the fabrication of this section of a missile.

Fig. 1-4. The metallic arc is used in fabricating many industrial products. (Fibre-Metals Products Co.)

Fig. 1-5. Welding is often indispensible in assembling the steel structure of a building.

Fig. 1-6. Welding plays an important role in the construction and repair of heavy machinery, trucks and road equipment. (Hobart Brothers Co.)

Fig. 1-7. Seam welding is being used here to weld aircraft sections. (Pratt & Whitney Aircraft)

Fig. 1-8. Manually operated inert-gas-shielded arc welding equipment is being used to weld this steel panel. (Miller Electric Manufacturing Co.)

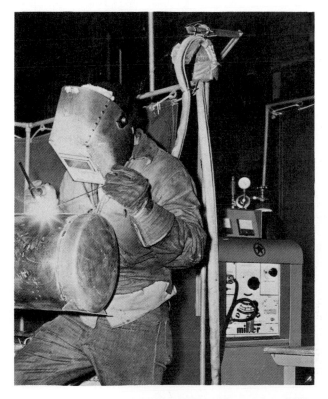

Fig. 1-9. This tank is being assembled by Tig welding. (Miller Electric Manufacturing Co.)

Fig. 1-10. A stainless steel restaurant kettle is completed by means of an inert-gas-shielded arc spot welder. (Linde Co.)

Types of Welding Processes

Of the many methods of welding in use today, gas, arc, and resistance dominate the field. We can best explain these processes from the standpoint of the operator's duties.

The principal duty of the operator employing gas welding equipment is to control and direct the heat on the edges of metal to be joined, while applying a suitable metal filler to the molten pool. The intense heat is obtained from the combustion of gas, usually acetylene and oxygen. For this reason this process is also called *oxyacetylene welding.*

The skills required for this job are adjustment of the regulators, selection of proper tips and filler rod, preparation of the metal edges to be joined, and the technique of flame and rod manipulation. The gas welder may also be called upon to do flame cutting with a cutting attachment and extra oxygen pressure. Flame or oxygen cutting is employed to cut various metals to a desired size or shape, or to remove excess metal from castings.

The arc welders perform their skill by first striking an arc at the starting point of a weld and maintaining this electric arc to fuse the metal joints. The molten metal from the tip of the electrode is then deposited in the joint, together with the molten metal of the edges, and solidifies to form a sound and uniform connection. The arc welding operator is expected to select the proper electrodes for the job or be able to follow

Fig. 1-11. One of the outstanding products of the welder's art is the mass-produced American automobile. (Chevrolet Div., General Motors Corp.)

instructions as stated in the job specifications, to read welding symbols and to weld any type seam using the technique required i.e., overhead, horizontal, etc.

In the field of arc welding, the gas-shielded arc processes have gained recognition as being superior to the standard metallic arc. With gas-shielded arc both the arc and molten puddle are covered by a shield of gas. The shield of gas prevents atmospheric contamination, thereby producing a sounder weld. The processes known as *Tig* and *Mig* welding are either manually or automatically operated.

Resistance welding operators are responsible for the control of machines which fuse metals together by heat and pressure, Fig. 1-11. If two pieces of metal are placed between electrodes which become conductors for a low voltage and high amperage current, the materials, because of their own resistance, will become heated to a plastic state. To complete the weld, the current is interrupted before pressure is released, thereby allowing the weld metal to cool for solid strength.

The operator's duty is to properly adjust the machine current, pressure, and feed settings suitable for the material to be welded. The welder usually will be responsible for the alignment of parts to be assembled and for controlling the passage of parts through the welding machine.

Selection of the Proper Welding Process

There are no hard and fast rules which govern the type of welding that is to be used for a particular job. In general, the controlling factors are kind of metals to be joined, costs involved, nature of products to be fabricated, and production techniques. Some jobs are more easily accomplished by the oxyacetylene welding process whereas others are more easily done by means of arc welding.

Gas welding is used in all metal working industries and in the field as well as for plant maintenance. Because of its flexibility and mobility, it is widely used in maintenance and repair work. The welding unit can be moved on a two-wheeled cart or transported by truck to any field job where breakdowns occur. Its adaptability makes the oxyacetylene process suitable for welding, brazing, cutting, and heat treating.

The chief advantage of arc welding is the rapidity with which a high quality weld can be made at a relatively low cost. Specific applications of this process are found in the manufacture of structural steel for buildings, bridges, and machinery. Arc welding is considered ideal for making storage and pressure tanks as well as for production line products using standard commercial metals.

Since the development of gas-shielded arc processes, there are indications that they will be used extensively in the future in welding all types of ferrous and nonferrous metals in both gage and plate thicknesses.

Resistance welding is primarily a production welding process. It is especially designed for the mass production of domestic goods, automobile bodies, electrical equipment, hardware, etc. Probably the outstanding characteristic of this type of welding is its adaptability to rapid fusion of seams.

Occupational Opportunities in Welding

The wide spread use of welding in American industry provides a constant source of employment for both skilled and semi-skilled operators. According to U.S. Department of Labor there are approximately 555,000 persons employed as welders. Three-fifths of these work in industries that manufacture durable goods, such as transportation equipment, machinery, and household products. Most of the others work for construction firms and repair shops.

Employment outlook. Employment of welders is expected to increase because of the development of newer and better welding processes. This is particularly true in ship building, tank and boiler fabrication, rail, automotive, and aircraft manufacture, building construction, piping, and many other metalworking industries. Although there is no uniform wage rate for welders, they are considered to be in one of the higher classifications of job wages.

Training. Learning the essential skills needed to fulfill the many welding job requirements varies from a few months of on-the-job training to several years of formal training. Most employers prefer applicants who have a high school education or vocational school training in welding. Courses in mathematics, mechanical drawing, blue-print reading and general metals are very helpful. See Fig. 1-12.

Young people planning careers as welders need manual dexterity, good eyesight and good eye-hand coordination. They should be able to concentrate on detailed work for long periods and must be free of any physical disabilities that would prevent them from bending, stooping or working in awkward positions.

Job classification. A beginner usually starts on simple production jobs and gradually works up to higher levels of skill as his experience and ability improves. Before being assigned to work

where the quality and strength of the weld are critical, a welder will generally have to pass a certification test given by the employer, government agency or some other inspection authority.

Welders are usually classified as skilled and semi-skilled. Skilled welders are those who have the ability to plan, lay out work from drawings or written specifications, and weld all types of joints in various positions, such as flat, vertical, horizontal and overhead. They also have a wide range of technical knowledge involving properties of metals, effects of heat on welded structures, control of expansion and contraction forces, reading welding symbols, and recognizing welding defects. The skilled welder may be proficient in several welding areas encompassing gas and arc welding processes. As a rule the skilled welder is always certified for the particular welding job he is required to perform.

The semi-skilled welder usually does repetitive work, that is, production work which generally does not involve critical safety and strength requirements. They primarily weld surfaces in only one position and may or may not have to be certified.

The following are some of the principal job titles of welders:

Welding engineer
Arc welder—shielded metal arc
      —gas shielded arc
Welding cutter
Submerged-arc welder
Welder—resistance, spot, automatic
Pipe welder
Boilermaker welder
Structural welder
Maintenance welder
Welding layout and set-up man
Welding inspector
Welding tester
Welding foreman-supervisor

Skilled welders may, by promotion, become inspectors, foremen, or supervisors. Actually there are unlimited opportunities for those who become thoroughly acquainted with the techniques, materials, designs, and new applications of welding processes.

Fig. 1-12. To attain the required welding skills, a person usually has to complete a formal course of instruction under a competent instructor.

CHAPTER 2 welding safety

Have you ever heard the saying "some people are accident-prone"? The implication is that accidents just seem to follow some individuals no matter what they do. They just seem plagued with bad luck. Actually, there is no such thing as being accident-prone. People have accidents simply because they are careless, or indifferent to safety regulations.

Each year thousands of people suffer the pain of injury because they have failed to use good judgment, Fig. 2-1. In many ways, safety can be considered a habit, a kind of behavior. A habit is acquired; you are not born with it. It is the result of repetition—doing something over and over again until it becomes part of you. Thus if you consistently follow good safety practices you subconsciously build within yourself a safety awareness that usually keeps you from making foolish mistakes. Taking a chance is another way of saying "I don't know any better." Thus safety simply means using a little common sense and in so doing avoid many serious accidents.

Finally safety is not something you read about or practice only on occasion. It has to be observed constantly. Industry places a high premium on safety—ask anyone in industry and they will tell you that a tremendous amount of time and effort is given to safety. So never take chances; you will enjoy your work more if you learn to become a safe worker.

Fig. 2-1. Being injured because of some foolish accident is no fun.

The primary purpose of this chapter is simply to alert you to some of the general precautions which should be followed while performing various welding operations. More definite safety practices are included throughout the text where they deal with specific welding situations.

9

Accident Reporting

Always report an accident regardless of how slight it may be. Even a little scratch might lead to an infection, or a minute particle could result in a serious eye injury. Prompt attention to any injury usually will minimize what may become serious if neglected.

Generally, where any physical work is performed, either in a learning situation or on an actual payroll job in industry, a definite accident reporting procedure is established. Since this reporting is for the best interest of the individual it is foolhardy to ignore it or go around it. Consequently, make it a practice to become fully informed about what should be done and then take immediate action if an accident occurs.

Work Behavior

On some occasion you may be tempted to engage in what might appear to be a harmless prank. Any form of horseplay in a shop is dangerous and can lead to an accident. There are many recorded incidents where foolish play ended in a serious injury. Most work areas are reasonably safe if proper work precautions are taken, but no one is safe if good work attitudes are ignored.

Welding Equipment Familiarization

No welding equipment of any kind should ever be used until exact instructions on how to operate it have been received. Manufacturer's recommended methods are very important and should be followed at all times. Attempting to operate a piece of equipment without instruction not only may damage the equipment but could result in a serious injury. Welding equipment of all kinds is safe to operate, providing it is used in the proper manner.

CAUTION: Equally important in operating welding equipment, is never try to remedy a malfunction without first consulting supervising personnel. This applies to a whole range of items from a leaking gas hose to a loose cable on a welding generator. The instructor or foreman is knowledgeable of what should be done and it is his responsibility to decide on the course of action to be taken.

Ventilation

All welding should be done in well ventilated areas. There must be sufficient movement of air to prevent accumulation of toxic fumes or possible oxygen deficiency. Adequate ventilation becomes extremely critical in confined spaces where dangerous fumes, smoke, and dust are likely to collect, Fig. 2-2.

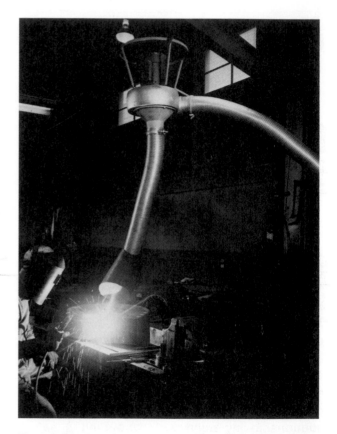

Fig. 2-2. Good ventillation is essential to the welder's continued health and functioning.

Where considerable welding is to be done, an exhaust system is necessary to keep toxic gases below the prescribed health limits. An adequate exhaust system is especially necessary when welding or cutting zinc, brass, bronze, lead, cadmium, or beryllium bearing metals. Fumes from these materials are toxic, and they are very hazardous to your health.

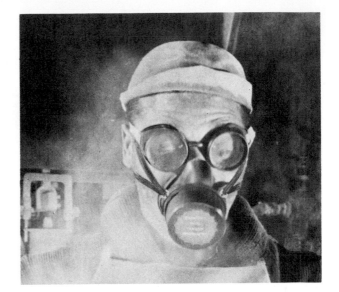

Fig. 2-3. A respirator should be worn when welding metals that produce highly toxic fumes. (American Optical Corp.)

CAUTION: Even with proper ventilation a respirator should be used. See Fig. 2-3.

Body Protection

In any welding or cutting operation sparks, dangerous ultra-violet and infrared rays are generated. Consequently, suitable clothing and proper eye protection are necessary. See Fig. 2-4. Sparks may lead to serious burns and rays are extremely dangerous to the eyes. More specific instructions concerning correct apparel and eye shields will be found in other sections dealing with various phases of welding and cutting. At the moment it is sufficient to point out that a welder must be aware of possible body dangers during any welding or cutting operation and learn the safe practices for his or her personal welfare.

Welding and Cutting Containers

CAUTION: No welding or cutting should ever be done on used drums, barrels, tanks, or other containers unless they have been thoroughly cleaned of all combustible substances that may produce flammable vapors or gases.

Flammable and explosive materials include gasoline, light oil, acids that react with metal to produce hydrogen, and non-volatile oils or solids that release vapors when exposed to heat.

Furthermore, sufficient precaution must be taken to insure that containers to be welded or cut are sufficiently vented. Any accumulation of air or gas in a confined area will expand when heated and the internal pressure may build up to cause an explosion.

Cleaning a container can be done by flushing several times with water, chemical solutions or steam. Water cleaning is satisfactory if the previous substance in the container is readily soluble in water, such as acetone or alcohol. For all less soluble substances, cleaning should be done by using a strong commercial caustic cleaning compound or by blowing steam into the container.

CAUTION: Never use oxygen to ventilate a container as it may start a fire or cause an explosion!

As a final precaution after cleaning, a container should be vented and filled with water before welding or cutting is undertaken. The container should be arranged so water can be kept filled to within a few inches of the point where the welding or cutting is to take place. See Fig. 2-5.

CAUTION: Be sure there is a vent or opening to provide for release of air pressure or steam.

Here are several more things which should be followed when working on containers:

Fig. 2-4. A welder should take full precaution to protect himself from rays that are generated during welding. (Fibre-Metal Products Co.)

SAFETY IN CUTTING

Fires often occur in a cutting operation simply because proper precautions were not taken. Too often a worker forgets that sparks and falling slag can travel as much as 35 feet and can pass through cracks out of sight of the goggled operator. Persons responsible for supervising or performing cutting of any kind should observe the following:

CAUTIONS:

1. Never use a cutting torch where sparks will be a hazard, such as near rooms containing flammable materials, especially dipping or spraying rooms.

2. If cutting is to be over a wooden floor, sweep the floor clean and wet it down before starting the cutting. Provide a bucket or pan containing water or sand to catch the dripping slag.

3. Keep a fire extinguisher nearby whenever any cutting is done, Fig. 2-6.

4. Whenever possible perform the cutting operation in wide open areas so sparks and slag will not become lodged in crevices or cracks.

5. If cutting is to be done near flammable materials, and the flammable materials cannot be moved, suitable fire-resisting guards, partitions, or screens must be used.

6. In plants where a greasy, dirty or gassy atmosphere exists, extra precaution should be taken to avoid explosions resulting from electric sparks or open fire during a cutting or welding operation.

7. Keep flame and sparks away from oxygen cylinders and hose.

8. Never do any cutting near ventilators.

9. Move combustible materials at least 40 feet (12.19 meters) away from any cutting or welding operation.

10. Where the risk of fire is great, always have stand-by watchers with fire extinguishers.

11. Never use oxygen to dust off clothing or work.

12. Never use oxygen as a substitute for compressed air.

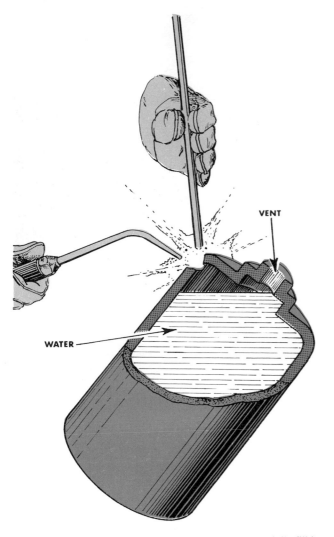

VENT

WATER

Fig. 2-5. When welding or cutting a container, partially fill it with water.

1. Never weld or cut drums, barrels, or tanks until you know there is no danger of fire or explosion.

2. Don't rely on your nose or eyes to determine if it is safe to weld or cut a closed container. First find out what was in the container. Remember, a very small amount of residual flammable liquid or gas may cause a serious explosion.

3. Don't clean a container where there is poor ventilation. Good ventilation is necessary to carry away harmful and explosive vapors.

4. Always use a spark resistive tool when heavy sludge or scale must be removed by scraping or hammering.

Fig. 2-6. Always keep a fire extinguisher handy. (Walter Kidde Co., Inc.)

SAFETY IN GAS WELDING

Specific instructions dealing with safety in gas welding are listed in the units involving oxyacetylene welding. These precautions cover proper handling of cylinders, operation of regulators, use of oxygen and acetylene, welding hose, testing for leaks and lighting a torch. All of these safety regulations are extremely important and should be followed with the utmost care and regularity.

In addition to the normal precautions to be observed in gas welding, a very significant safety procedure involves the piping of gas. All piping and fittings used to convey gases from a central supply system to work stations must withstand a minimum pressure of 150 psi. Oxygen piping can be of black steel, wrought iron, brass, or copper. Only oil-free compounds should be used on oxygen threaded connections. Piping for acetylene must be of wrought iron.

CAUTION: Under no circumstances should acetylene gas come in contact with unalloyed copper except in a torch. Any contact of acetylene with high alloyed copper piping will generate copper acetylide which is very reactive and may result in a violent explosion. After assembly all piping must be blown out with air or nitrogen to remove foreign materials.

There are five basic rules which contribute to the safe handling of oxy-acetylene equipment. These are:

1. Keep oxy-acetylene equipment clean, free of oil, and in good condition.
2. Avoid oxygen and acetylene leaks.
3. Open cylinder valves slowly.
4. Purge oxygen and acetylene lines before lighting torch.
5. Keep heat, flame and sparks away from combustibles.

SAFETY IN ARC WELDING

Arc welding includes shielded metal-arc, gas-shielded arc and resistance welding. Only general safety measures can be indicated for these areas because arc welding equipment varies considerably in size and type. Equipment may range from a small portable shielded metal-arc welder to highly mechanized production spot or gas-shielded arc welders. In each instance specific manufacturers' recommendations should be followed.

Safety practices which are generally common to all types of arc welding operations are as follows.

CAUTIONS:

1. Install welding equipment according to provisions of the National Electric Code.

2. Be sure a welding machine is equipped with a power disconnect switch which is conveniently located at or near the machine so the power can be shut off quickly, Fig. 2-7.

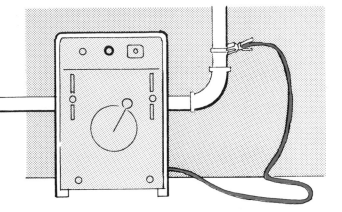

Fig. 2-8. Do not ground to pipelines carrying gases or flammable liquids.

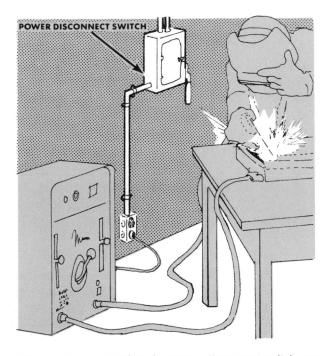

Fig. 2-7. Be sure that there is a power disconnect switch near at hand.

3. Don't make repairs to welding equipment unless the power to the machine is shut OFF. The high voltage used for arc welding machines can inflict severe and fatal injuries.

4. Don't use welding machines without proper grounding. Stray current may develop which can cause severe shock when ungrounded parts are touched. See Fig. 2-8. Do not ground to pipelines carrying gases or flammable liquids.

5. Don't use electrode holders with loose cable connections. Keep connections tight at all times. Avoid using electrode holders with defective jaws or poor insulation.

6. Don't change the polarity switch when the machine is under a load. Wait until the machine idles and the circuit is open. Otherwise, the contact surface of the switch may be burned and the person throwing the switch may receive a severe burn from the arcing.

7. Don't operate the range switch under load. The range switch which provides the current setting should be operated only while the machine is idling and the current is open. Switching the current while the machine is under a load will cause an arc to form between the contact surfaces.

8. Don't overload welding cables or operate a machine with poor connections. Operating with currents beyond the rated cable capacity causes overheating. Poor connections may cause the cable to arc when it touches metal grounded in the welding circuit. See Fig. 2-9.

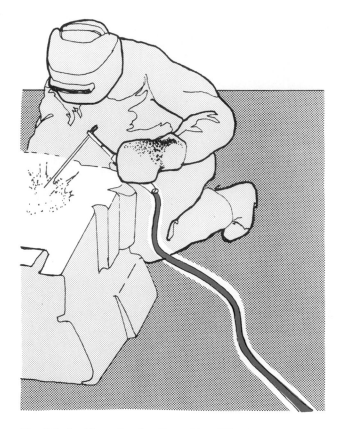

Fig. 2-9. Avoid overloading the cables. (Airco)

Fig. 2-10. Don't pick up hot objects. (Airco)

Fig. 2-11. Use safety glasses when chipping or grinding. (Airco)

9. *Don't weld in damp areas and keep hands and clothing dry at all times. Dampness on the body may cause an electric shock. Never stand or lie in puddles of water, on damp ground, or against grounded metal when welding without suitable insulation. Use a dry board or rubber mat to stand on.*

10. *Don't strike an arc if someone without proper eye protection is nearby. Arc rays are harmful to the eyes and skin. If other persons must work nearby, the welding area should be partitioned off with a fire-retardant canvas curtain to protect them from the arc welding flash.*

11. *Never pick up pieces of metal which have just been welded or heated. Fig. 2-10.*

12. *Always wear protective eye goggles when chipping or grinding. A small particle of slag or metal may cause a severe eye injury. Fig. 2-11.*

13. *Don't weld on hollow (cored) castings unless they have been properly vented, otherwise an explosion may occur. Fig. 2-12.*

14. *Be sure press-type welding machines are effectively guarded.*

15. *Be sure suitable spark shields are used around equipment in flash welding.*

Fig. 2-12. Be sure a hollow casting is vented before welding on it. (Airco)

16. When welding is completed, turn OFF the machine, pull the power disconnect switch and hang the electrode holder in its designated place.

Final Precaution

Remember, accidents do not just happen. Invariably they occur because of indifference to regulations, lack of information, or just plain carelessness.

Injury of any kind is painful and very often can incapacitate a person, or even produce a permanent deformity. If more thought were given to the consequences of injuries there would be less tendency to ignore safety precautions and thus, fewer accidents.

QUESTIONS FOR STUDY AND DISCUSSION

1. Why is there very little basis for saying some people are accident prone?

2. What are some of the main causes of accidents?

3. Why should all accidents be reported immediately?

4. How is it possible to become involved in an accident when playing around in the shop?

5. What may happen if you attempt to use welding equipment without proper instruction?

6. What should be done if some malfunction occurs in any welding equipment?

7. What general practice should be followed regarding ventilation during the performance of any welding operation?

8. Why should strict attention be given to proper clothing and eye protection?

9. Why should used containers be thoroughly cleaned before any welding or cutting is done on them?

10. Why do fires often occur during a cutting operation?

11. What are some of the precautions that should be taken when using a cutting torch?

12. Why is it dangerous for acetylene to come in contact with high alloyed copper piping?

13. What is the significance of having each welding machine equipped with a power disconnect switch?

14. Why should a welding machine never be overloaded?

15. Why should the polarity or range switch never be operated when the machine is under a load?

CHAPTER 3 welding metallurgy

In preparing to become a skillful welder you should become familiar with the effects of heat on the structure of metal and with what happens to metal when certain alloying elements are added to it.

You will also need to know what safeguards must be followed in welding metals because application of heat during the welding process may destroy the very elements which were originally added to improve the structure of the metal. For example, metals expand and contract, thereby setting up great stresses which often result in severe distortions. Improper welding of stainless steel may result in a complete loss of its corrosion-resistant qualities, and welding high-carbon steel in the same manner as low-carbon steel may produce such a brittle weld as to make the welded piece unusable.

This chapter deals with the metallurgy of welding; that is, the formation of impurities and the effects of heat on the chemical, physical, and mechanical properties of metals.

PROPERTIES OF MATERIALS

Chemical, physical, and mechanical properties have a very significant influence in any welding operation. This will become more apparent in later chapters dealing with specific welding techniques. These properties can be defined as follows:

Chemical properties. Chemical properties are those which involve corrosion, oxidation, and reduction. *Corrosion* is a wasting away of metal due to various atmospheric elements. *Oxidation* is the formation of metal oxides which occur when oxygen combines with a metal. *Reduction* refers to the removal of oxygen from the surrounding molten puddle to reduce the effects of atmospheric contamination.

In any welding situation, it is important to remember that oxygen is a highly reactive element. When it comes in contact with metal, especially at high temperatures, undesirable oxides and gases are formed, thereby complicating the welding process. Hence, the success of any welding operation depends on how well oxygen can be prevented from contaminating the molten metal.

Physical properties. Physical properties are those which affect metals when they are subject to heat generated by welding such as *melting point, thermal conductivity,* and *grain structure.* Solid metals change into a liquid state at different temperatures. When cooling from a liquid state the atoms will form various crystal patterns (lattices). The strength of a weld often depends on how these lattices are controlled and how much heat is necessary to produce proper fusion

19

of metal. Equally important is being aware that some metals have a high rate of heat conductivity while others have slower thermal conductivity. Also a welder needs to understand how heat will affect the grain structure of metals since the grain size of the crystalline structure has a direct bearing on the strength of a welded joint.

Mechanical properties. Mechanical properties are those which determine the behavior of metals under applied loads. These include a wide range of properties such as *tensile strength; ductility, toughness, brittleness* and others, all of which are extremely important in their relationship to welding.

STRUCTURE OF METALS

When you examine a polished piece of metal under a microscope, you will see small grains. Each of these grains is made up of smaller particles, called atoms, of which all matter is composed.

The grains, or crystals as they are often called, vary in shape and size. The arrangement of the atoms determines the shape of the crystalline structure. In general, the crystals of the more common types of metals arrange themselves in three different patterns. These are known as *space-lattices.*

A space-lattice is a visual representation of the orderly geometric pattern into which the atoms of all metals arrange themselves upon cooling from a liquid to a solid state.

The first type of space-lattice, illustrated in Fig. 3-1, is the *body-centered cube.* Here you will find nine atoms—one at each corner of the cube and one in the center. This crystal pattern is found in such metals as iron, molybdenum, chromium, columbium, tungsten, and vanadium.

The second crystal pattern is the *face-centered cube.* Notice in Fig. 3-2 how the atoms are arranged. Metals having this space-lattice pattern are aluminum, nickel, copper, lead, platinum, gold, and silver.

The third space-lattice is called the *close packed hexagonal form.* See Fig. 3-3. Among the

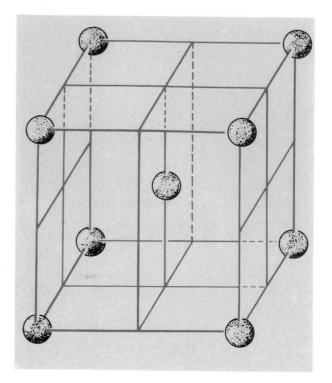

Fig. 3-1. Here is the arrangement of atoms in a body-centered cubic crystal.

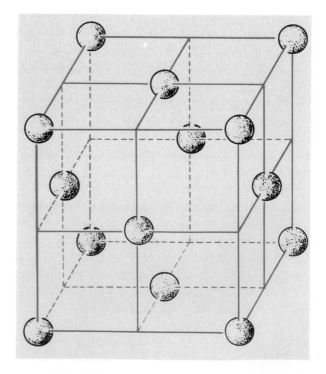

Fig. 3-2. The atoms in a face-centered cubic crystal assume this arrangement.

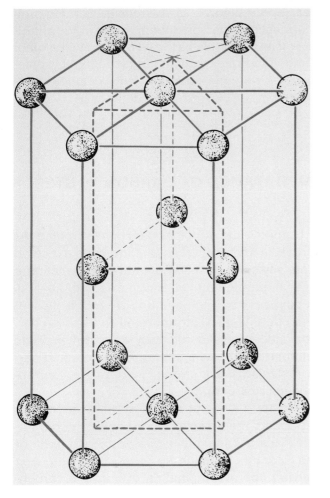

Fig. 3-3. This is the arrangement of the atoms in a hexagonal close-packed crystal.

metals having this type of crystalline structure are cadmium, bismuth, cobalt, magnesium, titanium, and zinc.

Metals with the face-centered lattice are generally *ductile;* that is, plastic and workable. Metals with close-packed hexagonal lattice lack plasticity and cannot be cold-worked, with the exception of zirconium and titanium. Metals with body-centered crystals have higher strength but lower cold working properties than those with the face-centered pattern.

Crystallization of Metals

All metals solidify in the form of crystals. Each metal has its own characteristic geometric pattern. Some metals may even change from one crystal structure to another crystal structure at various temperature levels. For example, iron when heated changes completely to a face-centered cubic structure at a temperature of 1670°F [910°C].

As liquid metal is cooled it loses thermal energy (heat) to the air and walls of the container. At the *solidification temperature* the atoms of the metal assume their characteristic crystal structure. Crystals begin growing at random in the melt at points of lowest energy. If the rate of cooling is fast, more crystals will form instantaneously than at slow rates of cooling. The more crystals that are growing simultaneously the finer will be the grain size of the metal.

Grain size is important since fine-grained steels have far superior mechanical properties than coarse-grained steels. Hence, it is important for a welder to preserve the grain size of the parent metal. The use of excessive heat leads to a slow rate of cooling, thus producing coarse grains and brittleness in a weldment.

Heating Effect on Grain Structure of Steel

When steel, which is carbon and iron, is heated from room temperature to above 1333° F [835°C], the pearlite grains change from a body-centered lattice to a face-centered structure. Such an arrangement of iron atoms is known as *gamma iron.*

What has happened is that while the steel went through its *critical temperature* (temperature above which steel must be heated so it will harden when quenched), the iron carbide separated into carbon and iron, with the carbon distributing itself evenly in the iron. The material is now called *austenite.*

If the heating is continued beyond the critical point, the grains grow larger or coarser until the melting point is reached. When the steel melts, the crystal structure is completely broken and the atoms float about without any definite relationship to one another.

Cooling Effect on Grain Structure of Steel

If you cool a metal from a molten state to room temperature, the change that takes place, under proper conditions, is exactly the opposite of what occurs while the metal is heating.

As the metal begins to cool, the crystals of pure iron start to solidify. This is followed by a crystallization of austenitic grains, and eventually the entire mass becomes solid.

During the range of temperatures at which various stages of solidification takes place, the metal passes from a mushy condition to a solid solution. While in a mushy stage the metal can be shaped easily. After it has reached a solid state, even though the alloy is still hot, it can be formed only by applying heavy pressure or hammering (forging).

With continued cooling of the solid metal, the austenite contracts evenly as the temperature falls. When it reaches its *transformation temperature,* the temperature drop stops for a time. At this point there occurs a rearrangement of *gamma iron* to *alpha iron* as well as a separation of iron carbide and pure iron into *pearlite* grains.

The transformation of the metal from a liquid to a solid is important because the proper rearrangement of the atoms depends on the rate of cooling. If, for example, a piece of 0.83 percent carbon steel is cooled rapidly after its critical temperature is reached, certain actions are arrested before the pearlitic structure can be formed. The result is a metal that is hard, but

Fig. 3-4. Structure of martensite.

very brittle, known as *martensite.* See Fig. 3-4. Martensite is the constituent found in fully hardened steel which is hard and brittle. On the other hand, if the rate of quenching (cooling) is somewhat slower, the structure will be much more ductile.

IMPORTANCE OF CARBON IN STEEL

Carbon is the principal element controlling the structure and properties that might be expected from any carbon steel. The influence that carbon has in strengthening and hardening steel is dependent upon the amount of carbon present and upon its microstructure. Slowly cooled carbon steels have a relatively soft iron pearlitic microstructure; whereas rapidly quenched carbon steels have a strong, hard, brittle, martensitic microstructure.

In carbon steel, at normal room temperature, the atoms are arranged in a body-centered lattice. This is known as *alpha iron.* Each grain of the structure is made up of layers of pure iron (ferrite) and a combination of iron and carbon. The compound of iron and carbon, or iron carbide, is called *cementite.* The cementite is very hard and has practically no ductility.

In a steel with 0.83 percent carbon, the grains are *pearlitic,* meaning that all the carbon is combined with iron to form iron carbide. This is known as a *eutectoid mixture* of carbon and iron. See Fig. 3-5.

If there is less than 0.83 percent carbon, the mixture of pearlite and ferrite is referred to as *hypoeutectoid.* An examination of such a mixture would show grains of pure iron and grains of pearlite as shown in Fig. 3-6.

When the metal contains more than 0.83 percent carbon, the mixture consists of pearlite and iron carbide and is called *hypereutectoid.* Notice in Fig. 3-7 how the grains of pearlite are surrounded by iron carbide. In general, the greatest percentage of steel used is of the hypoeutectoid type, that which has less than 0.83 percent carbon.

Fig. 3-5. Here is how the pearlite grains arrange themselves in a eutectoid mixture.

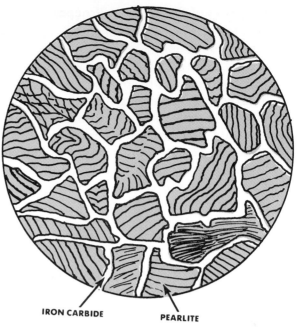

IRON CARBIDE   **PEARLITE**

Fig. 3-7. This is an example of a hypereutectoid structure.

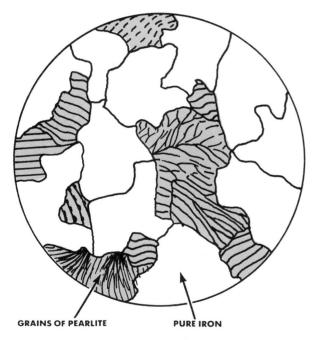

GRAINS OF PEARLITE   **PURE IRON**

Fig. 3-6. An example of hypoeutectoid grain structure.

Other Factors Altering Strength and Structure

When a metal is *cold-worked* (that is; hammered, rolled or drawn through a die) the ferrite and pearlite grains are made smaller and the metal becomes stronger and harder. If, after cold working, the metal is heated and allowed to cool, the grain size is again increased and the metal softened.

The grain size of some metals is reduced and the strength improved through a heating and quenching process. Thus, if a high-carbon steel is heated to a prescribed temperature and then immediately quenched in oil or water, followed by a tempering process, the grain size remains fine. But if you allow the same metal to heat for a long time or if you subject it to temperatures beyond the critical range, then the grain size increases and the metal is weakened. This point is particularly important to remember in welding various steel alloys. The problem of structural change is not too serious in welding mild steel. On the other hand, alloy steels are greatly dependent on space-lattice formation and grain size for their strength. Therefore, you must take extreme care during welding to avoid seriously altering a metal's space-lattice pattern through excessive application of heat or improper treatment of the weld during its cooling stages to avoid this problem.

Effects of Heat of the Welding Process

In welding you must realize, too, that one edge of the metal may cool rapidly, thereby resulting in the formation of hard spots which cause cracks or failure in the weld. Also, there will be conditions where the metal is in a molten state at one point while the surrounding areas may have a temperature ranging from near the molten point down to room temperature. This means that in some areas the crystal structure is completely broken down while elsewhere recrystallization is taking place.

Keep in mind that when hardenable steels are being fused, and you make no effort to control the structural changes either through preheating or by slowing down the cooling rate, the completed weld will be too brittle to be of any value. If a piece of steel, such as an automobile spring, is welded, the heat will remove the springiness from the metal. Moreover, you must remember that if a weld is made on a hardened structure, the act of welding will usually soften the steel and lower its strength. Such metals must then be heat treated to restore their original properties. It is evident then, that in welding any alloy steel, an understanding of the effects of heating and cooling is important.

Heat Treating Metals

Heat treatment is used to soften metal and relieve internal stresses (annealing), harden metal, and temper metal (to toughen certain parts). An understanding of these processes is important to a welder because often he must be aware of how welding heat will affect the structure which he is welding.

Annealing is a softening process which allows metal to be more readily machined and also eliminates stresses in metal after it has been welded. The steel is heated to a certain temperature and held at this temperature to allow the carbon to become evenly distributed throughout the steel. The degree of annealing temperature varies with different kinds of steel. After the metal has been heated for a sufficient period, it is allowed to cool slowly either in the furnace or by burying it in ashes, lime, or in some other insulating material.

For some metals, the *normalizing* treatment is used. It differs from standard annealing in that the steel is heated to a higher temperature for shorter periods and then air cooled.

Stress relieving is a means of removing the internal stresses which develop during the welding operation. The process consists of heating the structure to a temperature below the critical range (approximately 1100°F [594°C] and allowing it to cool slowly. Another method of relieving stresses is *peening* (hammering). However, peening must be undertaken with considerable care because there is always danger of cracking the metal.

Stress relieving is done only if there is a possibility that the structure will crack upon cooling and no other means can be used to eliminate expansion and contraction forces.

Hardening increases the strength of pieces after they are fabricated. It is accomplished by heating the steel to some temperature above the critical point and then cooling it rapidly in air, oil, water, or brine. Only medium, high, and very-high-carbon steels can be hardened by this method. The temperature at which the steel must be heated varies with the steel used.

The tendency of a steel to harden may or may not be desirable depending upon how it is going to be processed. For example, if it is to be welded, a strong tendency to harden will make a steel brittle and susceptible to cracking during the welding process. Special precautions such as preheating and a very careful control of heat input and cooling will be necessary to minimize this condition. During welding, an extremely high localized temperature difference exists between the molten metal of the weld and the metal being welded. The cold parent metal acts as a quench to the weld metal and the metal nearby which has been heated above the upper *critical temperature* (the metal's temperature of transformation). The resulting structure of these areas is hard, brittle martensite. The greater the hardenability of a steel, the less severe the rate of heat extraction necessary to cause it to harden. This is one of the reasons that alloy and high-carbon steels have to be welded with greater care than ordinary low-carbon steels.

Case Hardening

Case hardening is a process of hardening low-carbon or mild steels by adding carbon, nitrogen, or a combination of carbon and nitrogen to the outer surface, forming a hard, thin outer shell. The three principal case hardening techniques are known as carburizing, cyaniding, and nitriding.

Carburizing consists of heating low-carbon steel in a furnace containing a gas atmosphere with the desired amount of carbon monoxide. An alternate method is to heat the steel in contact with a carbon material such as charcoal, coal, nuts, beans, bone, leather or a combination of these. However, modern methods of carburizing use gas atmospheres almost exclusively.

The piece is heated to a temperature between 1650° and 1700°F [899° to 927°C] where steel in the austenitic condition readily absorbs carbon on its surface. The length of the heating period depends on the thickness of the hardened case desired. After heating, the steel is quenched, which produces a material with a hard surface and a relatively tough inner core.

Cyaniding involves heating a low-carbon steel in sodium cyanide or potassium cyanide. The cyanide is heated until it reaches a temperature of 1500°F [815°C] and then the steel is placed in the liquid bath. This produces a very thin outer case which is harder than that obtained by the carburizing process.

Nitriding is a case hardening method which produces the hardest surface of any hardening process. Hardness is obtained by the formation of hard, wear-resistant nitrogen compounds in certain alloy steels where distortion must be kept to a minimum. The alloy is heated to about 900° to 1000°F [482° to 538°C] in an atmosphere of dissociated ammonia gas.

MECHANICAL PROPERTIES OF METALS

Mechanical properties are measures of how materials behave under applied loads. Another way of saying this is how strong is a metal when it comes in contact with one or more forces. If you know the strength properties of a metal, you can build a structure that is safe and sound. Likewise, when a welder knows the strength of his weld as compared with the base metal, he can produce a weldment that is strong enough to do the job. Hence strength is the ability of a metal to withstand loads (forces) without breaking down.

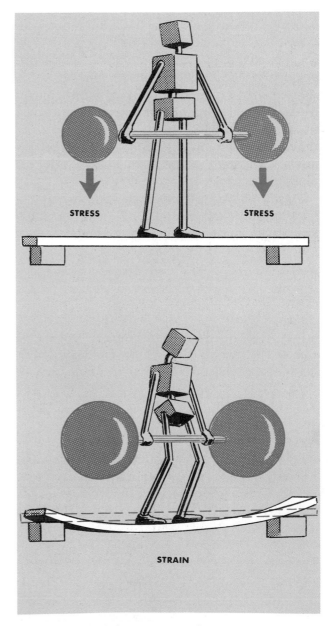

Fig. 3-8. Example of stress and strain.

Some of the basic terms that are associated with mechnical properties of metals are included in the paragraphs that follow. A welder should become familiar with them because they are often directly related to his ability to produce sound welds.

Stress is the internal resistance a material offers to being deformed and is measured in terms of the applied load over the area. See Fig. 3-8 top.

Strain is the deformation that results from a stress and is expressed in terms of the amount of deformation per inch. See Fig. 3-8 bottom.

Elasticity is the ability of a metal to return to its original shape after being elongated or distorted, when the forces are released. See Fig. 3-9. A rubber band is a good example of what is meant by elasticity. If the rubber is stretched, it will return to its original shape after you let it go. However, if the rubber is pulled beyond a certain point, it will break. Metals with elastic properties react in the same way.

Elastic limit is the last point at which a material may be stretched and still return to its undeformed condition upon release of the stress.

Modulus of elasticity is the ratio of stress to strain within the elastic limit. The less a material deforms under a given stress the higher the modulus of elasticity. By checking the modulus of elasticity the comparative stiffness of different materials can readily be ascertained. Rigidity or stiffness is very important for many machine and structural applications.

Tensile strength is that property which resists forces acting to pull the metal apart. See Fig. 3-10. It is one of the more important factors in the evaluation of a metal.

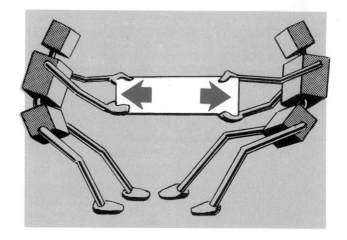

Fig. 3-10. A metal with tensile strength resists pulling forces.

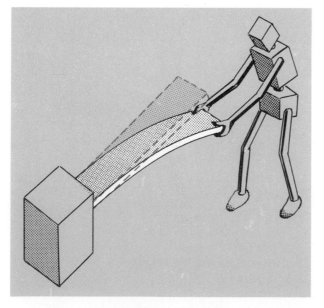

Fig. 3-9. A metal having elastic properties returns to its original shape after the load is removed.

Compressive strength is the ability of a material to resist being crushed. See Fig. 3-11. Compression is the opposite of tension with respect to the direction of the applied load. Most metals have high tensile strength and high compressive strength. However, brittle materials such as cast iron have high compressive strength but only moderate tensile strength.

Bending strength is that quality which resists forces from causing a member to bend or deflect in the direction in which the load is applied. Actually a bending stress is a combination of tensile and compressive stresses. See Fig. 3-12 top to grasp the idea.

Fig. 3-11. Compressive strength refers to the property of metal to resist crushing forces.

Torsional strength is the ability of a metal to withstand forces that cause a member to twist. See Fig. 3-12 middle.

Shear strength refers to how well a member can withstand two equal forces acting in opposite directions. See Fig. 3-12 bottom.

Fatigue strength is the property of a material to resist various kinds of rapidly alternating stresses. For example, a piston rod or an axle undergoes complete reversal of stresses from tension to compression.

Impact strength is the ability of a metal to resist loads that are applied suddenly and often at high velocity. The higher the impact strength of a metal the greater the energy required to break it. Impact strength may be seriously affected by welding since it is one of the most structure sensitive properties.

Ductility refers to the ability of metal to stretch, bend, or twist without breaking or cracking. See Fig. 3-13. A metal having high ductility, such as copper or soft iron, will fail or break gradually as

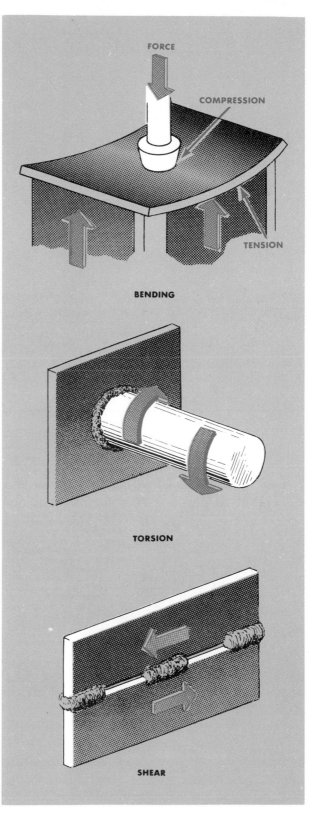

Fig. 3-12. Examples of bending, torsion, and of shearing stresses.

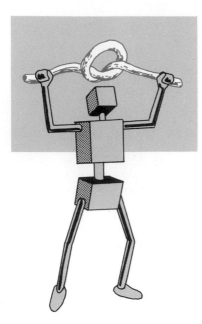

Fig. 3-13. A ductile metal can easily be shaped.

Fig. 3-14. Hardness resists penetration.

the load on it is increased. A metal of low ductility, such as cast iron, fails suddenly by cracking when subjected to a heavy load.

Hardness is that property in steel which resists indentation or penetration. See Fig. 3-14. Hardness is usually expressed in terms of the area of an indentation made by a special ball under a standard load, or the depth of a special indenter under a specific load.

Brittleness is a condition whereby a metal will easily fracture under low stress. It is a property which often develops because of improper welding techniques. Brittleness is a complete lack of ductility.

Toughness may be considered as. strength, together with ductility. A tough material or weld is one which may absorb large amounts of energy without breaking. It is found in metals which exhibit a high elastic limit and good ductility. Welding materials of this kind must be done with a great deal of care. For example, improper application of heat may change the grain size and carbon distribution in the metal so its inherent toughness will be completely destroyed.

Malleability is the ability of a metal to be deformed by compression forces without developing defects, such as encountered in rolling, pressing, or forging.

Creep is a slow but progressively increasing strain, usually at high temperatures, causing the metal to fail.

Cryogenic properties of metals represent behavior characteristics under stress in environments of very low temperatures. In addition to being sensitive to crystal structure and processing conditions, metals are also sensitive to low and high temperatures. Some alloys which perform satisfactorily at room temperatures may fail completely at low or high temperatures. The changes from ductile to brittle failure occurs rather suddenly at low temperatures.

Coefficient of expansion is the amount of expansion in one inch or one foot produced by a temperature rise of 1°F. The expansion rate of metals is always an important factor in welding.

CLASSIFICATION OF CARBON STEELS

A plain carbon steel is one in which carbon is the only alloying element. The amount of carbon in the steel controls its hardness, strength, and ductility. The higher the carbon content, the

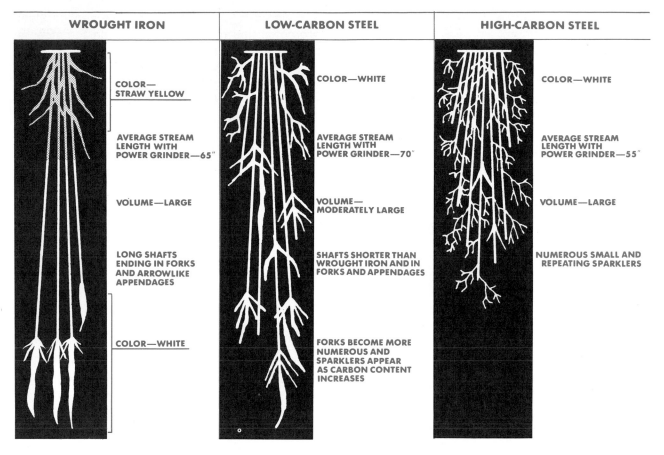

WROUGHT IRON	LOW-CARBON STEEL	HIGH-CARBON STEEL

COLOR—STRAW YELLOW

AVERAGE STREAM LENGTH WITH POWER GRINDER—65″

VOLUME—LARGE

LONG SHAFTS ENDING IN FORKS AND ARROWLIKE APPENDAGES

COLOR—WHITE

COLOR—WHITE

AVERAGE STREAM LENGTH WITH POWER GRINDER—70″

VOLUME— MODERATELY LARGE

SHAFTS SHORTER THAN WROUGHT IRON AND IN FORKS AND APPENDAGES

FORKS BECOME MORE NUMEROUS AND SPARKLERS APPEAR AS CARBON CONTENT INCREASES

COLOR—WHITE

AVERAGE STREAM LENGTH WITH POWER GRINDER—55″

VOLUME—LARGE

NUMEROUS SMALL AND REPEATING SPARKLERS

Fig. 3-15. Spark characteristics of carbon steels.

harder the steel. Conversely, the less the carbon the greater the ductility of the steel.

Carbon steels are classified according to the percentage of carbon they contain. They are referred to as low, medium, high, and very-high-carbon steels.

Low-carbon steels. Steels with a carbon range of 0.05 to 0.30 percent are called low-carbon steels. Steels in this class are tough, ductile, and easily machined, formed, and welded. Most of them do not respond to any heat treating process except case hardening. Low-carbon steel, when subjected to the spark test, will throw off long, white-colored streamers with very little or no sparklers. See Fig. 3-15.

Medium-carbon steels. These steels have a carbon range from 0.30 to 0.45 percent. They are strong and hard but cannot be worked or welded as easily as low-carbon steels. Because of their higher carbon content, they can be heat treated. Successful welding of these steels often requires special electrodes, but even then greater care must be taken to prevent formation of cracks around the weld area.

The spark test will show more numerous sparklers, beginning closer to the wheel, with the streamers much lighter in color.

High and very-high-carbon steels. Steels with a carbon range of 0.45 to 0.75 percent are classified as high-carbon and those with 0.75 to 1.7 percent carbon as very-high-carbon steels. Both of these steels respond well to heat treatment. As a rule, steels up to 0.65 percent carbon can be welded with special electrodes, although preheating and stress relieving techniques must often be used after the welding is completed.

Usually it is not practical to weld steels in the very-high-carbon range.

The spark test for high-carbon steels can easily be recognized by the numerous explosions or sparklers given off, which are practically white in color. See Fig. 3-15.

ALLOY STEELS

An alloy steel is a steel to which one or more of such elements as nickel, chromium, manganese, molybdenum, titanium, cobalt, tungsten, or vanadium have been added. The addition of these elements gives steel greater toughness, strength, resistance to wear, and resistance to corrosion.

Alloy steels are called by the predominating element which has been added. Most of them can be welded, provided special electrodes are used. The more common elements added to steel are:

Chromium. When quantities of chromium are added to steel the resulting product is a metal having extreme hardness and resistance to wear without making it brittle. Chromium also tends to refine the grain structure of steel, thereby increasing its toughness. It is used either alone in carbon steel or in combination with other elements such as nickel, vanadium, molybdenum, or tungsten.

Manganese. The addition of manganese to steel produces a fine grain structure which has greater toughness and ductility.

Molybdenum. This element produces the greatest hardening effect of any element except carbon and at the same time it reduces the enlargement of the grain structure. The result is a strong, tough steel. Although molybdenum is used alone in some alloys, often it is supplemented by other elements, particularly nickel or chromium or both.

Nickel. The addition of nickel increases the ductility of steel while allowing it to maintain its strength. When large quantities of nickel are added (25 to 35 percent), the steels not only become tough but develop high resistance to corrosion and shock.

Vanadium. Addition of this element to steel promotes fine grain structure when the steel is heated above its critical range for heat treatment. It also imparts toughness and strength to the metal.

Tungsten. This element is used mostly in steels designed for metal cutting tools. Tungsten steels are tough, hard, and very resistant to wear.

Cobalt. The chief function of cobalt is to strengthen the ferrite. It is used in combination with tungsten to develop red hardness; that is, the ability to remain hard when red hot.

STEEL CODE CLASSIFYING SYSTEMS

A uniform steel classification system has been adopted by the Society of Automotive Engineers (SAE) and the American Iron and Steel Institute (AISI). Identification is based on a four or five digit code. The first digit indicates the type of steel; thus 1 is a carbon steel, 2 a nickel steel, 3 a nickel-chromium steel, and so on. In the case of simple alloy steels the second number of the series indicates the approximate amount of the predominating alloying element. The last two or three digits refer to the carbon content and are expressed in hundredths of 1 percent. For example, a *2335* steel indicates a nickel steel of about 3 percent nickel and 0.35 percent carbon.

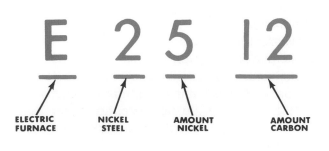

E	2	5	12
ELECTRIC FURNACE	NICKEL STEEL	AMOUNT NICKEL	AMOUNT CARBON

The following are the basic classification numerals for various steels:

Type of Steel	Series Designation
Carbon steels	1XXX
Plain carbon	10XX
Free machining, resulfurized (screw stock)	11XX
Free machining, resulfurized, rephosphorized	12XX
Manganese steels	13XX
High-manganese carburizing steels	15XX
Nickel steels	2XXX
3.50 percent nickel	23XX
5.00 percent nickel	25XX
Nickel-chromium steels	3XXX
1.25 percent nickel, 0.60 percent chromium	31XX
1.75 percent nickel, 1.00 percent chromium	32XX
3.50 percent nickel, 1.50 percent chromium	33XX
Corrosion and heat resisting steels	30XXX
Molybdenum steels	4XXX
Carbon-molybdenum	40XX
Chromium-molybdenum	41XX
Chromium-nickel-molybdenum	43XX
Nickel-molybdenum	46XX and 48XX
Chromium steels	5XXX
Low chromium	51XX
Medium chromium	52XXX
Corrosion and heat resisting	51XXX
Chromium-vanadium steels	6XXX
Chromium 1.0 percent	61XX
Nickel-chromium-molybdenum	86XX and 87XX
Manganese-silicon	92XX
Nickel-chromium-molybdenum	93XX
Manganese-nickel-chromium-molybdenum	94XX
Nickel-chromium-molybdenum	97XX
Nickel-chromium-molybdenum	98XX
Boron (0.0005% boron minimum)	XXBXX

AISI also uses a prefix to indicate the steel-making process. These prefixes are:

A—Open-hearth alloy steel
B—Acid Bessemer carbon steel
C—Basic open-hearth carbon steel
D—Acid open-hearth carbon steel
E—Electric furnace steel of both carbon and alloy steels

Examples:

C1078—Basic open-hearth carbon steel; carbon 0.72 to 0.85 percent

E50100—Electric furnace chromium steel 0.40 to 0.60 percent; chromium, 0.95 to 1.10 percent carbon.

E2512—Electric furnace nickel steel, 4.75 to 5.25 percent nickel; 0.09 to 0.14 percent carbon.

WELDING DEFECTS

In the process of welding various materials, precautions must be taken to prevent the development of certain defects in the weld metal otherwise these defects will severely weaken the weld. The following are some of the principal defects that are significant in any welding or brazing process.

Grain growth. A wide temperature differential will exist between the molten metal of the actual weld and the edges of the heat-affected zone of the base metal. This temperature may range from a point far above the critical temperature down to an area unaffected by the heat. Thus the grain size can be expected to be large at the molten zone of the weld puddle and gradually reducing in size until recrystallization is reached. Grain growth can be kept to a minimum by effective control of preheating and postheating.

Where heavy sections require successive passes, it is possible to use the heat of each successive pass to refine the grain of the previous pass. This can be done only if the metal is allowed to cool below the lower critical temperature between each pass. High-carbon and alloy steels are especially vulnerable to coarse growth if cooled rapidly. These metals usually require a certain amount of preheating before welding and then allowed to cool slowly after the weld is completed.

Blowholes. Blowholes are cavities caused by gas entrapment during the solidification of the weld metal. They usually develop because of improper manipulation of the electrode and failure to maintain the molten pool long enough to float out the entrapped gas, slag, and other foreign matter. When gas and other matter become trapped in the grains of the solid metal, small holes are left in the weld after the metal cools.

Blowholes can be avoided by keeping the molten pool at a uniform temperature throughout the welding operation. This can be done by using a constant welding speed so the metal solidifies evenly. Blowholes are most likely to occur during the stopping and starting of the weld along the seam, especially when the electrode must be changed.

Inclusions. Inclusions are impurities or foreign substances which are forced in a molten puddle during the welding process. Any inclusion tends to weaken a weld because it has the same effects as a crack. A typical example of an inclusion is slag which normally forms over a deposited weld. If the electrode is not manipulated correctly, the force of the arc causes some of the slag particles to be blown into the molten pool. When the molten metal freezes before these inclusions can float to the top, they become lodged in the metal, producing a defective weld.

Inclusions are more likely to occur in overhead welding, since the tendency is not to keep the molten pool too long to prevent it from dripping off the seam. However, if the electrode is manipulated correctly and the right electrodes are used with proper current settings, inclusion can be avoided, or at least kept to a minimum.

Segregation. Segregation is a condition where some regions of the metal are enriched with an alloy ingredient while surrounding areas are actually impoverished. For example, when metal begins to solidify, tiny crystals form along grain boundaries. These so-called crystals or dendrites tend to exclude alloying elements. As other crystals form, they become progressively richer in alloying elements leaving other regions without the benefits of the alloying ingredients. Segregation can be remedied by proper heat treating or slow cooling.

Porosity. Porosity refers to the formation of tiny pinholes generated by atmospheric contamination. Some metals have a high affinity for oxygen and nitrogen when in a molten state.

Unless an adequate protective shield is provided over the molten metal, gas will enter the metal and weaken it.

RESIDUAL STRESSES

The strength of a welded joint depends a great deal on the way you control the expansion and contraction of the metal during the welding operation. Whenever heat is applied to a piece of metal, expansion forces are created which tend to change the dimensions of the piece. Upon cooling, the metal undergoes a change again as it attempts to resume its original shape.

No serious consideration is given these factors when there are no restricting forces to prevent the free movements of the expansion and contraction forces or when welding ductile metal, because the flow of metal will usually relieve the stresses. When free movement is restricted there is likely to occur a warping or distortion if the metal is malleable or ductile, and a fracture if the metal is brittle, as with cast iron.

To better understand the effects of expansion and contraction, assume that the bar shown in Fig. 3-16 is thoroughly and uniformly heated. Since the bar is not restricted in its movements,

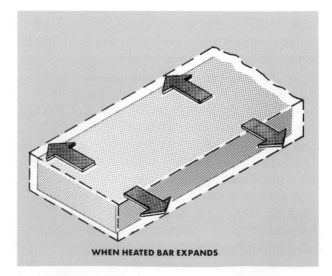

WHEN HEATED BAR EXPANDS

Fig. 3-16. This is what happens when a bar is heated.

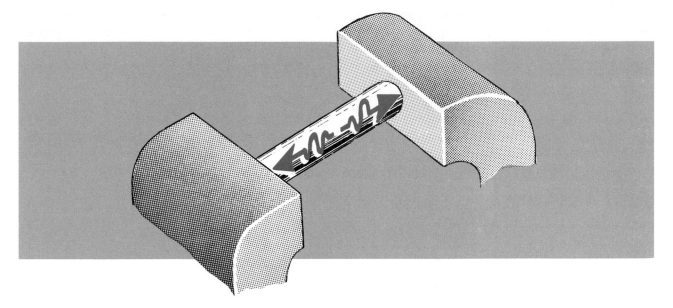

Fig. 3-17. The expansion forces are hindered when the bar has restricting forces like this.

expansion is free to take place in all directions. Consequently, the overall size of the bar is increased. If the bar is allowed to cool without restraint of any kind, it will contract to its original shape.

Suppose now that a similar bar is clamped in a vise, as shown in Fig. 3-17, and heated. Because the ends of the bar cannot move, expansion must take place in another direction. In this case the expansion occurs at the sides.

If heat is applied to one section only, the expansion becomes uneven. The surrounding cold metal prevents free expansion and the displacement of metal takes place only in the heated area. When this area starts to cool, contraction will also be uneven and some of the original displaced metal will become permanently distorted as illustrated in Fig. 3-18.

To show just how the expansion and contraction forces affect metal, study the results of welding two different pieces. In the first case, assume a break has occurred in the middle of a bar, as in Fig. 3-19. Upon welding the break, the heat naturally will cause the metal to expand. Since there are no obstructions on the ends of the bar the metal is permitted to move to what-

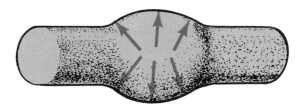

Fig. 3-18. This piece has been distorted because expansion forces were restricted.

ever limits it desires. When the piece begins to cool, there are still no forces to prevent the metal from assuming its original shape.

Suppose the break was in a center section as shown in Fig. 3-20. Note that in this case the ends of the bar are rigidly fastened to a solid frame. If the same procedure is used to weld the fracture as in the first case, something is bound to happen to the casting if no provisions are made for expansion and contraction. Since the vertical and horizontal sections (outside) of

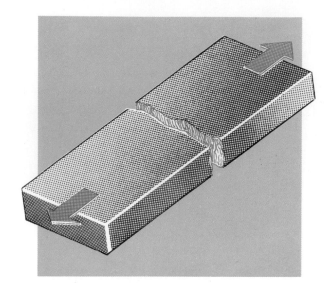

Fig. 3-19. In welding this break, expansion forces are free to move as heating and cooling occur.

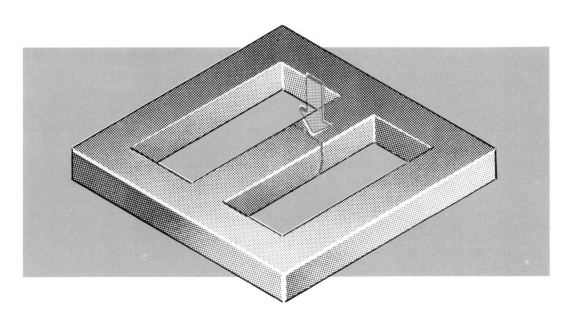

Fig. 3-20. Welding the frame in this confined portion will cause the frame to crack.

the frame will prevent expanding the ends of the center piece, there is only one direction in which this movement can go while the metal is being heated. That is at the point where fusion takes place. Now consider what will happen when the section begins to cool. The frame around the center section has not moved and, when contraction sets in, the center piece will be shortened. When the rigid frame resists this pull, a fracture or deformation at the line of weld or in some other place is bound to occur.

Controlling Residual Stresses

The following are a few simple procedures which will help control the forces caused by expansion and contraction:

Proper edge preparation and fit-up. Make certain that the edges are correctly beveled. Proper edge beveling will not only restrict the effects of distortion but will insure good weld penetration. See Fig. 3-21. Although sometimes the bevel angle can be reduced, care must be

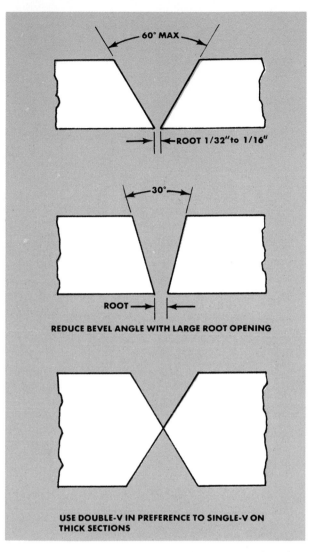

Fig. 3-21. Proper edge-preparation will minimize distortion.

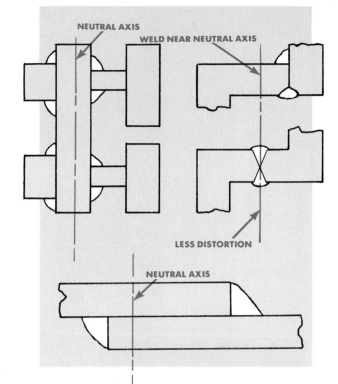

Fig. 3-22. Welding near the neutral axis helps to reduce distortion.

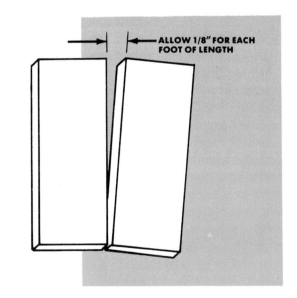

Fig. 3-23. Provide a space between the edges to be welded.

taken to insure that there is sufficient room in the joint to permit proper manipulation of the electrode when doing the weld.

Less distortion will occur if the welds are balanced around the center of gravity which is designated as the *neutral axis.* See Fig. 3-22 top, left. Furthermore, distortion is reduced if the joint nearest to the neutral axis is welded first, followed by welding the unit that is farthest from the neutral axis, Fig. 3-22 top, right, and bottom.

On long seams, especially on thin sections, the practice is to allow about 1/8″ [0.125″] at the end

for each foot in length of the weld for expansion. See Fig. 3-23 for example.

Tack welds are also used to control expansion on long seams as shown in Fig. 3-24. Tack welds

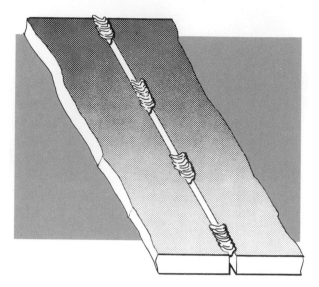

Fig. 3-24. Tacking the plates will hold them in position.

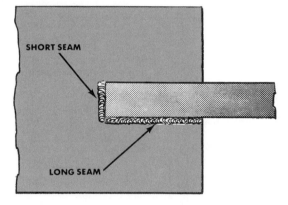

Fig. 3-25. Weld long longitudinal seam first.

are spaced about 12″ [305 mm] apart and run approximately twice as long as the thickness of the weld. When tack welds are used, progressive spacing is not necessary. The plates are simply spaced an equal amount throughout the seam. Also, a long longitudinal (end-ways) seam is welded before a short transverse (side-ways) seam. See Fig. 3-25.

Minimizing heat input. Controlling the amount of heat input is somewhat more difficult for the beginner. An experienced welder is able to join a seam with the minimum amount of heat by rapid welding.

A technique often employed to minimize the heat input is the *intermittent,* or *skip weld.* Instead of making one continuous weld, a short weld is made at the beginning of the joint. Next a few inches is welded at the center of the seam, and then a short length is welded at the end of the joint. Finally you return to where the first weld ended and proceed in the same manner, repeating the cycle until the weld is completed. See Fig. 3-26.

The use of the *back-step,* or *step-back,* welding method also minimizes distortion. With this technique, instead of laying a continuous bead from left to right, you deposit short sections of the beads from right to left as illustrated in Fig. 3-27, then complete the pass.

Preheating. On many pieces, particularly alloy steels and cast iron, expansion and contraction forces can be better controlled if the

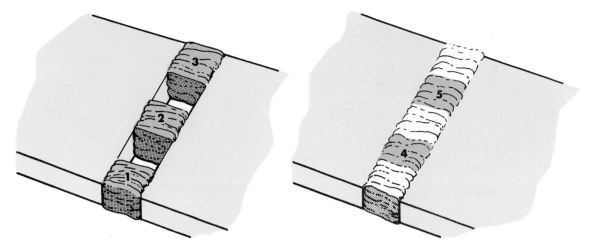

Fig. 3-26. The intermittent weld, sometimes referred to as the skip weld, will prevent distortion.

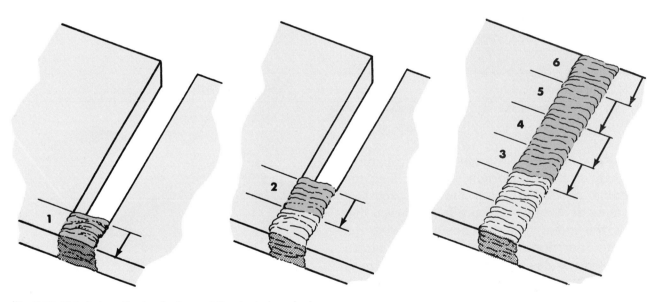

Fig. 3-27. This is how the back-step welding technique is done.

entire structure is preheated before the welding is started. To be effective, preheating must be kept uniform throughout the welding operation, and after the weld is completed the piece must be allowed to cool slowly. Preheating can be done with an oxyacetylene or carbon flame. Usually for work of this kind a second operator manipulates the preheating torch.

Peening. To help a welded joint stretch as it cools, a common practice is to peen it lightly with the round end of a ball peen hammer. However, peening should be done with care because too much hammering will add stresses to the weld or cause the weld to work-harden and become brittle. See Fig. 3-28.

Stress relieving. A common stress relieving method is heat treating. The welded component is placed in a furnace capable of uniform heating and temperature control. The metal must be kept in a soaking temperature until it is heated

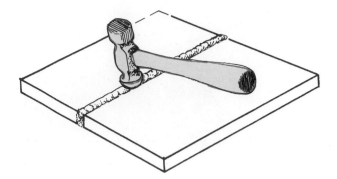

Fig. 3-28. Peening a weld helps to release the locked-up stresses.

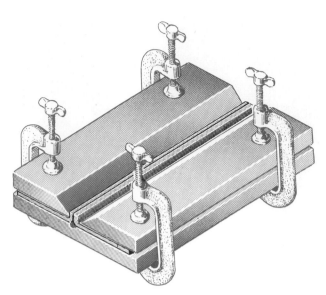

Fig. 3-29. Clamping pieces between heavy blocks will keep them straight.

throughout. Correct temperatures are important to prevent injury to the metal being treated. For example, mild steels require temperatures of 1100° to 1200°F (595° to 650°C) while other alloy steels must be heated to temperatures of 1600°F (870°C) or more.

After the proper soaking period the heat must be reduced gradually to nearly atmospheric temperature.

Jigs and fixtures. The use of jigs and fixtures will help prevent distortion, since holding the metal in a fixed position prevents excessive movements. A jig or a fixture is any device that holds the metal rigidly in position during the welding operation. Fig. 3-29 illustrates a simple way to hold pieces firmly in a flat position. These heavy plates not only prevent distortion but they also serve as *chill blocks* to avoid excessive heat building up in the work. Special chill plates made of copper or other metal having good conductivity are particularly effective in dissipating heat away from the weld area. See Fig. 3-30.

Jigs and fixtures are used extensively in production welding since they permit greater welding speed while reducing to a minimum any form of distortion. By and large, industrial jigs and fixtures are designed to accommodate the specific production work being done. Fig. 3-31 shows such a device.

Number of passes. Distortion can be kept to a minimum by using as few passes as possible

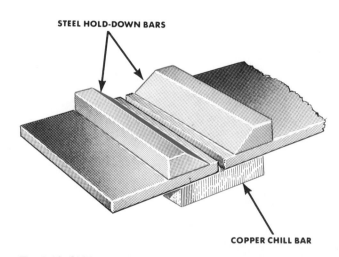

STEEL HOLD-DOWN BARS

COPPER CHILL BAR

Fig. 3-30. Chill plates help to reduce heat and warpage in the weld area. (Republic Steel Corp.)

over the seam. Two passes made with large electrodes are often better than three or four with smaller electrodes. See Fig. 3-32.

Parts out of position. When a single-V butt joint is welded, the greater amount of hot metal at the top than at the root of the V will cause more contraction across the top of the welded joint. The result is a distortion of the plate as shown in Fig. 3-33.

Fig. 3-31. Example of an industrial type of welding jig. (Lincoln Electric Co.)

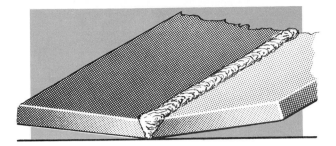

Fig. 3-33. Unless the plates are clamped, a butt joint will be distorted when welded.

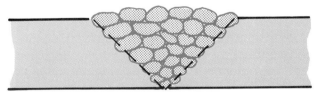

INCORRECT

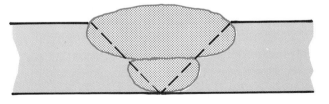

CORRECT

Fig. 3-32. Use few passes to reduce distortion.

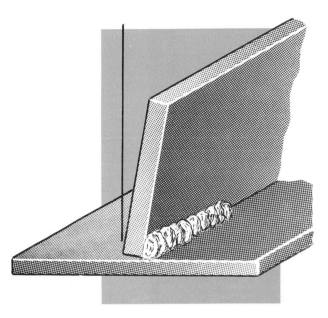

Fig. 3-34. A T-joint is likely to be distorted in this manner.

In a T-joint, the weld along the seam will bend both the upright and flat piece. See Fig. 3-34.

To minimize these distortions, the simplest thing to do is to angle the pieces slightly in the opposite direction in which contraction is to take place. Then, upon cooling, the contraction forces will pull the pieces back into position. Thus, the distortion shown in Figs. 3-33 and 3-34 can be prevented by placing the pieces out of alignment as illustrated in Fig. 3-35.

Final note. A skillful welder is one who possesses a considerable amount of technical information. Merely being able to lay a good bead is not enough, because in the process of making a weld he may, from lack of understanding, jeopardize the strength of the welded structure. Consequently, such factors as properties of metals, expansion and contraction, grain growth, effects of heat, and others should definitely be considered essential knowledge for any welder.

As a final consideration a welder should always remember that the higher the carbon con-

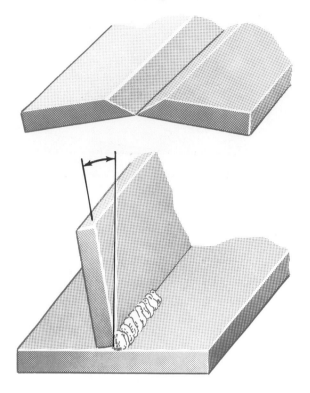

Fig. 3-35. To minimize distortion, set the pieces of the joint slightly out of alignment.

tent of a metal the more difficult it is to weld and the more difficult the weld the more extensive should be the planning, selection, and execution of the welding process. Additional details on the application of this kind of knowledge in specific welding situations will be covered in subsequent chapters.

Points to Remember

1. Be sure you know the kind of metal you are welding and the effects heat may have on the welded structure.

2. When welding alloy steels make certain that the piece is not subjected to prolonged periods of heat that are beyond the critical point.

3. The greater the carbon content in steel, the more difficult it is to weld.

4. Be sure provisions are made for expansion and contraction forces in any welding job.

5. For most butt welds, allow about 1/8″ space at the end of the joint for each foot in length of the weld.

6. Keep the heat as low as possible on a piece being welded.

7. Heat can be controlled on a weld by using intermittent or back-step welding techniques.

8. Use as few passes as possible to minimize distortion.

9. Always guard against blowholes and inclusions in a weld.

10. Select the type of joint that best meets the load requirements of the welded structure.

QUESTIONS FOR STUDY AND DISCUSSION

1. What is the difference between a stress and a strain?

2. Why is elasticity an important property in metals?

3. How does tensile strength differ from compressive strength?

4. What is meant by torsional strength?

5. What is shear strength?

6. Why is fatigue strength in some structures very important?

7. Impact strength refers to what particular quality in metals?

8. What do cryogenic properties refer to?

9. Why is ductility important in some metals?

10. How are carbon steels classified?

11. What is meant by an alloy steel?

12. What are some of the alloying elements that are added to steel?

13. How would you identify a steel labeled C1024?

14. Why are space-lattice formations important in steels?

15. Ductile metals have what kind of space-lattice pattern?

16. How does grain size affect the strength of steel?

17. What is alpha iron?

18. What is meant by cementite?

19. What is meant by pearlite?

20. What is gamma iron?

21. What effects does heating beyond the critical range have on the grain size?

22. What is martensite?

23. What may cause brittleness in a weld area?

24. What are the principal functions of the annealing process?

25. What is stress relieving?

26. What is the difference between hardening and tempering?

27. Why are some metals case hardened?

28. What are the principal case hardening processes?

29. How can grain growth in a weld be controlled?

30. What are blowholes?

31. What causes inclusions in a weld?

32. How can porosity in a weld be prevented?

33. Why must a welder take into account expansion and contraction forces when preparing to make a weld?

34. What means can be used to minimize distortion?

35. What is meant by an intermittent weld?

CHAPTER 4 joint design and welding terms

Joint design is greatly influenced by the cost of preparing the joint, the accessibility of the weld, its adaptability for the product being designed or welded, and the type of loading the weld is required to withstand.

The five basic joints used in welding are butt, T, lap, edge, and corner. See Fig. 4-1. Each has certain advantages and limitations. A welder should be especially concerned with their limitations since the effectiveness of a weld is often contingent on the type of joint that is used as well as his skill in welding.

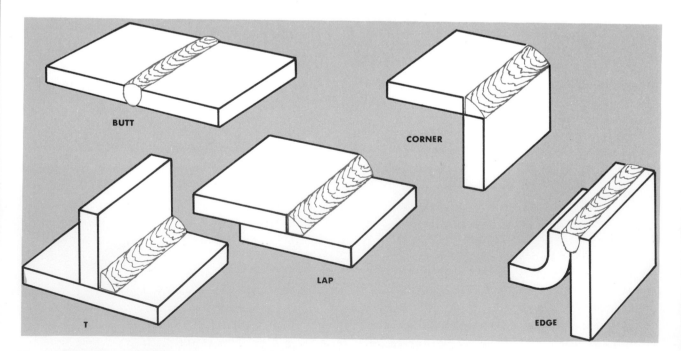

Fig. 4-1. Basic types of joints.

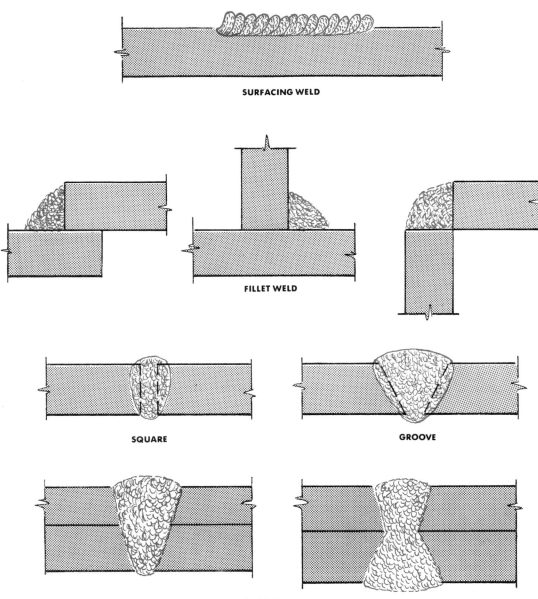

SURFACING WELD

FILLET WELD

SQUARE

GROOVE

GROOVE WELD

Fig. 4-2. Types of welds.

Weld Types

The various joint configurations are used with the following types of welds: *surfacing, fillet, groove, plug* and *slot*. See Fig. 4-2.

Surfacing weld. A type of weld composed of one or more stringer or weave beads deposited on an unbroken surface to obtain desired properties or dimensions.

Fillet weld. A fillet weld is approximately a triangle in cross-section, joining two surfaces at right angles to each other in a lap, T or corner joint.

Groove weld. A groove weld is a weld made in the groove between two members to be joined. The weld is adaptable for joints classified as square butt, single-V, double-V, single-U, double-U, single-J and double-J.

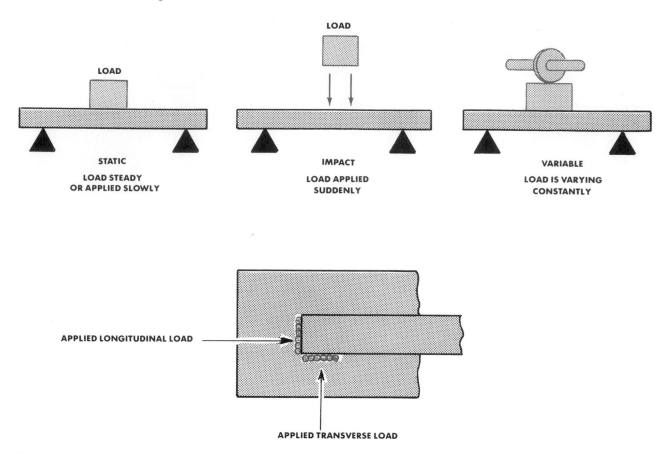

LOAD

LOAD

LOAD

STATIC

**LOAD STEADY
OR APPLIED SLOWLY**

IMPACT

**LOAD APPLIED
SUDDENLY**

VARIABLE

**LOAD IS VARYING
CONSTANTLY**

APPLIED LONGITUDINAL LOAD

APPLIED TRANSVERSE LOAD

Fig. 4-3. Nature of loads applied to welded joints.

Plug and slot-weld. These welds are used to join two overlapping pieces of metal by welding through circular holes or slots. Such welds are often used instead of rivets.

Joint Selection

Just what type of joint is best suited for a particular job depends on many factors. Although the designer or engineer is primarily responsible for determining the kind of joint that is to be used, nevertheless if a welder knows something about joint design he usually produces welds that will better meet the established specifications for the job. In general there are five basic considerations in the selection of any welding joint. These five considerations are:

1. Whether a load is in tension or in compression and if bending, fatigue, or impact stresses will be encountered.

2. How a load is applied, that is, whether the load is steady, sudden, or variable. See Fig. 4-3.

3. Direction of the load as applied to the joint.

4. Thickness of the plates.

5. Cost of preparing the joint.

Joint Geometry[1]

Proper joint geometry is based on the following principles:

1. *Fit-up must be consistent for the entire joint.* Sheet metal and most fillet and lap joints should

---

1. The Lincoln Electric Co.

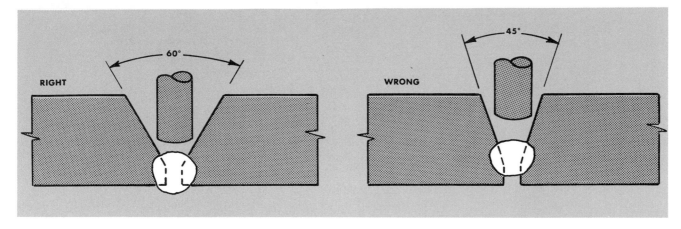

Fig. 4-4. Correct groove bevel is essential for a good weld.

Fig. 4-5. Grooves that are too wide usually result in greater welding costs.

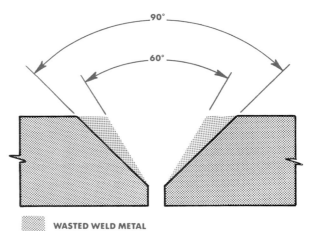

WASTED WELD METAL

be clamped tight for their entire length. Gaps or bevels must be accurately controlled over the entire joint. Any variation in a given joint will force the operator to slow the welding speed to avoid burnthrough and to handle the different electrode manipulation required by the fit-up variation.

2. *Sufficient bevel is required for good bead shape and penetration.* See Fig. 4-4. Insufficient bevel prevents getting the electrode into the joint. A deep narrow bead may lack penetration and has a strong tendency to crack.

3. *Excess bevel wastes weld metal.* See Fig. 4-5. Since filler metal in the form of electrodes and wire is expensive, any variation from the

recommended groove angle size simply contributes to the cost of making a weld both in terms of material and time.

4. *Sufficient gap is needed for full penetration.* See Fig. 4-6. Unless there is adequate penetration, a welded joint will not withstand the loads imposed on it. Although proper penetration depends to some extent on electrode manipulation, the first essential is providing a correct root opening.

5. *Either a ⅛″ [0.125″] land (root face) or a back-up strip is required for fast welding and good quality.* See Fig. 4-7 and Fig. 4-17. Feather-edge preparations require a slow costly seal (root) bead. However, double-V butt joints

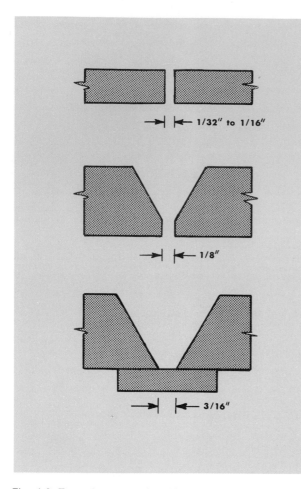

Fig. 4-6. To make a sound weld a correct root opening is important.

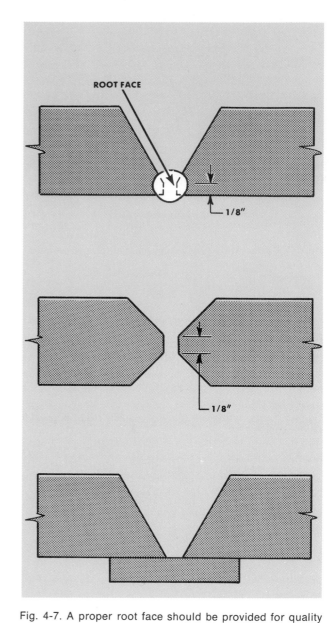

Fig. 4-7. A proper root face should be provided for quality welds.

without a land are practical when the root bead is offset by easier edge preparations and the gap can be limited to about ³/₃₂″.

Butt Joints

In a butt joint the weld is made between the edge surfaces of the two sections to be fused. The joint may be either of the square or grooved type. See Fig. 4-8.

Square butt joint. The square butt joint is intended primarily for materials that are ³/₁₆″ or lighter in thickness, with adequate root opening, and require full and complete fusion for optimum strength. For submerged-arc welding, ma-

terials up to ³/₈″ with a minimum gap of ¹/₈″ can be welded. The joint is reasonably strong in static tension but is not recommended when it is subjected to fatigue or impact loads, especially at low temperatures. The preparation of the joint is relatively simple since it requires only matching the edges of the plates, consequently the cost of making the joint is low. See Fig. 4-8A for this type of joint.

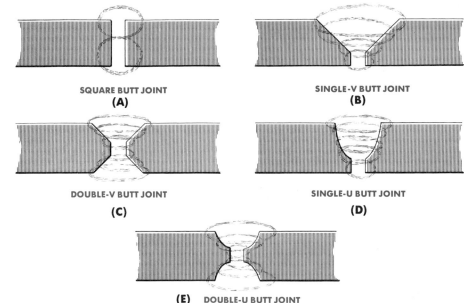

SQUARE BUTT JOINT
(A)

SINGLE-V BUTT JOINT
(B)

Fig. 4-8. Types of butt joint designs.

DOUBLE-V BUTT JOINT
(C)

SINGLE-U BUTT JOINT
(D)

(E) **DOUBLE-U BUTT JOINT**

Single-V butt joint. This joint is used on plate ³/₈″ or greater in thickness, but not to exceed ³/₄″. Preparation is more costly because a special beveling operation is required, and more filler material is necessary. The joint is strong in static loading but as in the square joint it is not particularly suitable when subject to bending at the weld root. See Fig. 4-8B.

Double-V butt joint. The double-V butt joint is best for all load conditions. It is often specified for stock that is heavier than metal used for a single-V (up to ³/₄″). For maximum strength the penetration must be complete on both sides. The cost of preparing the joint is higher than the single-V, but often less filler material is required because a narrower included angle can be used. To keep the joint symmetrical and warpage to a minimum the weld bead must be alternated, welding first on one side and then the other. See Fig. 4-8C.

Single-U butt joint. A joint of this type readily meets all ordinary load conditions and is used for work requiring high quality. It has greatest applications for joining plates ¹/₂″ to ³/₄″. See Fig. 4-8D. The joint needs less filler metal than the single or double-V groove, and less warpage is likely to occur.

Double-U butt joint. The double-U butt joint is intended for metals ³/₄″ or more in thickness where welding can readily be accomplished on both sides. The joint meets all regular load conditions although the preparation cost is higher than the single-U butt joint. See Fig. 4-8E.

T-Joints

A T-joint is made by placing the edge of one piece of metal on the surface of the other piece at approximately a 90° angle and is used for all ordinary plate thicknesses. The basic T-joints are classified as square, single bevel, double bevel, single-J and double-J.

Square T-joint. The square T-joint requires a fillet weld which can be made on one or both sides. It can be used for light or reasonably thick materials where loads subject the weld to longitudinal shear. Since the stress distribution of the joint may not be uniform this factor should be considered where severe impact or heavy transverse loads are encountered. For maximum strength, considerable weld metal is required. See Fig. 4-9A.

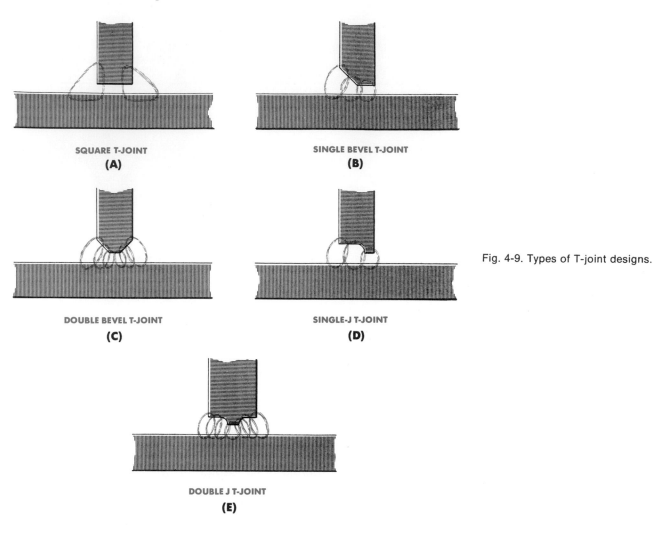

SQUARE T-JOINT
(A)

SINGLE BEVEL T-JOINT
(B)

DOUBLE BEVEL T-JOINT
(C)

SINGLE-J T-JOINT
(D)

DOUBLE J T-JOINT
(E)

Fig. 4-9. Types of T-joint designs.

Single bevel T-joint. This joint will withstand more severe loadings than the square T-joint due to better distribution of stresses. It is generally confined to plates $1/2''$ or less in thickness where welding can be done from one side only. See Fig. 4-9B for example.

Double bevel T-joint. The double bevel T-joint is intended for use where heavy loads are applied in both longitudinal and transverse shear and where welding can be done on both sides. See Fig. 4-9C.

Single-J T-joint. The single-J T-joint is used on plates one inch or more in thickness where welding is limited to one side. It is especially suitable where severe loads are encountered. See Fig. 4-9D.

Double-J T-joint. A joint of this kind is particularly suitable for heavy plates $1^1/2''$ or more in thickness where unusually severe loads must be absorbed. Joint location should permit welding on both sides. See Fig. 4-9E.

Lap Joints

A lap joint as the name implies is made by lapping one piece of metal over another. It is one of the strongest joints, despite the lower unit strength of the filler metal. For joint efficiency, an overlap greater than three times the thickness of the thinnest member is recommended. The two basic lap joints are known as single fillet and double fillet lap joints.

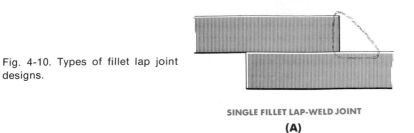
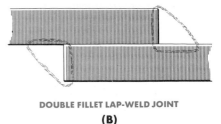

Fig. 4-10. Types of fillet lap joint designs.

SINGLE FILLET LAP-WELD JOINT
(A)

DOUBLE FILLET LAP-WELD JOINT
(B)

Single fillet lap joint. The single fillet lap joint is very easy to weld. Filler metal is simply deposited along the seam. Actually the strength of this weld depends on the size of the fillet. Metal up to $1/2''$ in thickness can be welded satisfactorily with a single fillet if the loading is not too severe. See Fig. 4-10A.

Double fillet lap joint. This joint can withstand greater loads than the single fillet lap joint. It is one of the more widely used joints in welding. As a rule if the weld is properly made its strength is very comparable to that of the parent metal. See Fig. 4-10B.

Corner Joints

Corner joints have wide applications in joining sheet and plate metal sections where generally severe loads are not encountered. The common corner joints are classified as flush, half open and full open.

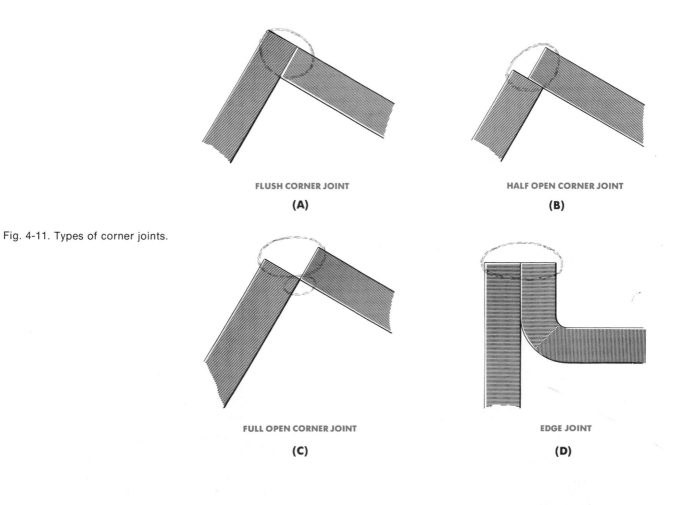

FLUSH CORNER JOINT
(A)

HALF OPEN CORNER JOINT
(B)

Fig. 4-11. Types of corner joints.

FULL OPEN CORNER JOINT
(C)

EDGE JOINT
(D)

Flush corner joint. This joint is designed primarily for welding sheet 12 gage and lighter. It is restricted to lighter materials because deep penetration is sometimes difficult and it supports only moderate loads. See Fig. 4-11A.

Half-open corner joints. The half-open corner joint is usually more adaptable for materials heavier than 12 gage. It is suitable for loads where fatigue or impact are not too severe and where the welding can be done from only one side. Since the two edges of the pieces are shouldered together there is less tendency to burn through the plates at the corner. See Fig. 4-11B for example.

Full-open corner joint. Since this joint permits welding on both sides it produces a strong joint capable of carrying heavy loads. Plates of all thicknesses can be welded. It is recommended for fatigue and impact applications because of good stress distribution. See Fig. 4-11C.

Edge Joint

The edge joint is suitable for plate 1/4″ or less in thickness and can sustain only light loads. See Fig. 4-11D.

Basic Welding Terms

Before proceeding with various welding operations, an understanding of the following terms is important:

Welding positions. The four positions assumed in welding are flat, overhead, vertical and horizontal. See Fig. 4-12. The flat position is the most widely used because welding can be done faster and easier. Overhead welding is somewhat more difficult because the molten metal has a tendency to sag and considerable skill is required to secure a uniform bead with proper penetration. Vertical welding is done in a vertical line from the bottom to the top or top to bottom

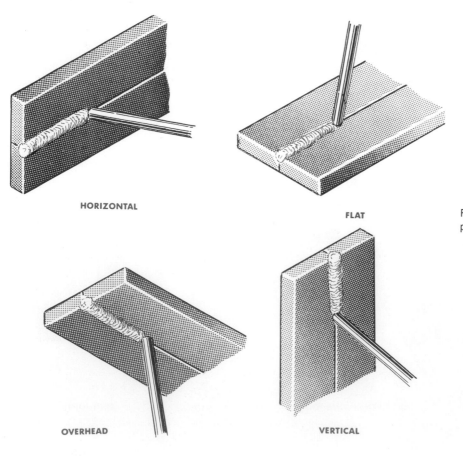

HORIZONTAL

FLAT

OVERHEAD

VERTICAL

Fig. 4-12. There are four main welding positions.

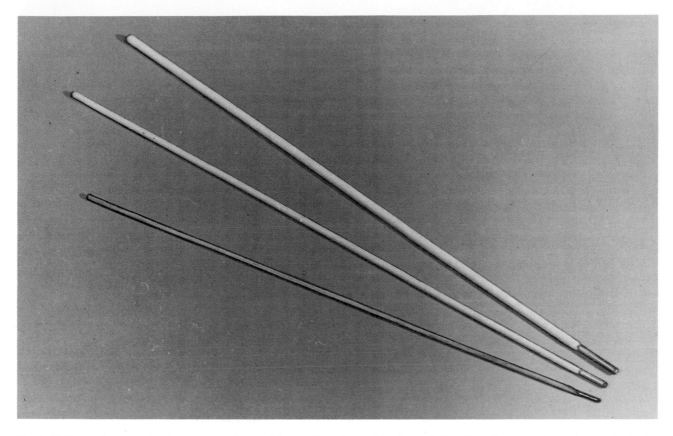

Fig. 4-13. These three electrodes are coated except for small portions at the lower right, where they are gripped by the electrode holder.

of the plates. On thin material a down-hand or downhill welding technique is usually more applicable. Horizontal welding is also difficult to perform because as in overhead welding the molten puddle has a tendency to sag.

Electrode. Thin metal wire, as shown in Fig. 4-13, coated with a special substance and used as a filler to join the metal to be welded.

Base metal or parent metal. Metal to be welded. See Fig. 4-14.

Bead. Narrow layer or layers of metal, as shown in Fig. 4-14, deposited on the base metal as the electrode melts.

Ripple. Shape of the deposited bead that is caused by movement of the electrode is shown in Fig. 4-14.

Pass. Each layer of beads deposited on the base metal as in Fig. 4-14.

Crater. Depression in the base metal, shown in Fig. 4-15, made by the arc as the electrode comes in contact with the base metal.

Penetration. Depth of fusion with the base metal shown in Fig. 4-16.

Reinforcement. Refers to the amount of weld metal that is piled up above the surface of the pieces being joined. It is particularly applicable to butt welds. See Fig. 4-17.

Toes. The points where the base metal and weld metal meet. See Fig. 4-17.

Face. The exposed surface of the weld bounded by the toes of the weld. The face may be either concave or convex. See Fig. 4-17.

Root. The point of the weld triangle opposite the face of the weld. See Fig. 4-17.

Root face. The bottom lip near the slanted surface of the groove. See Fig. 4-17.

Throat. The distance through the center of the weld from the face to the root. See Fig. 4-17.

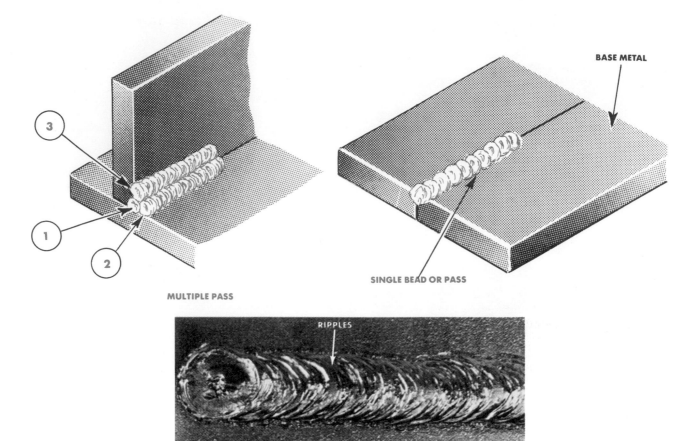

Fig. 4-14. These views illustrate bead, ripple, pass, and the base metal.

Fig. 4-15. Crater appearance in the base metal. (The Lincoln Electric Co.)

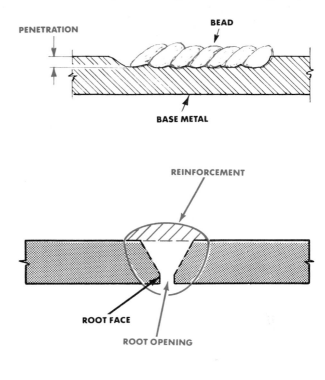

Fig. 4-16. The depth of the surface of the base metal to the bottom of the bead is called penetration.

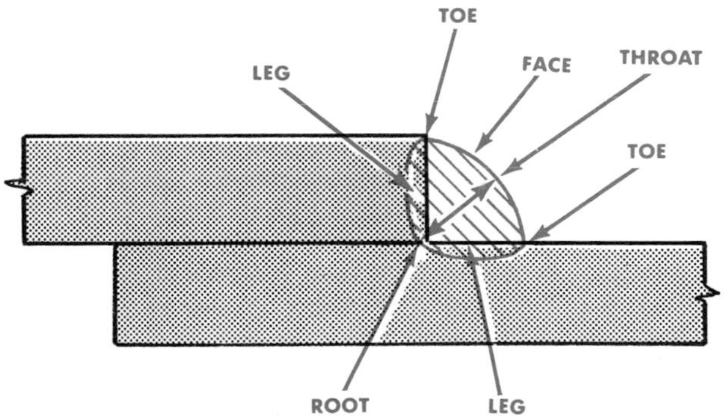

Fig. 4-17. Terms used in indicating the various parts of a weld, as defined in the text.

Weld width. Distance from toe to toe across the face of the weld. See Fig. 4-17.

Weld legs. Size of fillet welds made in lap or T-joints. See Fig. 4-17.

QUESTIONS FOR STUDY AND DISCUSSION

1. What are some of the factors that must be considered in the type of joint that should be used in welding any structural unit?

2. When is a surfacing weld used?

3. What is a fillet weld?

4. In what type of joints are groove welds made?

5. What is a plug weld?

6. Why are grooved butt joints usually superior to square butt joints for welding thick plates?

7. What are the basic types of T-joints?

8. How would you describe a double fillet lap joint?

9. Of the various types of corner joints, which is the strongest?

10. Why is the flat position the most practical for welding?

11. Why is overhead welding more difficult to perform?

12. What is meant by downhill welding? By uphill welding?

13. What does it mean when a weld requirement calls for three passes?

14. What do we mean when we speak of the toes of a weld?

15. What is meant by the root of a weld?

16. What are some of the basic principles which contribute to good joint geometry?

17. When are double bevel T-joints normally used?

CHAPTER 5 machines and accessories

Shielded metal-arc welding, sometimes referred to as metallic-arc welding, or just stick welding, is widely used in the construction of many products, ranging from steamships, tanks, locomotives, and automobiles to small household appliances. Arc welding machines today are designed to join light and heavy gage metals of all kinds. The process of arc welding not only simplifies the maintenance and manufacture of goods and machines, but it permits the skilled operator to perform welding operations quickly and easily.

Welding Current

When an electrical current moves through a wire, heat is generated by the resistance of the wire to the flow of electricity. The greater the current flow the greater the resistance and the more intense the heat.

The heat generated for welding comes from an arc which develops when electricity jumps across an air gap between the end of an electrode and the base metal. The air gap produces a high resistance to the flow of current and this resistance generates an intense arc heat which may be anywhere from 6000°F to 10,000°F (approximately 3300°C to 5500°C).

Welding current is provided by an AC or DC machine. The primary current (input) to a weld-

ing machine is either 220 or 440 volts. *Since voltage of this magnitude is always dangerous, extreme care must be taken to insure that the motor and frame are well grounded.*

The actual voltage used to provide welding current is low (18 to 36V) whereas high amperage is necessary to produce the heat required for welding. However, low voltage and high amperage for welding are not particularly dangerous if there is adequate grounding and proper insulation. Although they need not be feared, both should be treated with care to avoid any electrical accident.

Electrical Terms

To understand the correct operation of an electric arc welding machine, you must know a few basic electrical terms and electric principles. The following are especially important:

Alternating current (AC). An electrical current having alternating positive and negative values. In the first half cycle the current flows in one direction then reverses direction and for the next half cycle flows the opposite way. See Fig. 5-1. The rate of change is referred to as frequency. This frequency is indicated as 25, 40, 50 and 60 cycles per second. In the United States, alternating current is usually established at 60 cycles per second.

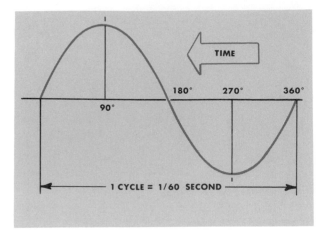

Fig. 5-1. Alternating current has alternate positive and negative values. (Miller Electric Manufacturing Co.)

Fig. 5-3. An *ammeter* shows the amount of current that is flowing. (Weston Electrical Instrument Corp.)

Direct current (DC). Electrical current which flows in one direction only.

Conductor. A conductor is any material in the form of a wire, cable or bus bar which allows a free passage of an electrical current.

Electrical circuit. Path taken by an electric current in flowing through a conductor from one terminal of the source of supply to the other. It starts from the negative terminal of the power supply where the current is produced, moves along the wire or cable to the load or working source, and then returns to the positive terminal. See Fig. 5-2.

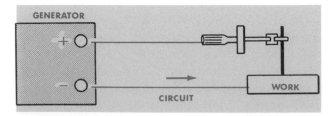

Fig. 5-2. An example of a simple electrical circuit.

Ampere. Amperes (abbreviated amp), or amperage, refers to the amount or rate of current that flows in a circuit. The instrument that meas-

ures this rate is called an ammeter is shown in a diagram, Fig. 5-3.

Volt. The force (emf, or electro-motive force) that causes current to flow in a circuit is known as voltage. This force is similar to the pressure used to make water flow in pipes. In a water system, the pump provides the pressure, whereas in an electrical circuit the power supply produces the force that pushes the current through the wires. Voltage does not flow; only current flows. The force is measured in volts and the instrument used to measure voltage is called a voltmeter. See Fig. 5-4.

Resistance. Resistance is the opposition of the material in a conductor to the passage of an electric current causing electrical energy to be transformed into heat.

Static electricity. Static electricity refers to electricity at rest or electricity that is not moving.

Dynamic electricity. Dynamic electricity is electricity in motion in an electrical current.

Constant potential. Potential is synonymous with voltage. It refers to the generation of a stable voltage regardless of the amperage output produced by the welding power supply. This

Fig. 5-4. A *voltmeter* measures the force of electricity flowing in a circuit. (Weston Electrical Instrument Corp.)

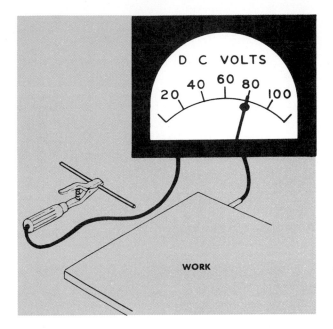

Fig. 5-5. When the machine is running and no welding is being done, you have an *open circuit voltage.*

characteristic is particularly important in Mig welding. See Chapter 19.

Voltage drop.   Just as the pressure in a water system drops as the distance increases from the water pump, so does the voltage lessen as the distance increases from the generator. This fact is important to remember in using a welding machine because if the cable is too long, there will be too great a voltage drop. When there is too much drop, the welding machine cannot supply enough current for welding.

Open-circuit voltage and arc voltage. Open-circuit voltage is the voltage produced when the machine is running and no welding is being done. This voltage varies from 50 to 100V. After the arc is struck, the voltage drops to what is known as the *arc or working voltage,* which is between 18 and 36V. An adjustment is provided to vary the open circuit voltage so welding can be done in different positions. See Figs. 5-5 and 5-6.

Variable voltage.   A control which spans a range of voltages is used to set the open circuit voltage on a welding machine.

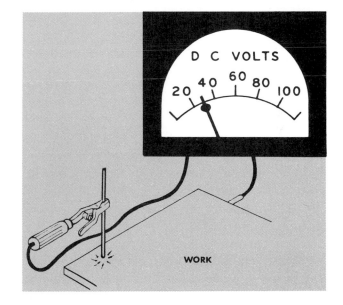

Fig. 5-6. *Arc voltage* is the voltage used when welding is in process.

Polarity.   Polarity indicates the direction of the current in that circuit. Since the current moves in one direction only in *DC welders,* polarity is important because for some welding operations the flow of current must be changed. When the electrode holder is fastened to the

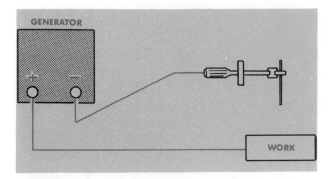

Fig. 5-7. This is how the circuit is arranged for *straight polarity*.

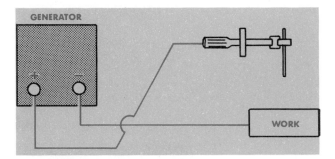

Fig. 5-8. This is how the cables are connected for *reversed polarity*.

negative pole of the generator and the work to the positive pole, the polarity is negative, or more commonly referred to as *straight polarity*. See Fig. 5-7. If the electrode holder is attached to the positive pole of the generator and the cable leading to the work to the negative pole, the circuit is called *reversed polarity*. See Fig. 5-8.

Polarity has a direct relationship to the location of the liberated heat since it is possible to control the amount of heat going into the base metal. By changing polarity the greatest heat can be concentrated where it is most needed.

For some types of welding situations, it is preferable to have more heat at the workpiece because the area of the work is greater and more heat is required to melt the metal than the electrode. Thus, if large, heavy deposits are to be made, the work should be hotter than the electrode. For this purpose, straight polarity would be more effective.

On the other hand, in overhead welding it is necessary to quickly freeze the filler metal to help hold the molten metal in position against the force of gravity. By using reverse polarity, less heat is generated at the workpiece thereby giving the filler metal greater holding power for out-of-position welding.

In other situations, it may be expedient to keep the workpiece as cool as possible, such as in repairing a cast iron casting. With reverse polarity less heat is produced in the base metal and more heat at the electrode. The result is that the deposits can be applied rapidly while the base metal is prevented from overheating.

On early DC welders, the change of polarity involved reversing the cables. Modern machines equipped with a *polarity switch* eliminate disconnecting the cables. Moving the switch to straight or reverse, see Fig. 5-15, changes the polarity.

Inasmuch as the current is constantly reversing in AC welders, polarity is of no consequence.

WELDING MACHINES

To supply the current for welding, three types of units are available: transformers, motor generators, and rectifiers. The source of power to run these welding machines may come from regular electrical lines, or gasoline or diesel engines. The gasoline or diesel run types are especially useful for field work where electrical power is not available. See Figs. 5-9 and 5-10.

Sizes of machines. Sizes of welding machines are rated according to their approximate amperage capacity at *60 percent duty cycle,* such as 150, 200, 250, 300, 400, 500 or 600. This amperage is the rated current output at the working terminal. *Thus a machine rated at 150 amperes can be adjusted to produce a range of power up to 150 amperes.* The duty cycle for welding machines is based on a ten minute period of time. Every welding machine is rated at a certain amperage output and voltage output for a given period of time. The National Electrical Manufacturers Association (NEMA) has set a standard based on the ten-minute period. Thus a

Fig. 5-9. DC welder with power supplied by gasoline engine. (Hobart Brothers Co.)

Fig. 5-10. AC welder driven by gasoline engine. (Miller Electric Manufacturing Co.)

welder rated at 300 amperes, 32 volts, 60 percent duty cycle will put out the rated amperage at the rated voltage for six minutes out of every ten. The machine must idle and cool the other four of every ten minutes.

Some machines used for automatic welding are rated at 100 percent duty cycle and as such can be run continuously without overheating. The size of the welding machine to be used is

governed largely by the kind of welding that is to be done. The following serves as a general guide to size and service.

150–200 Ampere. For light to medium duty welding. Excellent for all fabrication purposes, and rugged enough for continuous operation on light or medium production work.

250–300 Ampere. For average welding requirements. Used in plants for production, maintenance, repair, tool room work and all general shop welding.

400–600 Ampere. A machine for heavy duty welding with large capacity, and for a wide range of purposes. It is used extensively in heavy structural work, fabricating heavy machine parts, heavy pipe and tank welding, and for cutting scrap and cast iron.

CLASSIFICATION OF WELDING MACHINES

Welding machines are classified into two main groups—*constant current* and *constant potential* (voltage). Constant current machines are designed primarily for manual stick welding, whereas constant potential machines are used mostly for Mig (gas-metal arc) welding. Constant potential machines will be described in Chapter 19, dealing with Mig Welding.

Constant Current Machines

Constant current simply means that a steady supply of current is produced over a wide range of welding voltages regardless of changes in arc length. In manual stick welding, whether an AC or DC machine is used, it is difficult to hold the arc length truly constant. However, with a constant current machine there will be relatively small changes in current with changes in arc length. The result is that welding heat and burn-off rate of the electrode are influenced very little, thereby permitting the welder to maintain good control of the weld puddle.

Constant current welding machines have a sloping volt-amp characteristic. The volt-amp characteristic is actually a curve which shows how the voltage varies in its relationship to amperage between the open-circuit (where there is static electrical potential but no current is flowing) and short circuit (when the electrode touches the work). See Fig. 5-11.

Under normal welding operations, the open-circuit is between 50 and 100 volts where the output voltage is between 18 to 36 volts. By having a high open-circuit voltage, easier arc starting is possible with all types of electrodes.

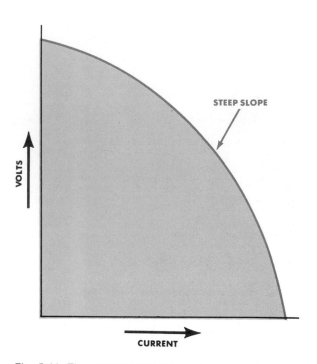

Fig. 5-11. The conventional DC welder has a steep sloping volt-amp curve to control the arc and welding heat.

As the welding proceeds, the high voltage drops to the arc (welding) voltage but regardless of the arc length, due to raising or lowering the electrode, the current output will not fluctuate appreciably. The actual arc voltage will vary, depending on the length of the arc. Thus to strike an arc, the electrode must be shorted to the work. At the moment of contact (short circuit) the amperage shoots up while the voltage drops. Then, as the electrode moves away from the work, the voltage rises to maintain the arc while the amperage drops to the required working level.

During welding if the arc length increases, the voltage increases. Conversely, if the arc length decreases the arc voltage decreases. This enables the operator to vary the heat by lengthening or shortening the arc.

Transformers

The transformer type of welding machine produces AC current and is considered to be the least expensive, lightest, and smallest machine. It takes power directly from a power supply line and transforms it to the voltage required for welding.

Fig. 5-12. Welding current output on some AC machines is regulated by plugging leads into sockets. (Miller Electric Manufacturing Co.)

Fig. 5-13. An AC arc welder with both a tap type and a moving core adjustment. Coarse adjustments are made by moving the electrode selector from the high to the low range tap. Fine adjustments are made by rotating the crank. (Miller Electrical Manufacturing Co.)

The welding current output may be adjusted by plugging leads of the electrode holder into sockets on the front of the machine in various locations, or by rotating a handwheel or crank. See Figs. 5-12 and 5-13 and Fig. 5-10.

Some AC transformers also have an arc booster switch which supplies a burst of current for easy arc starting as soon as the electrode comes in contact with the work. After the arc is struck, the current automatically returns to the amount set for the job. See Fig. 5-14. The arc booster switch has several settings to permit quick arc starting for welding either thin sheets or heavy metal plates.

One outstanding advantage of the AC welder is the freedom from *magnetic arc blow* which often occurs when welding with DC machines. Arc blow is a condition that causes the arc to wander while welding in corners on heavy metal or if using large electrodes.

Since the current in a DC welder flows in one direction, metal being welded becomes magnetized. When this happens, the arc is often deflected, resulting in excessive spatter. Moreover, arc blow breaks the continuity of the deposited metal, usually referred to as the bead, making it necessary to refill the crater, or concave surface, which results from this arc blow. The process of refilling the crater not only slows down the welding but very often leaves weak spots in the weld.

Another feature of the AC welder is its low operating and maintenance cost, its high overall electrical efficiency, and noiseless operation.

DC Welding Machines

Motor generators are designed to produce DC current in either straight or reverse polarity. The polarity selected for welding depends on the kind of electrode used and the material to be

Fig. 5-14. AC transformer type general service welding unit. (Hobart Brothers Co.)

Fig. 5-16. Constant-current DC welding machine. (The Lincoln Electric Co.)

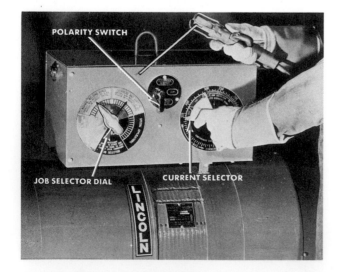

Fig. 5-15. DC motor-generator type for stationary or mobile use. (The Lincoln Electric Co.)

welded. A switch on the machine can be turned for straight or reverse polarity. See Fig. 5-15.

Present day motor generators for manual stick welding are usually of the constant current, dual control type.

With a dual control machine welding current is adjusted by two controls. One control provides an approximate or coarse setting. The second control is usually a rheostat that can be turned to provide a fine adjustment of the welding current, and increase or decrease the heat. See Fig. 5-16.

On dual control machines the slope of the output current can be varied to produce a soft or harsh arc. By flattening the volt-amp curve (increasing the amperage) a digging arc can be obtained for a deeper penetration. With a steeper curve (reduced amperage in relation to volt-

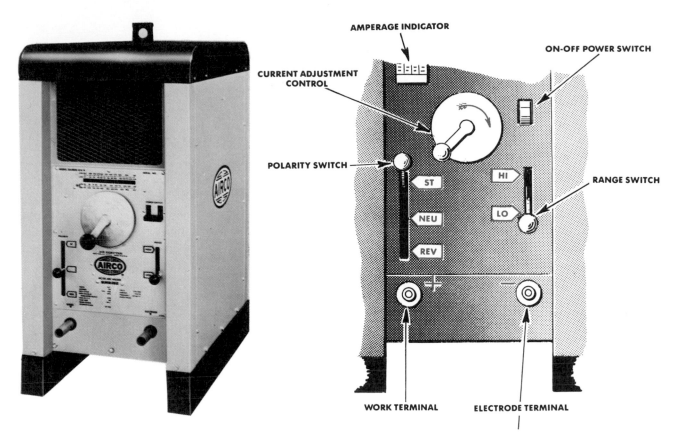

AMPERAGE INDICATOR

ON-OFF POWER SWITCH

CURRENT ADJUSTMENT
CONTROL

POLARITY SWITCH

ST

NEU

REV

HI

LO

RANGE SWITCH

+

−

WORK TERMINAL

ELECTRODE TERMINAL

Fig. 5-17. DC/AC rectifier type adaptable to a variety of welding applications. (Airco)

age) a soft or quiet arc results which is useful for welding light gage materials. In other words, a machine with dual control allows the greatest flexibility for welding materials of different thicknesses.

Rectifiers

Rectifiers are essentially transformers containing an electrical device which changes alternating current into direct current. Some types are designed to provide a choice of low voltage for metal inert-gas welding and submerged-arc welding, and others are designed for a high open circuit with drooping voltage characteristics for tungsten inert-gas welding and metallic arc (stick) electrode welding.

Rectifier welding machines are also available to produce both DC as well as AC current. By turning a switch the output terminals can be changed to the transformer or to the rectifier and produce either AC or DC straight or reverse polarity current.

The transformer rectifier is usually considered more efficient electrically than the generator and provides quiet operation. Current control is achieved by a switching arrangement where one switch can be set for the desired current range and a second dial for securing the fine adjustment desired for welding. See Fig. 5-17.

PERSONAL EQUIPMENT

Helmet and hand shield. An electric arc not only produces a brilliant light but it also gives off invisible ultraviolet and infrared rays which are extremely dangerous to the eyes and skin.

Fig. 5-18. Types of head shields used for arc welding.

CAUTION: You should never look at an arc with the naked eye!

To protect yourself from harmful rays, you must use either a helmet or hand shield as illustrated in Fig. 5-18.

The *helmet* fits over the head and can be swung upward or the lenses opened when you are not welding. The chief advantage of the helmet is that it leaves both hands free, thereby making it possible to hold the work and weld at the same time.

The *hand shield* provides the same protection as the helmet except that it is held in position by a handle. This shield is frequently used by an observer or a person who welds for only a short period. Both types of shields are equipped with special colored lenses that reduce the brilliancy of the light and screen out the infrared and ultraviolet rays. Lenses come in different shades, and the type used depends on the kind of welding done. In general, the recommended practice is as follows:

Shade 2 for resistance welding and stray light.

Shade 5 for light oxygen cutting and gas welding.

Shade 6 and 7 for arc welding up to 30 amperes, oxygen cutting and medium gas welding.

Shade 8 for arc welding beyond 30 amperes and up to 75 amperes and heavy gas welding.

Shade 10 for arc cutting and welding beyond 75 amperes and up to 200 amperes.

Shade 12 for arc welding and cutting beyond 200 amperes and up to 400 amperes.

Shade 14 for arc welding and cutting over 400 amperes.

During the welding process, small particles of metal fly upward from the work and may lodge on the lens. Colored lenses are protected by use of *clear glass* or *plastic cover plates*. Although the methods of inserting the cover plates vary on different makes of shields, you will find that the change can be made easily and quickly.

These clear glasses are inexpensive and can be purchased from any welding supply dealer. *Since you must have clear vision at all times during welding, always replace the cover glass when enough spatter has accumulated on it to interfere with your vision.*

Goggles. Goggles, as shown in Fig. 5-19, are worn when chipping slag from a weld. In arc welding a thin crust forms on the deposited bead. This substance, known as slag, must be removed. While removing slag, tiny particles are often deflected upward. Without proper eye protection, these particles may cause a serious eye injury. Therefore, *always wear goggles when removing slag from a weld.*

Gloves. Another important item you will need for arc welding is a pair of gloves. *You must wear gloves to protect your hands from the ultraviolet rays and spattering hot metal.*

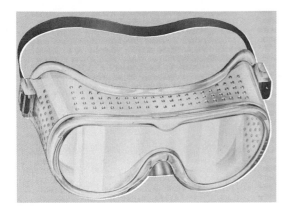

Fig. 5-19. Goggles must always be used when removing slag from a weld. (Purity Cylinder Gasas Inc.)

Fig. 5-21. An apron provides protection for your clothes. (The Lincoln Electric Co.)

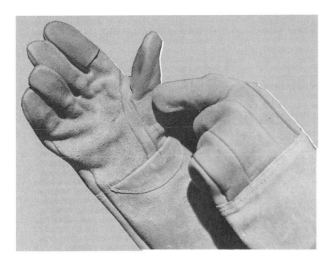

Fig. 5-20. Always wear gloves when arc welding. (The Lincoln Electric Co.)

Several different kinds of gloves are available. As a rule, the leather gauntlet gloves shown in Fig. 5-20 provide ample protection. Regardless of the type used, they should be flexible enough to permit proper hand movement, yet not so thin as to allow the heat to penetrate easily.

Apron. It is a good idea to wear an apron when learning to arc weld. Otherwise, spattering metal might ruin your clothes. Since the spattering particles are hot, a leather apron offers the best protection. A typical welding apron is illustrated in Fig. 5-21.

Most experienced welders seldom wear an apron on the job except in situations where there may be an excessive amount of metal spatter resulting from awkward welding positions. Usually they wear suitable coveralls (fire retardant) to protect their clothing.

Coveralls or ordinary clothing should be sufficiently heavy to prevent infrared and ultraviolet rays from penetrating to the skin. Cuffs on overalls should be turned down or eliminated, and pockets removed so they will not serve as lodging places for falling globules of molten metal. Sleeves and collars should be kept buttoned. Ankle-type shoes are preferable over Oxford type.

SHOP EQUIPMENT

Electrode holder. To do a good welding job, a properly designed electrode holder is essential. The holder is a handle-like tool attached to the cable that holds the electrode during welding. See Fig. 5-22.

A well designed holder can be identified by these features:

1. It is reasonably light, to reduce excessive fatigue while welding.

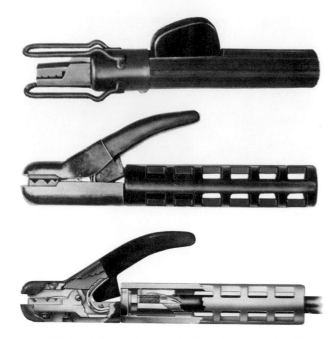

Fig. 5-22. Insulated and ventillated handles on rod holders. (The Lincoln Electric Co.)

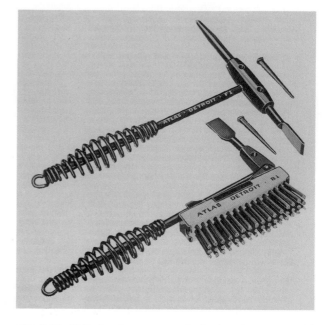

Fig. 5-23. Chipping hammers and wire brushes are used to clean a weld. (Hobart Brothers Co.)

2. It does not heat too rapidly.
3. It is well balanced.
4. It receives and ejects the electrodes easily.
5. All exposed surfaces, including the jaws, are protected by insulation.

The jaws of some holders are not insulated. *When using a holder with uninsulated jaws, never lay it on the bench plate while the machine is running because it will cause a flash.*

Cleaning tools. To produce a strong welded joint, the surface of the metal must be free of all foreign matter such as rust, oil, and paint. A steel brush is used for cleaning purposes.

After a bead is deposited on the metal, the slag which covers the weld is removed with a chipping hammer, which is pictured in Fig. 5-23. The chipping operation is followed by additional wire brushing. Complete removal of slag is especially important when several passes must be made over a joint. Otherwise, gas holes will form in the bead, resulting in porosity which weakens the weld.

Welding screen. Whenever welding is done in areas where other people may be working, the welding operation should be enclosed with screens so the ultraviolet rays will not injure those nearby. These screens can easily be constructed from heavy fire-resistant canvas painted with black or gray ultraviolet protective paint.

When welding is to be done in a permanent location such as a school or shop, a booth is desirable as illustrated in Fig. 5-24.

Cables. The cables carry the current to and from the work. One cable runs from the welding machine to the holder and the other cable is attached to the work or bench. The cable connected to the work is called the *ground cable.* Thus when the welding machine is turned on and the electrode in the holder comes in contact with the work, a circuit is formed, allowing the electricity to flow.

It is important to use the correct diameter cable specified for the welding machine. If the cable is too small for the current, it overheats and a lot of power is lost. Furthermore, a larger cable is necessary to carry a required voltage any distance from the machine. Otherwise, there will be an excessive voltage drop. Even with larger diameter cables, you must take precaution

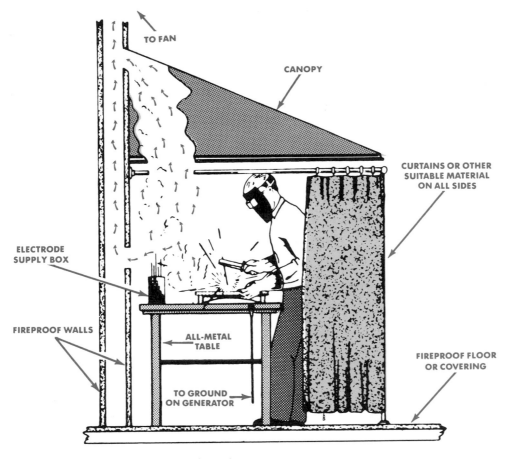

Fig. 5-24. A permanent welding booth provides a safe work area.

not to exceed the recommended lengths, because the voltage drop will lower the efficiency of your welding.

All cable connections should be tight because any loose connection will cause the cable or clamp to overheat. A loose connection may even produce arcing at the connection.

Ground connections. Proper ground connections can be made in several ways. The ground cable can be fastened to the work or bench by a *C* clamp, a special ground clamp, or by bolting or welding the lug on the end of the cable to the bench. See Fig. 5-25 and 5-26.

Providing Ventilation

Electrodes used in arc-welding give off a great deal of smoke and fumes. These fumes are not harmful if the welding area is properly ventilated. *Unless there is sufficient movement of air in the room, arc welding should not be done.* There should be either a suction fan or other good fresh air circulation.

Permanent welding booths should be equipped with a sheet metal hood mounted directly above the welding table, and a suction fan to draw out the smoke and fumes, (as illustrated in Fig. 5-24.)

Where there is considerable welding an exhaust system is necessary to keep toxic gases within the prescribed health limits. The general recommendation for adequate ventilation is a minimum of 2000 cubic feet of air flow per minute per welder. If individual movable exhaust hoods can be placed near the work the rate of air

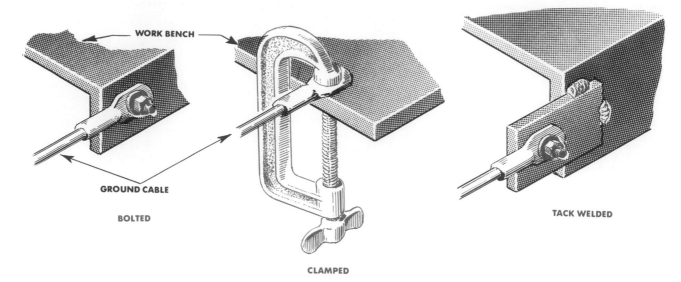

Fig. 5-25. Several connections can be used to provide a proper ground.

Fig. 5-26. Newly designed spring-loaded clamp. (The Lincoln Electric Co.)

flow in the direction of the hood should be approximately 100 linear feet per minute in the welding zone.

Points to Remember

1. Never look at a welding arc without a shield.

2. Always replace the clear cover glasses when they become spattered.

3. Examine the colored lenses in the helmet. Replace cracked ones at once.

4. Wear goggles when chipping slag off a finished weld.

5. Always wear gloves and an apron when welding.

6. Use a holder that is completely insulated. If the jaws of your holder are not insulated, never lay it on the bench while the machine is running.

7. Do welding only in areas where there is enough ventilation.

8. If you weld outside of a permanent booth, be sure to have screens so the arc will not harm persons nearby.

9. Prevent welding cables from coming in contact with hot metal, water, oil, or grease. Avoid dragging the cables over or around sharp corners.

10. Make sure that you have a good ground connection.

11. Keep cables in an orderly manner to prevent them from becoming a stumbling hazard. Fasten the cables overhead whenever possible.

12. Do not weld near inflammable materials.

13. Always turn off the machine when leaving the work.

14. Never stand in water or on a wet floor or

use wet gloves when welding. Water is an electrical conductor and any wet surface will carry current. Always dry out the work pieces or bench if there is any evidence of moisture.

15. Keep electrodes off the floor since they are a serious slipping hazard. Accidents often happen when an operator steps on an electrode causing him to slip and lose his balance.

QUESTIONS FOR STUDY AND DISCUSSION

1. What is a circuit?

2. What is voltage? What instrument is used to measure voltage?

3. What term is used to indicate rate of current flow in a circuit?

4. What is voltage drop? What effect does it have on welding current?

5. What is the difference between AC and DC current?

6. What is the difference between static and dynamic electricity?

7. What is meant by open-circuit voltage and arc voltage?

8. What is polarity?

9. What determines whether the machine is to be set for straight or reversed polarity?

10. How are sizes of welding machines rated?

11. How does a DC motor generator differ from a transformer type unit?

12. What is meant by a constant current, dual control machine?

13. What is meant when a welding machine has a sloping volt-amp curve?

14. What is one main advantage of an AC welder?

15. Why is a rectifier type welding unit often preferred?

16. Why should you never look at an electric arc without eye protection?

17. Why is a helmet more suitable than a hand shield for continuous and long term welding operations?

18. What determines the correct shade of lens to use for welding?

19. Why should shaded lenses be covered with clear glass?

20. Why are goggles important when chipping slag off a weld?

21. Why should welder's gloves be worn when arc welding?

22. What are some of the requirements for a good electrode holder?

23. Why is it important to weld only where there is a good circulation of air?

24. What is likely to happen if cable connections are loose?

CHAPTER 6 selecting the electrode

There are many different kinds and sizes of electrodes and unless the correct one is selected you will have difficulty in doing a good welding job.

In general, all electrodes are classified into five main groups: *mild steel, high-carbon steel, special alloy steel, cast iron, and non-ferrous.* The greatest range of arc welding is done with electrodes in the *mild steel* group. Special alloy steel electrodes are made for welding various kinds of steel alloys. Cast iron electrodes are used for welding cast iron, and non-ferrous electrodes for welding such metals as aluminum, copper, and brass. In this chapter we will discuss mild steel electrodes. Other types of electrodes will be covered in subsequent chapters dealing with the welding of special metals.

WHAT IS AN ELECTRODE?

An electrode is a coated metal wire having approximately the same composition as the metal to be welded. When the current is produced by the generator or transformer and flows through the circuit to the electrode, an arc is formed between the end of the electrode and the work. The arc melts the electrode and the base metal. The melted metal of the electrode flows into the molten crater and forms a bond between the two pieces of metal being joined.

Electrodes are not only manufactured to weld different metals, but they are also designed for DC or AC welding machines. A few electrodes work equally well on either DC or AC. Then too, electrode usage depends on the welding position. Some electrodes are best suited for flat position welding, others are intended primarily for horizontal and flat welding, and some types are used for welding in any position.

The two kinds of mild steel electrodes are known as *bare* and *shielded*. Originally, bare electrodes were uncoated metal rods; today they have a very light coating. Their use for welding is very limited because such electrodes are difficult to weld with and they produce brittle welds with low strength. Practically all welding is done with shielded electrodes.

Shielded electrodes have heavy coatings of various substances such as cellulose sodium, cellulose potassium, titania sodium, titania potassium, iron oxide, iron powder as well as several other ingredients. Each of the substances in the coating is intended to serve a particular function in the welding process.

1. Act as a cleansing and deoxidizing agent in the molten crater.

2. Releases an inert gas to protect the molten metal from atmospheric oxides and nitrides. See Fig. 6-1. Since oxygen and nitrogen weakens a weld if allowed to come in contact with the molten metal, the exclusion of these contaminants is important.

70

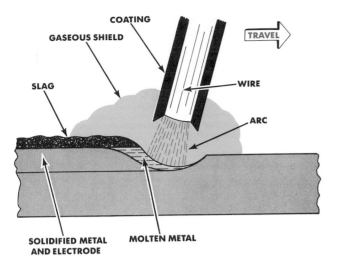

Fig. 6-1. Cross-section of a coated electrode in the process of welding. (The Lincoln Electric Co.)

3. Form a slag over the deposited metal which further protects the weld until the metal cools sufficiently so it is no longer affected by atmospheric contamination. The slag also slows the cooling rate of the deposited metal thereby permitting a more ductile weld to form.

4. Provide easier arc starting, stabilize the arc better, and reduce splatter.

5. Permit better penetration and improve the X-ray quality of the weld.

The coating of some electrodes contains powdered iron which converts to steel and becomes a part of the weld deposit. The powdered iron also helps to increase the speed of welding and improve the weld appearance.

A group of electrodes known as low-hydrogen electrodes have coatings high in limestone and other ingredients with low-hydrogen content, such as calcium fluoride, calcium carbonate, magnesium-aluminum-silicate, and ferrous alloys. These electrodes are used to weld high-sulfur and high-carbon steels that have a great affinity for hydrogen which often causes porosity and underbead cracking in a weld.

Identifying Electrodes

You will find that electrodes are referred to by a manufacturer's trade name. To insure some degree of uniformity in manufacturing electrodes, the American Welding Society and the American Society for Testing of Materials have set up certain requirements for electrodes. Thus different manufacturers' electrodes which are within the classification established by the AWS and ASTM, may be expected to have the same welding characteristics.

In this classification, each type of electrode has been assigned specific symbols, such as E-6010, E-7010, E-8010, etc. The *prefix E* identifies the electrode for electric arc welding as illustrated in Fig. 6-2. The *first two digits* in the symbol designate the minimum allowable tensile strength of the deposited weld metal in thousands of pounds per square inch (psi). For example, the 60 series electrodes have a minimum pull strength of 60,000 psi; the 70 series, a strength of 70,000 psi.

The *third digit* of the symbol indicates possible welding positions. Three numbers are used for this purpose: 1, 2 and 3. Number 1 is for an electrode which can be used for welding in any position. Number 2 represents an electrode restricted for welding in horizontal and flat positions only. Number 3 represents an electrode to be used in the flat position only.

The *fourth digit* of the symbol simply shows some special characteristic of the electrode, such as type of coating, weld quality, type of arc, and amount of penetration. The fourth digit may be 0, 1, 2, 3, 4, 5, 6, 7, or 8. Their meanings are as follows: (see Table 6-1.)

0. Direct current with reverse polarity only.

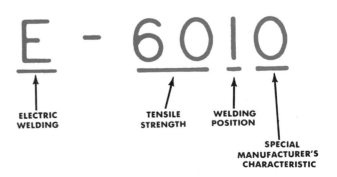

Fig. 6-2. The letter and each number used to classify electrodes has a specific meaning.

TABLE 6-1. FOURTH DIGIT ELECTRODE CHARACTERISTICS.

FOURTH DIGIT	COATING	WELD CURRENT	WELD CHARACTERISTICS
0	Cellulose sodium	DCR	Deep penetration, flat or concave beads, fast fill
1	Cellulose potassium	AC, DCR	Deep penetration, flat or concave beads, fast fill
2	Titania sodium	AC, DCS	Medium penetration, convex beads, full freeze
3	Titania potassium	AC, DCR, DCS	Shallow penetration, convex beads, full freeze
4	Titania iron powder	AC, DCR, DCS	Medium penetration, fast deposit, full freeze-fast freeze
5	Low-hydrogen sodium	DCR	Moderate penetration, convex beads, welding high sulfur, low-carbon steels
6	Low-hydrogen potassium	AC, DCR	Moderate penetration, convex beads, welding high-sulfur, high-carbon steels
7	Iron powder, iron oxide	AC, DCR, DCS	Medium penetration, flat beads, fast freeze
8	Iron powder, low-hydrogen	AC, DCR	Shallow to medium penetration, convex bead, fill freeze

Produces high quality deposits with deep penetration and flat or concave beads, cellulose sodium coating.

1. Alternating current or direct current with reverse polarity. Produces high quality deposits with deep penetration and flat to slightly concave beads, cellulose potassium coating.

2. Direct current with straight polarity only or AC current. Medium quality deposits, medium arc, medium penetration and convex heads, titania sodium coating.

3. Alternating current or direct current with either polarity. Medium to high quality deposits, soft arc, shallow penetration and slightly convex beads, titania potassium coating.

4. AC or DC either polarity, fast deposition rate, deep groove butt, fillet and lap welds, medium penetration, easy slag removal, iron powder, titania coating.

5. Direct current with reverse polarity, high quality deposits, soft arc, moderate penetration, flat to slightly convex bead, low-hydrogen content in weld deposits, low-hydrogen sodium coating.

6. AC or DC reverse polarity with qualities similar to number 5. Low-hydrogen, potassium coating.

7. AC or DC straight polarity, iron powder iron oxide, fast fill, fast deposition rate medium penetration, low spatter, flat beads.

8. DC reverse polarity, iron powder, low hydrogen, fill freeze, shallow to medium penetration, high deposition, easy slag removal, convex beads.

Thus for mild steel, the complete classification number E-6010 would signify an electrode that (a) has a minimum tensile strength of 60,000 for the as-welded deposited weld metal, (b) is usable in all welding positions, and (c) can be used with DC reverse polarity only. In the same way, E-7024 designates an electrode with 70,000 psi minimum tensile strength usable in the flat and horizontal positions only, and operates on DC, either polarity, or AC, and its coating contains iron powder.

It is significant to note that the fourth digit cannot be considered individually; it must be associated with the third digit since in this way it

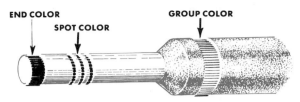

LOCATION OF COLOR MARKINGS FOR END-GRIP ELECTRODES

Fig. 6-3. NEMA Color Code for covered electrodes. Notice that the color markings for end-grip electrodes are quite different from those of the center-grip type. The center grip variety may be read from either end.

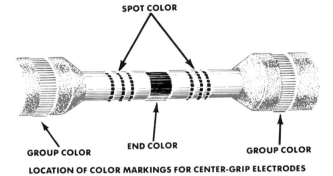

LOCATION OF COLOR MARKINGS FOR CENTER-GRIP ELECTRODES

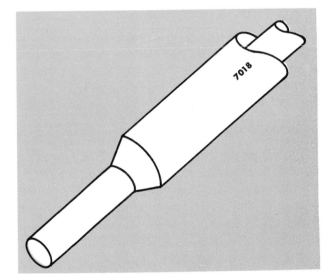

Fig. 6-4. AWS numerical classification. Numerical classification requires careful reference to a specification chart.

identifies both the polarity and position of the electrode.

In addition to the number classification, electrodes are often identified by a *color code* established by the National Electrical Manufacturers Association (NEMA) See Fig. 6-3.

Some manufacturers do not use the color code but simply print the AWS classification on each electrode. See Fig. 6-4.

The NEMA color markings of the more commonly used covered electrodes are given in Table 6-2. The location is shown in Fig. 6-3.

TABLE 6-2. COLOR IDENTIFICATION FOR COVERED MILD-STEEL AND LOW-ALLOY-STEEL ELECTRODES.

AWS ELECTRODE CLASSIFICATION	GROUP COLOR	END COLOR	SPOT COLOR
E6010	no color	no color	no color
E6011	no color	no color	blue
E6012	no color	no color	white
E6013	no color	no color	brown
E6020	no color	no color	green
E6027	no color	no color	silver
E7010-A1	no color	blue	white
E7010-G	no color	blue	no color
E7011-A1	no color	blue	yellow
E7011-G	no color	blue	blue
E7020-G	no color	blue	green
E7027-A1	no color	blue	orange
E7014	no color	black	brown
E7024	no color	black	yellow
E8010-G	no color	white	no color
E8011-G	no color	white	blue
E9010-G	no color	brown	no color
E9011-G	no color	brown	blue
E10010-G	no color	green	no color
E10011-G	no color	green	blue

Selecting the Correct Electrode

The ideal electrode is one that will provide good arc stability, smooth weld bead, fast deposition, minimum spatter, maximum weld strength, and easy slag removal. To achieve these characteristics seven factors should be considered in selecting an electrode. These are:

Properties of the base metal. A top quality weld should be as strong as the parent metal. This means that the electrode to be used must produce a weld metal with approximately the same mechanical properties as the parent metal.

Electrodes are available for welding different classifications of metal. Thus some electrodes are designed to weld carbon steels, others are best suited for low-alloy steels and some are intended specifically for high-strength alloy steels. Therefore, in undertaking any welding operation, the first consideration is to check the chemical analysis of the metal and then select an electrode that is recommended for that metal. Most welding supply distributors are able to provide this information.

Electrode diameter. As a rule, an electrode is never used that has a diameter larger than the thickness of the metal to be welded. Some operators prefer larger electrodes because they permit faster travel along the joint and thus speed up the welding operation; but this requires considerable skill. For example, it takes approximately half the time to deposit a quantity of weld metal from $1/4''$ coated mild steel electrodes than from $3/16''$ electrodes of the same type. The larger sizes not only make possible the use of higher current but require fewer stops to change the electrode. Therefore, from the standpoint of economy it is always a good practice to use the largest size electrode that is practical for the work at hand.

When making vertical or overhead welds, 3/16″ is the largest diameter electrode that should be used regardless of plate thickness. Larger electrodes make it too difficult to control the deposited metal. Ordinarily, a *fast-freeze* type of electrode is best for vertical and overhead welding. See Table 6-3.

The diameter of the electrode is also influenced by the factors of the joint design. Thus, in a thick metal section with a narrow V, a small diameter electrode is always used to run the first weld bead or root pass. This is done to insure thorough penetration at the root of the weld. Successive passes are then made with larger diameter electrodes.

Joint design and fit-up. Joints with insufficient beveled edges require deep penetrating, fast-freeze electrodes. Some electrodes have this particular digging characteristic and may require more skillful electrode manipulation by the operator. On the other hand, joints with open gaps need a mild penetrating fill-freeze electrode that rapidly bridges gaps. See Table 6-3 for variety of electrode characteristics.

Welding position. The position of the weld joint is an important factor in the type of electrode to be used. Some electrodes produce better results when the welding is done in a flat position. Other electrodes are designed for vertical, horizontal, and overhead welding. See Fig. 6-5 for these.

Welding current. Electrodes are made for use with either AC current or DC current reverse polarity or DC current straight polarity, although some electrodes function as well on both AC or DC current.

Production efficiency. Deposition rate is extremely significant in any production work. The faster a weld can be made the lower the cost. Not

TABLE 6-3.   ELECTRODE CHARACTERISTICS.

TYPE	AWS CLASS	CURRENT TYPE	WELDING POSITION	WELD RESULTS
Mild steel	E-6010	DCR	F, V, OH, H	Fast freeze, deep penetrating, flat beads,
	E-6011	DCR, AC	F, V, OH, H	all-purpose welding
	E-6012	DCS, AC	F, V, OH, H	Fill-freeze, low penetration, for poor fit-up,
	E-6013	DCR, DCS, AC	F, V, OH, H	good bead contour, minimum spatter
	E-6014	DCS, AC	F, V, OH, H	
	E-6020	DCR, DCS, AC	F, H	Fast-fill, high deposition, deep groove
	E-6024	DCR, DCS, AC	F, H	welds, single pass
	E-6027	DCR, DCS, AC	F, H	Iron powder, high deposition, deep penetration
	E-7014	DCR, DCS, AC	F, V, OH, H	Iron powder, low penetration, high speed
	E-7024	DCR, DCS, AC	F, H	Iron powder, high deposition, single and multiple pass
Low hydrogen	E-7016	DCR, AC	F, V, OH, H	Welding of high-sulfur and high-carbon
	E-7018	DCR, AC	F, V, OH, H	steels that tend to develop porosity and
	E-7028	DCR, AC	F, H	crack under weld bead

DCR—Direct Current Reverse Polarity
DCS—Direct Current Straight Polarity
AC —Alternating Current
F —flat, V—vertical, OH—overhead, H—horizontal

Fig. 6-5. Be sure to use the right electrode for the job being done—auto shop repair. (Hobart Brothers Co.)

all electrodes have a high-speed high-current rating and still produce smooth, even bead ripples. Unless electrodes are noted for a fast deposition rate they may prove very difficult to handle when used at high speed travel.

Service conditions. The service requirements of the part being welded may demand special weld deposits. For example, high-corrosion resistance, or ductility, or high strength may be important factors. In such cases electrodes must be selected that will produce these specific characteristics.

Conserving and Storing Electrodes

Most electrodes are costly; therefore, every bit of the electrode should be consumed. Do not discard stub ends until they are down to only $1\frac{1}{2}''$ to $2''$ long. See Fig. 6-6.

Always store electrodes in a dry place at a normal room temperature and 50 percent maximum relative humidity. When exposed to moisture, the coating has a tendency to disintegrate. In storing electrodes, be sure they are not bumped, bent, or stepped on, since this will remove the coating and make the electrode useless.

Fig. 6-6. Do not discard electrodes of this length.

Deposition Classification of Electrodes

Electrodes for welding mild steel are sometimes classified as fast-freeze, fill-freeze, and fast-fill[1] (See Table 6-3.) The *fast-freeze* electrodes are those which produce a snappy, deep penetrating arc and fast-freezing deposits. They are commonly called reverse polarity electrodes even though some can be used on AC. These electrodes have little slag and produce flat beads. They are widely used for all types of

---

1. The Lincoln Electric Company.

all-position welding for both fabrication and repair work.

Fill-freeze electrodes have a moderately forceful arc and deposit rate between those of the fast-freeze and fast-fill electrodes. They are commonly called straight-polarity electrodes even though they may be used on AC. These electrodes have complete slag coverage and weld beads with distinct, even ripples. They are the general-purpose electrode for production shop and are also widely used for repair work. They can be used in all positions, though the fast-freeze electrodes are preferred for vertical and overhead welding.

The *fast-fill* group includes the heavy coated, iron powder electrodes with soft arc and fast deposit rate. These electrodes have a heavy slag and produce exceptionally smooth weld beads. They are generally used for production welding where all work can be positioned for downhand (flat) welding.

Types of Mild Steel Electrodes

E-6010. This is an all-position, fast-freeze electrode. It is suitable only on DC machines with reversed polarity, and is designed primarily for welding mild and low-alloy steels. It should be used only where there is an absolutely good fit-up. The E-6010 electrode has wide applications in ship construction, buildings, bridges, tanks, and piping. Table 6-4 shows the amperage settings for different sizes of this electrode.

TABLE 6-4. CURRENT SETTINGS FOR E-6010 ELECTRODES.

ELECTRODES DIA (INCHES)	AMPERES*
3/32	60–90
1/8	80–120
5/32	110–160
3/16	150–200
7/32	175–250
1/4	225–300
5/16	250–450

*These ranges may vary slightly for electrodes made by different manufacturers

TABLE 6-5. CURRENT SETTINGS FOR E-6011 ELECTRODES.

ELECTRODES DIA (INCHES)	AMPERES*
3/32	50–90
1/8	80–130
5/32	120–180
3/16	140–220
7/32	170–250
1/4	225–325

*These ranges may vary slightly for electrodes made by different manufacturers

E-6011. The E-6011 electrode is similar to the E-6010 except that it is made especially for AC machines. Although the electrode can be used on DC machines with reversed polarity, it does not work quite as well as the E-6010. Its amperage setting is slightly lower than for E-6010. See Table 6-5.

E-6012. This is a fill-freeze electrode and may be used on either DC or AC welders. When employed on DC welders the current must be set for straight polarity. The electrode provides medium penetration, a quiet type arc, slight spatter, and dense slag. Although it is considered an all-position electrode, it is used in greater quantities for flat and horizontal position welds. This electrode is especially useful to bridge gaps under conditions of poor *fit-up work;* that is, joints where the edges do not fit closely together. Higher currents can be used with the E-6012 electrodes than with any other type of all-position electrodes. See Table 6-6.

E-6013. Electrodes of this type are very similar to E-6012 with a few slight exceptions. Slag removal is better and the arc can be maintained easier, especially with small diameter electrodes. This permits better operation with lower open-circuit voltage. The bead deposited is noticeably flatter and smoother but with shallower penetration than the E-6012 class. Although the electrode is used particularly for welding sheet metal, it has many other applications. It works well in all positions and it functions very well on AC

TABLE 6-6.   CURRENT SETTINGS FOR E-6012
ELECTRODES.

ELECTRODE DIA (INCHES)	AMPERES*
3/32	40–90
1/8	80–120
5/32	120–190
3/16	140–240
7/32	180–315
1/4	225–350

*These ranges may vary slightly for electrodes made by different manufacturers

TABLE 6-7.   CURRENT SETTINGS FOR E-6013
ELECTRODES.

ELECTRODE DIA (INCHES)	AMPERES*
1/16	20–40
5/64	25–50
3/32	30–80
1/8	80–120
5/32	120–190
3/16	140–240
7/32	225–300
1/4	250–350

*These ranges may vary slightly for electrodes made by different manufacturers

TABLE 6-8.   CURRENT SETTING FOR E-6027
ELECTRODES.

ELECTRODE DIA (INCHES)	AMPERES*
3/16	225–300
7/32	275–375
1/4	350–450

*These ranges may vary slightly for electrodes made by different manufacturers

TABLE 6-9.   CURRENT SETTING FOR E-7014
ELECTRODES.

ELECTRODE DIA (INCHES)	AMPERES*
3/32	80–110
1/8	110–150
5/32	140–190
3/16	180–260
7/32	250–325
1/4	300–400
5/16	400–500

*These ranges may vary slightly for electrodes made by different manufacturers

welders. When used with DC machines the polarity may be straight or reverse. Current settings for E-6013 electrodes are shown in Table 6-7.

Iron Powder Electrodes

Iron powder electrodes are those which contain a high content of iron powder. They are designed for welding mild steels where high speed and fast deposition rate are required. The three principal types are E-6027, E-7014, and E-7024. All of them produce low spatter with easy slag removal. Typical application includes railroad cars, earth-moving equipment, posi-tioned welds in pressure vessels, piping, and ships. The E-7014 and E-7024 are often used where higher strength joints are necessary.

E-6027. Produces high quality welds for high-speed deposition of 1/4" and 5/16" horizontal fillets and for butt and fillet welds in the flat position and for cover beads on butt welds where complete coverage and good bead appearance are required. Current may be AC or DC with either polarity. A drag welding technique is recommended to keep the cover over both legs of fillet welds. See Table 6-8 for approximate current settings.

E-7014. This is a fast-fill and fast-freeze electrode where high speed is necessary. May be used in all positions with AC or DCR and DCS current. The E-7014 electrode deposits much more metal than an E-6012 or E-6013 type. It is particularly effective in vertical downhill welding.

TABLE 6-10. CURRENT SETTING FOR E-7024 ELECTRODES.

ELECTRODE DIA (INCHES)	AMPERES*
3/32	90–120
1/8	120–150
5/32	180–230
3/16	250–300
7/32	300–350
1/4	350–400
5/16	400–500

*These ranges may vary slightly for electrodes made by different manufacturers

TABLE 6-11. CURRENT SETTINGS FOR E-7016 ELECTRODES.

ELECTRODE DIA (INCHES)	AMPERES*
3/32	75–105
1/8	100–150
5/32	140–190
3/16	190–250
7/32	250–300
1/4	300–375

*These ranges may vary slightly for electrodes made by different manufacturers

See Table 6-9 for approximate current settings.

E-7024. The E-7024 is a fast-fill electrode which provides exceptional economy for single or multiple pass welds and is excellent on build-up applications because of the high deposition rate and easy slag removal. It is recommended only for flat and horizontal positions with AC or DC straight or reverse polarity. See Table 6-10 for approximate current settings.

Low-Hydrogen Electrodes

Low-hydrogen electrodes are designed for welding high-sulfur and high-carbon steels. When such steels are welded they tend to develop porosity and cracks under the weld bead because of the hydrogen absorption from arc atmospheres. Low-hydrogen electrodes were developed to prevent the introduction of hydrogen in the weld.

The basic low-hydrogen electrodes are E-7016, E-7018, and E-7028.

E-7016. This is an all-position electrode suitable for AC or DC reverse polarity current. It is especially recommended for welding hardenable steels where no preheat is used, and where stress-relieving normally would be required but cannot be effected. See Table 6-11 for approximate current settings.

E-7018. The E-7018 is a low-hydrogen type electrode but also contains iron powder. It is a high-speed fast-deposition rate electrode de-

TABLE 6-12. CURRENT SETTINGS FOR E-7018 ELECTRODES.

ELECTRODE DIA (INCHES)	AMPERES*
3/32	70–120
1/8	100–150
5/32	120–200
3/16	200–275
7/32	275–350
1/4	300–400

*These ranges may vary slightly for electrodes made by different manufacturers

signed to pass the most severe X-ray requirements when applied in all welding positions, using either AC or DC reverse polarity current. Its puddle fluidity permits gases to escape when the lowest currents are used for out-of-position welding. See Table 6-12 for approximate current settings for low-hydrogen electrodes.

E-7028. The E-7028 is a low-hydrogen electrode with a heavy iron powder type covering and is considered the counterpart of E-7018 but for flat and horizontal positions only. See Table 6-13 for approximate current settings.

TABLE 6-13.   CURRENT SETTINGS FOR E-7028 ELECTRODES.

ELECTRODE DIA (INCHES)	AMPERES*
5/32	175–250
3/16	250–325
7/32	300–400
1/4	375–475

*These ranges may vary slightly for electrodes made by different manufacturers

Variables Governing Electrode Selection

Although there are a variety of electrode classification charts which list the basic characteristics or differences in electrodes, many of the variables encountered in production will often require the use of tests to determine the suitability of an electrode for a specific application. By first analyzing the variables in term of their importance in a welding situation, considerable time and effort can be saved.

The variables normally associated with most types of welding are listed in Table 6-14. These are shown with a relative rating ranging from 1 to 10, with 10 as the highest value and 1 the lowest. These variables and their corresponding ratings are based on experience and intended primarily as an aid in the electrode selection process. For example, notice in Table 6-14 that if high-sulfur steel is to be welded, either a E-7016 or E-7018 electrode should be used. If poor fit-up is the problem, electrode E-6012 is considered best for this condition. On the other hand, if deposition rate is the primary factor then either E-6027 or E-7024 is the most suited for this purpose.

Special Electrodes

Standard electrodes are used for most general types of welding. With these electrodes fusion is

TABLE 6-14.   MILD STEEL ELECTRODE SELECTION CHART.*

	ELECTRODE CLASS										
	E6010	E6011	E6012	E6013	E7014	E7016	E7018	E6020	E7024	E6027	E7028
Groove butt welds, flat (< 1/4″)	5	5	3	8	9	7	9	10	9	10	10
Groove butt welds, all positions (< 1/4″)	10	9	5	8	6	7	6	(b)	(b)	(b)	(b)
Fillet welds, flat or horizontal	2	3	8	7	9	5	9	10	10	9	9
Fillet welds, all positions	10	9	6	7	7	8	6	(b)	(b)	(b)	(b)
Current(c)	DCR	AC DCR	DCS AC	AC DC	DC AC	DCR AC	DCR AC	DC AC	DC AC	AC DC	DCR AC
Thin material (1/4″)	5	7	8	9	8	2	2	(b)	7	(b)	(b)
Heavy plate or highly restrained joint	8	8	8	8	8	10	9	8	7	8	9
High-sulfur or off-analysis steel	(b)	(b)	5	3	3	9	9	(b)	5	(b)	9
Deposition rate	4	4	5	5	6	4	6	6	10	10	8
Depth of penetration	10	9	6	5	66	7	7	8	4	8	7
Appearance, undercutting	6	6	8	9	9	7	10	9	10	10	10
Soundness	6	6	33	5	7	10	9	9	8	9	9
Ductility	6	7	44	5	6	10	10	10	5	10	10
Low-temperature impact strength	8	8	4	5	8	10	10	8	9	9	10
Low spatter loss	1	2	6	7	9	6	8	9	10	10	9
Poor fit-up	6	7	10	8	9	4	4	(b)	8	(b)	4
Welder appeal	7	6	8	9	10	6	8	9	10	10	9
Slag removal	9	8	6	8	8	4	7	9	9	9	8

(a) Rating is on a comparative basis of same size electrodes with 10 as the highest value. Ratings may change with size.
(b) Not recommended
(c) DCR—direct current reverse, electrode positive; DCS—direct current straight, electrode negative; AC—alternating current; DC—direct current, either polarity.
*AWS

achieved by changing the solid state of metal into a molten mass, which combines with the metal deposit of the electrode, to form a permanent bond. The success of the bond depends on the generation of sufficient heat to produce a completely molten puddle. However, metals often undergo unfavorable changes when subjected to high heat, and precautions must be taken to avoid stresses, distortion, warpage, and metallurgical structural changes.

A wide range of low-heat-input electrodes is available under the name *Eutectic* for welding where high heat may be a critical factor. These electrodes are designed so a strong joint can be made without heating the parent metal to its full melting point. Such welding is done with considerably lower current values without necessarily sacrificing the strength of the joint. For example, a $1/8''$ high nickel content electrode may be used to cold weld castings only five thirty-seconds of an inch thick. An arc of only one-sixteenth of an inch instead of the conventional one-eighth inch arc gap is required.

Low-heat-input electrodes may be procured to weld all kinds of ferrous and non-ferrous metals. Special information concerning these electrodes can be obtained from the Eutectic Welding Alloys Corporation.

Points to Remember

1. Use the correct type electrode for the welding to be done.

2. Remember, some electrodes can be used only on DC machines and others on AC machines.

3. If welding is to be done on a DC machine, check whether straight or reverse polarity is needed for the particular electrode to be used.

4. Select an electrode with a diameter that is about the same as the thickness of the plate to be welded.

5. Always use up the electrode until the stub is down to $1^1/2''$ to $2''$ long.

6. Store electrodes in a dry place where the coating cannot be damaged.

QUESTIONS FOR STUDY AND DISCUSSION

1. What is the difference between bare and shielded electrodes?

2. Why are bare electrodes rarely used?

3. What are the functions of the heavy coating on shielded electrodes?

4. What has been done to insure uniformity of electrode specifications?

5. What symbols have been adopted to identify different types of electrodes?

6. In addition to identification symbols, what other means are used to mark electrodes? Where are these markings located?

7. Explain the identifying symbols of the electrode classification E-6010.

8. What is an all-position electrode?

9. How is it possible to determine if an electrode is designed for an AC or DC welder?

10. What factors should be taken into consideration when selecting an electrode for the job to be done?

11. Why are smaller diameter electrodes used for overhead welding?

12. What precautions must be taken in storing electrodes?

13. What is the specific feature of electrodes with coatings containing powdered iron?

14. Why are low-hydrogen electrodes used?

15. What are some of the specific characteristics of electrodes designated as fast-freeze?

16. Some electrodes are classified as fill-freeze. What does this mean?

17. Fast-fill electrodes are intended for what types of welding?

CHAPTER 7 *striking an arc*

Learning to arc weld involves mastery of a specific series of operations. Skill in performing these operations requires practice. Once this skill has been acquired, the operations can be applied on any welding job. The first basic operation is learning to strike an arc and run a straight bead.

BASIC PRINCIPLES OF SUSTAINING A WELDING ARC

Before proceeding with the first welding operation, it is well to review several basic principles for maintaining an arc since the success of any welding operation depends upon a stable arc. To sustain a stable arc, three elements will be necessary:

1. arc gap (arc length)
2. arc voltage
3. load current (amperage)

A proper *arc length* between the electrode and the work is essential to generate the heat needed for welding. A long arc unfortunately produces an unstable welding arc, reduces penetration, increases spatter, causes flat and wide beads, and prevents the gas shield from protecting the molten puddle from atmospheric contamination. Too short an arc will not create enough heat to melt the base metal, the electrode will have a tendency to stick, penetration will be poor, and uneven beads with irregular ripples will result.

Arc voltage or load voltage as it is sometimes called, is the voltage measured across the arc during welding. This voltage is affected by the arc length and therefore must fall within a certain range for each welding condition. Large changes of voltage resulting from wide changes in arc gap will cause uneven heating of the base metal and filler metal, thereby influencing weld penetration and bead shape. In addition to arc length, arc voltage is also affected by the coating of the electrode, composition of the filler metal, electrode diameter, and current density (amount of current per square inch of electrode).

Load current represents the actual flow of electricity and is regulated by the current setting of the power supply. As with arc voltage, load current must also fall within a certain range to provide a stable arc. Otherwise, uneven heating will occur and cause uneven melting of the base metal and filler metal.

Although the experienced welder can usually minimize the effects of arc length by his or her skill in manipulating the electrode, a stable arc also depends on the design of the machine and the proper setting of the controls for the type of welding to be done. See Chapter 5. Therefore, one of the first requirements in carrying out a successful welding operation is the selection of the right electrode and the proper adjustment of the welding current.

Checking and Adjusting the Equipment

To start your first welding operation, proceed as follows:

1. Inspect the cable connections to make certain that they are all tight.

2. Make sure the bench top and metal to be welded are dry and free from dirt, rust, and grease.

3. If you are using a DC welder, set the polarity switch for the desired kind of current—straight or reversed.

4. Adjust the control unit for the amperage and voltage needed for the selected electrode. *Remember,* the recommended current setting specified for the electrode is only approximate. Final adjustment of current value is made as you proceed with the welding operation. For example, the amperage range for the electrode may be 90–100. It is best for the beginner to set the control midway between the two limits, which in this case is 95 amperes. If after the welding is started the arc is too hot, turn the control to reduce the amperage. Increase the current setting if the arc is not hot enough for penetration. No specific rules can be given for the final

setting because many factors are involved, such as skill of the operator, welding position, type of metal, and the nature of the welding job. Ability to make the final adjustment comes as you gain experience.

Gripping the Electrode Holder

Place the bare end of the electrode in the holder as shown in Fig. 7-1. By gripping the electrode near the end, most of the coated portion can be used. *Always keep the jaws of the holder clean* to insure good electrical contact with the electrode. *Be careful not to touch the welding bench with an uninsulated holder, as this will cause a flash. When not in use, hang the holder in the place provided for it.*

Grip the holder lightly in your hand. If you hold it too tightly, your hand and arm will tire quickly. Whenever possible, drape the cable over the shoulder or knee to lessen its drag on the holder.

Striking the Arc

1. Obtain a piece of 1/8″ or 1/4″ steel plate and lay it flat on the bench top. Insert a 1/8″ or 5/32″ E-6010, E-6011, or E-6012 electrode in the holder and set the machine for the correct current.

2. There are two methods which can be used to start, or strike, the arc—the *tapping* and the *scratching motion*. The tapping method is the one preferred by experienced welders, whereas

Fig. 7-1. How to grip the electrode. (Hobart Brothers Co.)

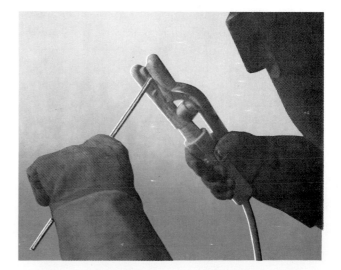

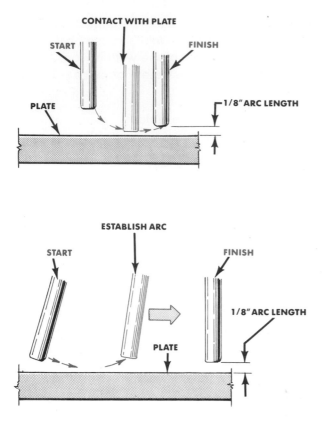

Fig. 7-2. There are two methods of starting, or striking, a welding arc.

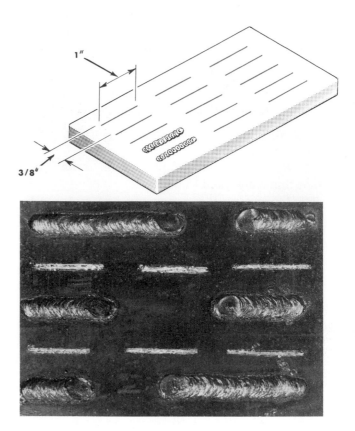

Fig. 7-3. Lay out guide lines with a soapstone for running short beads.

the scratching motion is found to be easier for the beginner.

In the tapping motion, the electrode is brought straight down and withdrawn instantly, as shown in top of Fig. 7-2. With the scratching method, the electrode is moved at an angle to the plate in a scratching motion much as in striking a match shown in bottom of Fig. 7-2. Regardless of which motion you use, upon contact with the plate, promptly raise the electrode a distance equal to the diameter of the electrode. Otherwise, the electrode will stick to the metal. If it is allowed to remain in this position with the current flowing, the electrode will become red hot. *Should the electrode weld fast to the plate, break it loose by quickly twisting or bending the holder. If it should fail to dislodge, disengage the electrode by releasing it from the holder.*

3. Practice starting the arc until this operation can be performed quickly and easily.

Running Short Beads

With a soapstone, which is a marking chalk used to draw lines on metal, draw a series of lines on a steel plate, each line to be approximately 1″ in length and ³/₈″ apart, as illustrated in Fig. 7-3. Run a continuous bead over each line, moving the electrode from left to right. Hold the electrode in a vertical position or slant it slightly away from you as shown in Fig. 7-4.

Move the electrode just rapidly enough so deposited metal has time to penetrate into the base plate. If the current is set properly and the arc is maintained at the correct length, there will be a continuous crackling or frying noise. Learn to recognize this sound. An arc that is too long will have somewhat of a humming sound. Too short of an arc makes a popping sound. Notice the action of the molten puddle and how the trailing edge of the puddle solidifies as the electrode travels forward.

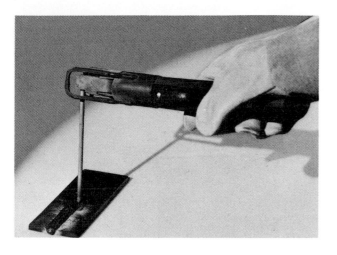

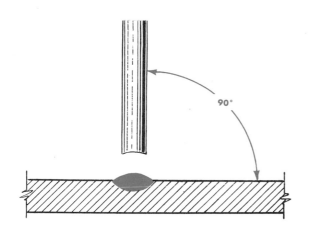

END VIEW

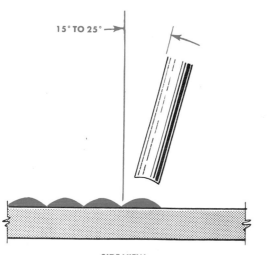

SIDE VIEW

Fig. 7-4. Hold the electrode in this position for running straight beads.

The appearance of the puddle is often an indication of how good a weld is being made. If the molten metal is clear and bright it means that no molten slag is mixing with the puddle. Slag is brittle and when it flows in the molten metal the weld is weakened. Normally, if the edges of the weld bead have a dull irregular appearance it means that slag is being trapped into the puddle.

Checking the Welding Heat

After you have become accustomed to striking an arc and running short weld beads, vary the welding current to see how it affects the welding heat. First turn the machine down about five amperes and check if there is any difference when you run a bead. Then turn it down another five amperes and again try to run a bead. As you reduce the amperage it soon becomes apparent that there is insufficient heat to melt the base metal. Furthermore, you will find that as the electrode burns off, it does not fuse with the base metal but lies on the surface as spatter which easily scrapes off.

Now reverse the process by gradually raising the amperage. Turn the machine up five amperes in several steps and each time run short beads. It will soon become obvious that as the amperage is increased the arc gets hotter and the electrode melts faster.

From this experiment you can appreciate the importance of having the correct welding heat to make a sound weld. However, as you gain experience in welding, proper adjustment of welding current becomes relatively easy.

Points to Remember

1. Inspect the equipment before starting to weld.

2. See that the polarity switch is set in the right position.

3. See that no combustible materials are near where the welding is to be done, as flying sparks from the spatter of the arc may easily ignite the materials.

4. Do not lay the holder on the bench while the current is flowing.

5. Release the electrode if it sticks to the plate.

6. Always shut off the machine when leaving the welding bench.

QUESTIONS FOR STUDY AND DISCUSSION

1. How does an arc gap affect a weld?

2. What happens if arc voltage fluctuates?

3. Why must the load current be correct for a particular welding operation?

4. What are some of the items to be checked before proceeding to weld?

5. Why should the electrode be clamped at its extreme end?

6. Why should the holder never be placed on the work bench while the current is on?

7. What two methods may be used in striking an arc?

8. In striking an arc, why should the electrode be withdrawn instantly?

9. What should be done if the electrode welds fast to the plate?

10. The arc should be maintained at approximately what length?

11. If the arc length and current are correct, what is the characteristic noise that is heard by the experienced welder?

12. How does the appearance of the molten puddle and weld bead indicate a good or poor weld?

CHAPTER 8 *running continuous beads*

To produce good welds you must not only know how to manipulate the electrode, but you need to know certain weld characteristics. Especially important is a knowledge of what constitutes a good weld and what causes a poor weld. Some of the more important elements affecting good welds are discussed in this chapter.

FIVE ESSENTIALS OF ARC WELDING

To secure a weld that has proper penetration, you must keep in mind the following five factors: (1) Correct electrode, (2) Correct arc length, (3) Correct current, (4) Correct travel speed, and (5) Correct electrode angle.

Correct electrode. The choice of an electrode involves such items as position of the weld, properties of the base metal, diameter of the electrode, type of joint, and current value. Since many different kinds of electrodes are manufactured, you must know the results that can be expected from different electrodes. If the characteristics of the electrodes are known, then you have greater assurance that a correct weld will be made. Without the right kind of electrode it is almost impossible to get the results desired, regardless of the welding technique used.

Correct arc length. The significance of *arc length* was briefly mentioned in Chapter 7. Since it is one of the essentials for good welding,

further amplification of correct arc length is included here.

If the arc is too long, the metal melts off the electrode in large globules which wobble from side to side as the arc wavers. This produces a wide, spattered, and irregular bead without sufficient fusion between the original metal and the deposited metal. An arc that is too short fails to generate enough heat to melt the base metal properly. Furthermore, the electrode sticks frequently, producing a high, uneven bead with irregular ripples.

The length of the arc depends on the size of electrode used and the kind of welding done. Thus for small diameter electrodes, a shorter arc is necessary than for larger electrodes. As a rule, the length of the arc should be approximately equal to the diameter of the electrode. For example, an electrode $1/8''$ in diameter should have an arc length of about $1/8''$. You will find, too, that a shorter arc is required for vertical and overhead welding than for most flat position welds because it gives better control of the puddle.

The use of a short arc is important because it prevents impurities from entering a weld. A long arc allows the atmosphere to flow into the arc stream, thereby permitting impurities of nitrides and oxides to form. Moreover, when the arc is too long, heat from the arc stream is dissipated too rapidly, causing considerable metal spatter. Study Fig. 8-1 and notice the condition of the three sample welds shown. The top sample illus-

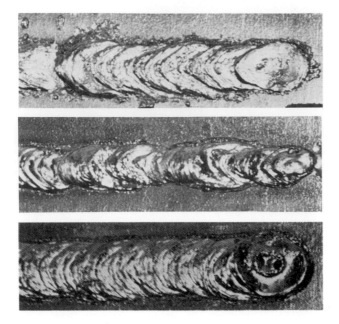

Fig. 8-1. This is how the beads appear when (top) the arc is too long; (middle) when the arc is too short; and (bottom) when the arc is the correct length.

trates a weld formed when the arc is too long. This is indicated by the large amount of spatter and the fact that the beads are coarse. The middle sample shows a weld made with an arc that is too short. Notice the excessive height of the beads. Such beads are a sign of improper penetration. The bottom sample is an example of a good weld. In this case the beads have the proper height and width, and the ripples are uniformly spaced.

Correct current. If the current is too high, the electrode melts too fast and the molten pool is large and irregular. When the current is too low, there is not enough heat to melt the base metal and the molten pool will be too small. The result is not only poor fusion but the beads will pile up and be irregular in shape. See Fig. 8-2.

Correct travel speed. Where the speed is too fast, the molten pool does not last long enough and impurities are locked in the weld. The bead

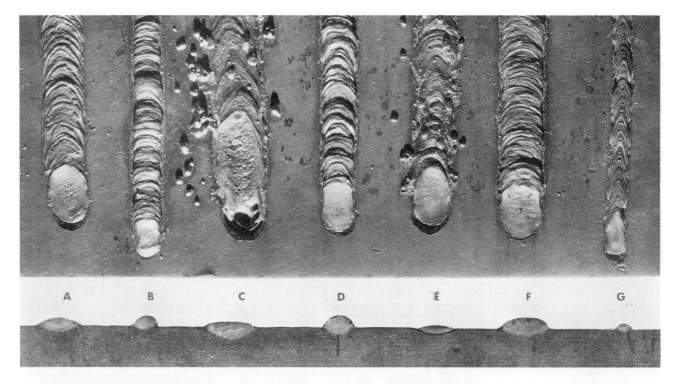

Fig. 8-2. Examples of properly and improperly formed beads. (The Lincoln Electric Co.)

A. Current, voltage, and speed normal
B. Current too low
C. Current too high
D. Voltage too low
E. Voltage too high
F. Speed too slow
G. Speed too fast

is narrow and the ripples pointed. If the rate of travel is too slow, the metal piles up excessively and the bead is high and wide with straight ripples as illustrated in Fig. 8-2.

Correct electrode angle. The angle at which the electrode is held will greatly affect weld bead shape and is particularly important in fillet and deep groove welding. Electrode angle involves two positions—incline and side angles. *Incline angle* is the angle in the line of welding and may vary from 5° to 30° from the vertical, depending on welder preference and welding conditions. *Side angle* is the angle from horizontal, measured at right angles to the line of welding, which normally splits the angle of the weld joint. See Fig. 8-3 for examples.

Ordinarily, a variation of 15° in either direction from the horizontal will not affect weld appearance or quality. However, whenever under-cuts occur in the vertical member of a fillet weld the angle of the arc should be lowered and the electrode directed more toward the vertical member. Side angle is especially important in multiple pass welding. See Fig. 8-3.

Crater Formation

As the arc comes in contact with the base metal, a pool, or pocket, is formed. As previously stated, this pool is known as a *crater*. The size and depth of a crater indicate the amount of penetration. In general, the depth of penetration should be from one-third to one-half the total

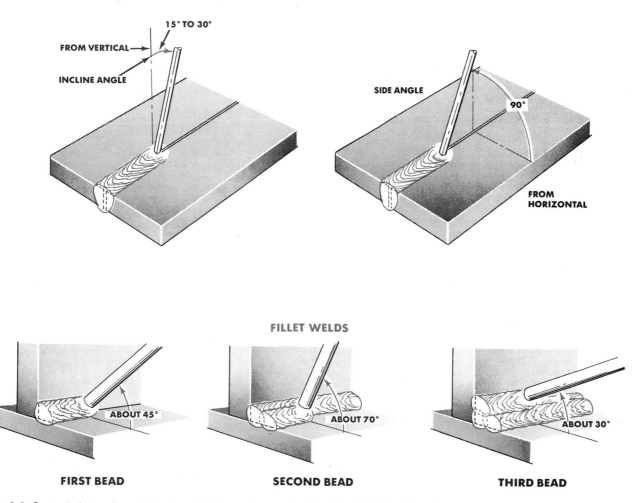

Fig. 8-3. Correct electrode angle is important to make good welds. (Republic Steel Corp.)

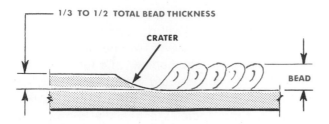

Fig. 8-4. This crater and bead show proper penetration.

thickness of the bead, depending upon the size of the electrode, as pictured in Fig. 8-4.

To secure a sound weld, the metal deposited from the electrode must fuse completely with the base metal. Fusion will result only when the base metal has been heated to a liquid state and the molten metal from the electrode readily flows into it. Thus, if the arc is too short, there will be insufficient *spread* of heat to form the correct size crater of molten metal. When the arc is too long, the heat is not centralized or intense enough to form·the desired crater.

Remelting and controlling a crater. An improperly filled crater may cause a weld to fail when a load is applied on a welded structure. Therefore, be sure to fill a crater properly. There is always a tendency when starting an electrode for a large globule of metal to fall on the surface of the plate with little or no penetration. This is especially true when beginning a new electrode

at the crater left from a previously deposited weld. To fill the crater and secure proper fusion, strike the arc approximately 1/2″ in front of the crater as shown at *A* in Fig. 8-5. Then bring it back through the crater to point *B* just beyond the crater, and weld back through the crater.

Occasionally, you will find that the crater is getting too hot and the fluid metal has a tendency to run. When this happens, lift the electrode slightly and quickly, and shift it to the side or ahead of the crater. Such a movement reduces the heat, allows the crater to solidify momentarily, and stops the deposit of metal from the electrode. Then return the electrode to the crater and shorten the arc.

Another method used by welders to control the temperature of the molten puddle is a *whipping* action of the electrode. This technique is especially helpful when welding pieces that do not have a tight fit and large openings have to be filled. It is also used in overhead and vertical welding to better control the weld puddle.

In a whipping action the electrode is struck and held momentarily. Then it is moved forward about 1/4″ or 3/8″ and raised a similar distance at the same time. Raising the electrode temporarily reduces the heat. Then just as the puddle begins to freeze the electrode is moved back into the center of the puddle and the sequences repeated. The movement of the electrode should be done by pivoting the wrist and not moving the arm while making the pass.

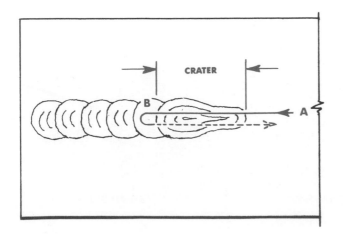

Fig. 8-5. Fill the crater by moving the electrode from A to B, and then back again through the crater to A.

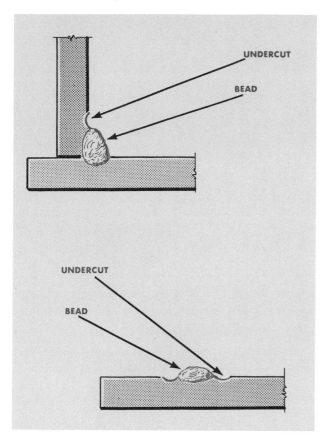

Fig. 8-6. Undercutting greatly weakens a weld.

Undercutting and Overlapping

Undercutting is a condition that results when the welding current is too high. The excessive current leaves a groove in the base metal along both sides of the bead which greatly reduces the strength of a weld as illustrated in Fig. 8-6. Undercutting may also occur when there is insufficient deposition of metal on the vertical plate. This can be corrected by slightly changing the electrode angle.

Overlapping occurs when the current is set too low. In this instance, the molten metal falls from the electrode without actually fusing with the base metal as shown in Fig. 8-7.

Cleaning a Weld

It has already been mentioned that when a weld is made, a layer of slag covers the deposited bead. If additional layers of weld metal are deposited, this slag must be removed; otherwise it will flow in with the deposited metal and cause a weak weld. To remove the slag, strike the weld with a chipping hammer. Hammer the bead so the chipping is away from the body, and away from the eyes and face as pictured in Fig. 8-8.

CAUTION: Always wear safety glasses when chipping. Do not pound the bead too hard; otherwise the structure of the weld may be damaged. After the slag is loosened, drag the

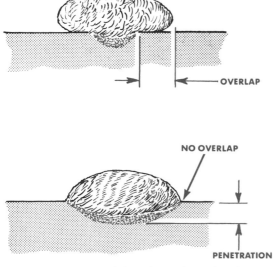

Fig. 8-7. When not enough heat is used, overlapping occurs as shown in top illustration. A satisfactory weld is shown at the bottom.

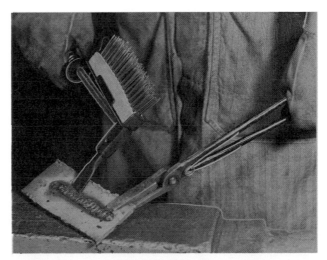

Fig. 8-8. Slag is removed from the weld by chipping.

Fig. 8-9. After chipping, brush the weld with a wire brush. (Hobart Brothers Co.)

pointed end of the hammer along the weld where it joins the plate. This will remove the remaining particles of slag. Follow the chipping with a good, hard brushing, using a stiff wire brush as illustrated in Fig. 8-9.

Welding Continuous Beads

Plate no. 1. After you have mastered the operation of depositing short beads, secure another steel plate $1/4'' \times 4'' \times 6''$. With a soapstone, draw a number of lines approximately $3/4''$ apart as shown in Fig. 8-10. Use a $1/8''$ diameter E-6010 or E-6011 electrode and run continuous beads on these lines, starting from the left edge and working to the right. After the plate is filled, remove the slag and examine the beads.

Plate no. 2. This exercise provides practice in moving the electrode in several directions.

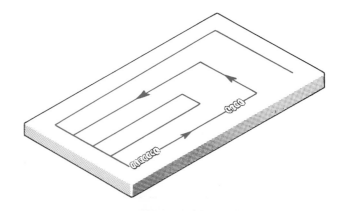

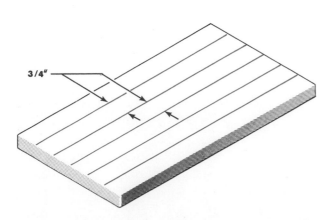

Fig. 8-10. Using a steel plate with lines $3/4''$ apart, deposit continuous beads from left to right.

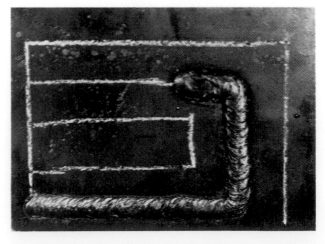

Fig. 8-11. To practice moving the electrode in several directions, deposit beads along the lines as indicated by the arrows (above) and the finished work (below).

Draw the lines on the plate as shown in Fig. 8-11. Then deposit a continuous bead, moving the electrode from left to right, bottom to top, right to left, and top to bottom.

Plate no. 3. The purpose of this plate is to develop skill in re-striking an arc while making a continuous bead. Draw a series of straight lines on the plate and divide these lines into 2-inch sections as shown in Fig. 8-12. Run a bead over the first line but break the arc when reaching the end of the 2-inch mark. Re-start the arc and continue the deposit for another two inches; then repeat the practice of breaking the arc and refilling the crater. Follow this procedure until skill is mastered in depositing uniform and continuous beads with properly filled craters.

Weaving the Electrode

Weaving is a technique used to increase the width and volume of the bead. Enlarging the size of the bead is often necessary on deep groove or fillet welds where a number of passes must be made. Fig. 8-13 illustrates several weaving patterns. The pattern used depends to some extent

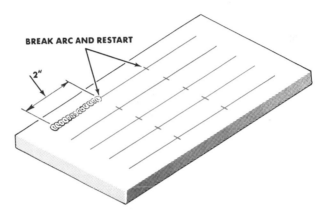

Fig. 8-12. To develop skill in re-starting an arc, break the arc and re-start every two inches.

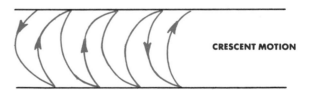

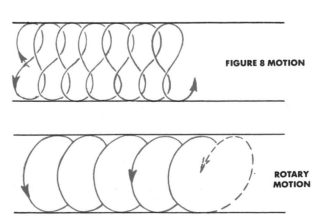

Fig. 8-13. The crescent, figure-8, and rotary motions are three typical weaving patterns.

on the position of the weld. In subsequent chapters additional instructions will be given as to the most suitable weaving pattern for specific weld joints.

Plate no. 4. Lay out a series of straight lines on a 1/4" × 4" × 6" plate. Run continuous beads over these lines and then clean each bead. Now

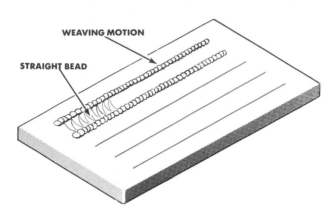

Fig. 8-14. Practice weaving by depositing a weld from left to right between the straight beads.

proceed to practice weaving by depositing a weld back and forth between the first pair of continuous beads as illustrated in Fig. 8-14. Use one type of weaving motion to fill the first space and then try several other weaves on the remaining sections. Make certain that the short beads are fused into the long, straight beads. Continue the weaving practice on several plates until a workmanlike job is accomplished.

Plate no. 5, depositing a weld metal pad. *Padding* or surfacing is a process for building worn surfaces of shafts, wheels, and other machine parts. The operation consists of depositing several layers of beads, one on top of the other as pictured in Fig. 8-15. For this exercise divide the standard 4" × 6" plate into three sections as shown in Fig. 8-16. Fill the first section completely with beads, clean the weld thoroughly, and then deposit a second layer of weaving beads about 1/2" wide at right angles to the first layer. After the weld is cleaned, deposit a third layer at right angles to the second layer. Clean

Fig. 8-15. This shaft is being built up by padding. (Hobart Brothers Co.)

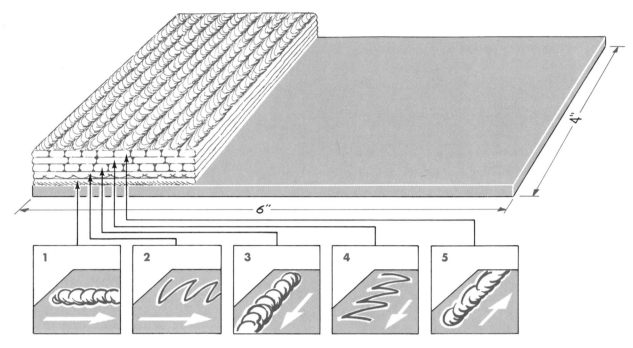

Fig. 8-16. A weld metal pad is built up by depositing successive layers of beads on top of each other.

this surface and deposit a fourth and a fifth layer. Follow a similar procedure to build up the two remaining sections on the plate.

Points to Remember

1. Use the correct type of electrode.

2. Use an arc that is about as long as the diameter of the electrode.

3. Maintain the correct welding current.

4. Move the electrode just fast enough to produce evenly spaced ripples.

5. Keep the penetration to a depth equal to one-third to one-half the total thickness of the weld bead.

6. Be sure the molten metal from the electrode fuses completely with the base metal.

7. Always re-start the electrode $1/2''$ in front of the previously made crater.

8. Avoid undercutting and overlapping.

9. Clean the slag from a weld. Remember, chip away from your body to prevent the slag from flying up into your face.

QUESTIONS FOR STUDY AND DISCUSSION

1. What causes a bead to overlap the base metal? Why does this usually indicate a poor weld?

2. What causes undercutting? How can undercutting be avoided?

3. What is meant by a crater?

4. What should be the depth of penetration? How thick should the bead be?

5. How is a crater affected when the arc is too short? What happens when the arc is too long?

6. What are the five essentials for securing a sound weld?

7. What factors must be taken into consideration when selecting an electrode?

8. When an arc is too long what happens to the metal as it melts from the electrode?

9. How is it possible to identify a weld that has been made with an arc that is too long?

10. What is likely to happen to the electrode when the arc is too short?

11. What are some of the characteristics of a weld made with an arc that is too short?

12. What are some of the factors that you must think of in deciding the length of an arc?

13. In what way does the amount of current affect a weld?

14. What determines the speed at which an electrode should be moved?

15. How should an electrode be restarted to fill a crater left from a previously deposited weld?

16. What should be done when the crater gets too hot and the metal has a tendency to run over the surface?

17. How should slag from a weld be removed from a workpiece?

18. What is meant by weaving?

19. When is a weaving motion used?

20. What is the purpose of padding?

shielded metal-arc welding

CHAPTER 9 the flat position

Although welding can be done in any position, the operation is simplified if the joint is flat. When placed in this way, the welding speed is increased, the molten metal has less tendency to run, better penetration can be secured, and the work is less fatiguing. Some structures may appear at first glance to require horizontal, vertical, or overhead welding, but upon more careful examination you may be able to change them to the easier and more efficient flat position. See Fig. 9-1.

Fig. 9-1. Welding with work in the flat position is easier and results in better quality at greater economy. (The Lincoln Electric Co.)

WELD PASSES

In carrying out some welding operations, very often the pieces have to be tack-welded. Tack welds are simply short sections of weld beads $1/4''$ to $1/2''$ long used to maintain the proper root opening between the two sections of metal being welded. See Chapter 3. These tack welds are spaced along the seam and must be firmly fused into the joint.

Once the joint is tacked, the remaining weld passes are deposited. The first pass, known as the *root pass* or *stringer bead* is a narrow bead laid in the bottom of the V. It is made with a small diameter electrode by moving it straight down into the groove without any weaving motion. See Fig. 9-2. Its principal function is to fill the root opening and join the two metal sections. Since it serves as the base for the other passes it is very important that it produces complete penetration. Complete penetration is more assured if the root pass penetrates the bottom surface of the groove but not more than $1/16''$, and fuses all tack welds previously made.

The next layer is called the *fill* or *filler pass*. One or more filler passes may be needed to fill the groove, depending on the thickness of the metal. In depositing the filler passes a slight weaving motion is generally advisable to insure proper fusion to the previously laid beads and sides of the groove joint.

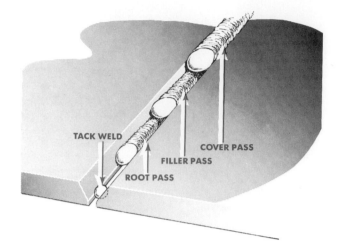

Fig. 9-2. Types of weld passes.

The final pass called the *cap* or *cover pass* (sometimes referred to as the *wash bead*), is intended to provide additional reinforcement to the weld and give it a nice appearance. The cover pass should not extend more than $1/16''$ above the plate surface. Inasmuch as it has to extend completely over and beyond the filler passes, a weaving motion is necessary to cover the wider area.

How to Weld a Butt Joint

The butt joint is often used when structural pieces have flat surfaces, such as in tanks, boilers, and a variety of machine parts. See Fig. 9-3. The joint may be opened, closed, or the edges beveled, as in Fig. 9-4. A *closed butt joint* has the edges of the two plates in direct contact with each other. This joint is suitable for welding steel plates that *do not exceed $1/8''$ to $3/16''$ in thickness.* Heavier metal can be welded but only if the machine has sufficient amperage capacity and if heavier electrodes are used. *Remember that on thick metal it is difficult to secure ample penetration to produce a strong weld by a single pass bead.*

In the *open butt joint* the edges are placed slightly apart, usually $3/32''$ to $1/8''$, to allow for expansion. As a rule, a back-up strip or block of scrap steel, or copper, is placed under this joint as shown in Fig. 9-5. A back-up strip prevents the bottom edges from burning through.

Fig. 9-3. The narrow section in this tank is welded with a butt joint. (Hobart Brothers Co.)

When the thickness of the metal exceeds $1/8''$ the edges of a butt joint should be beveled. The beveling can be done by cutting the edges with a flame torch or by grinding them on an emery wheel. The included angle of the V should not exceed 60°, to limit the amount of contraction that usually results when the metal cools. The edges may be shaped in several ways as shown

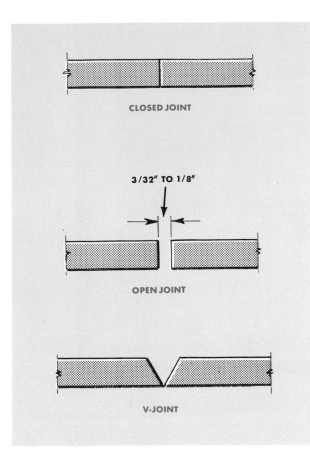

Fig. 9-4. There are three types of butt joints.

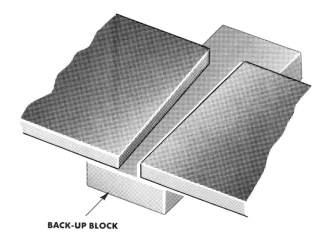

Fig. 9-5. A back-up strip or block should be used on an open butt joint.

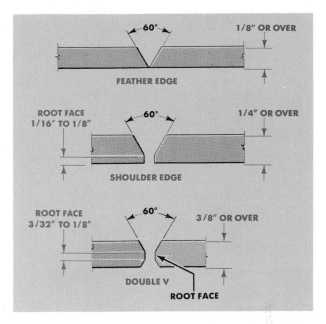

Fig. 9-6. Here are the three methods of preparing a V-joint.

in Fig. 9-6. Notice that on heavy metal ³/₈″ or more in thickness the edges are beveled on both sides. Beveling in this manner will insure better penetration, requires less weld metal, and contraction forces are better equalized.

Welding procedure. Fig. 9-7 illustrates the position of the electrode for welding a butt joint. However, when the joint consists of two pieces of different thicknesses, adjust the position of the electrode so the greatest portion of the heat is concentrated on the thickest plate.

In welding thin stock with a single pass, as in a closed or open butt joint, simply allow the electrode to travel along the seam without any weaving motion. Move the electrode slow enough to give the arc sufficient time to melt the metal. Take care that the travel is not too slow because the arc will burn through the metal.

When a multiple pass is to be made in a grooved joint, be sure to hold the electrode down in the groove so it almost touches both sides of the joint while depositing the root pass. Move the electrode fast enough to keep the slag flowing back on the finished weld. If the electrode is not moved rapidly enough, the slag may become trapped in the bottom of the weld, thereby preventing proper fusion.

After completing the root pass proceed with whatever filler passes are required and then

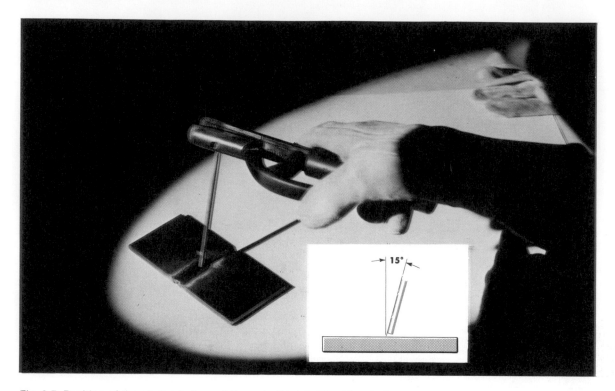

Fig. 9-7. Position of the electrode for welding a butt joint. (The Lincoln Electric Co.)

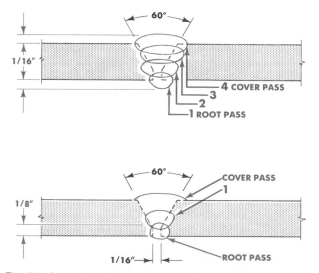

Fig. 9-8. Sequence of passes in a multiple pass butt joint.

Making a Single-Pass Fillet Lap Weld

The lap joint is one of the most frequently used joints in welding. It is a relatively simple joint, since no beveling or machining is necessary. One standard requisite is to have clean, evenly aligned surfaces. As shown in Fig. 9-9, the joint

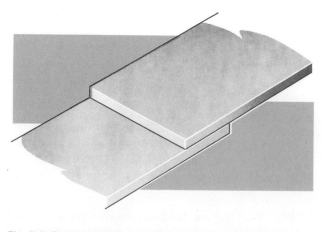

Fig. 9-9. The lap joint is frequently used in welding because it provides great strength.

complete the weld with a cover pass. See Fig. 9-8. Remember, always remove the slag completely after each pass. If any slag particles are allowed to remain they will weaken the weld.

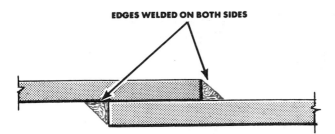

Fig. 9-10. For a strong joint weld both edges.

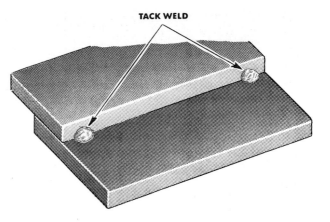

Fig. 9-11. Tack the plates at each end before starting to weld a lap joint.

consists of lapping one edge over another and joining. The amount the edges should overlap depends upon the thickness of the plates and the strength required of the welded piece. Usually the thicker the plates the greater the amount of overlap.

When the structure is subjected to heavy bending stresses, it is advisable to weld the edges of both sides of the joint as illustrated in Fig. 9-10.

The lap joint is adaptable for a variety of new construction work as well as for making numerous types of repairs. For example, such a joint can be employed when joining a series of metal plates together or in reinforcing another structural member. Since a lap weld stiffens the structure where the plates are lapped, this joint is used a great deal in tank and ship building.

To practice welding a single-pass lap joint, obtain two pieces of ³/₁₆″ or ¹/₄″ steel plate. Remember, a *single pass* simply means depositing one layer of beads. When a weld is built up of more than one layer it is known as a *multiple pass* weld. Use ¹/₈″ electrodes and adjust the machine for the correct current. Then tack the plates at each end. See Fig. 9-11.

With the plates properly tacked, run a ¹/₄″ fillet along the edge. Hold the electrode at a 45° angle and point it toward the weld as shown in Fig. 9-12. Weave the electrode slightly, maintaining the arc for a little longer period on the lower plate. See Fig. 9-13. Be sure to get complete fusion at the *root,* or joining point, of the joint and avoid overlaps on the top surface. Watch the crater carefully to prevent an undercut on the bottom plate. Fig. 9-14, *top,* illustrates a *properly made fillet weld* on a lap joint. A weld made as in

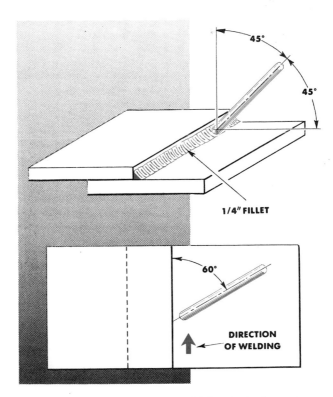

Fig. 9-12. For a sound lap weld, hold the electrode as shown in the above two views.

Fig. 9-14, *center,* usually is too weak because it lacks sufficient reinforcing material, while the weld shown in Fig. 9-14, *bottom,* has too much waste metal, which is of no value to the joint.

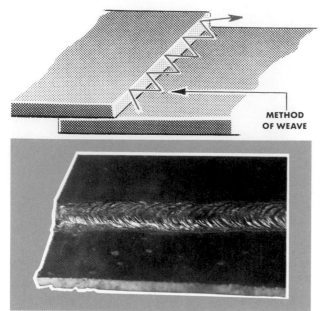

Fig. 9-13. Use this weave for a flat lap weld.

Making a Multiple-Pass Fillet Lap Weld

When an exceptionally strong lap joint is required, especially on heavy plates $3/8''$ and over in thickness, a multiple-pass fillet weld is recommended. This joint has two or more layers of beads along the seam, with each bead lapping over the other.

To make such a weld, deposit the first bead as shown in Fig. 9-15 by moving the electrode straight down the seam without any weaving motion. Clean the weld carefully and lay the second pass over this *stringer bead*. During the second pass, weave the electrode, pausing for an instant at the top of the weave to favor or deposit extra metal on the vertical edge of the upper plate.

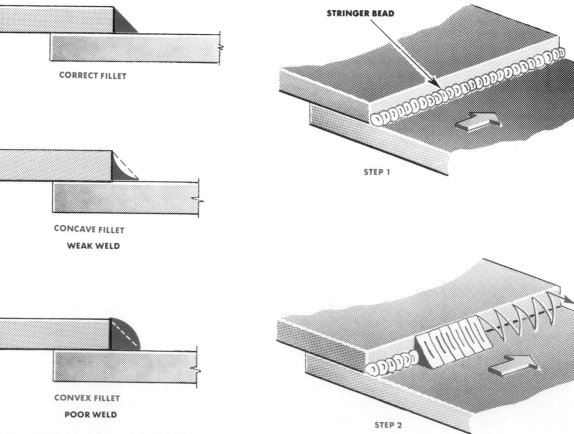

Fig. 9-14. How a good lap weld and two poor lap welds appear from a side view.

Fig. 9-15. Making a multiple pass fillet weld on a lap joint.

Making a Single-Pass T-Fillet Joint

The T-fillet joint is frequently used in fabricating straight and rolled shapes. See Figs. 9-16 and 9-17. The strength of this joint depends considerably on having the edges of the joint fit close together. The T-joint should not be used if

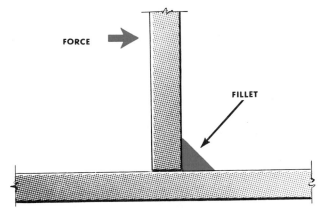

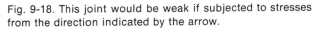

Fig. 9-18. This joint would be weak if subjected to stresses from the direction indicated by the arrow.

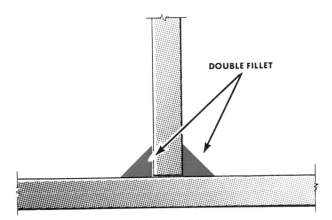

Fig. 9-19. A double fillet T-joint is stronger.

Fig. 9-16. Single pass T-fillet joint. (The Lincoln Electric Co.)

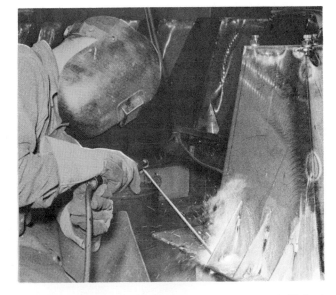

Fig. 9-17. Example of work where a T-joint is used. (Hobart Brothers Co.)

it is subjected to heavy stresses from the opposite direction of the welded seam. This weakness can be partially overcome by using a double fillet—that is, welding both sides of the joint. See Figs. 9-18 and 9-19.

To practice welding a T-fillet joint, obtain two $3/16''$ or $1/4''$ plates. Set the vertical plate on the middle of the horizontal plate and tack weld each end. Deposit a $1/4''$ fillet bead along the edge. Hold the electrode as shown in Fig. 9-20 and advance it in a straight line without any weaving motion. Point the tip of the electrode toward the completed portion of the weld and travel rapidly enough to stay ahead of the molten pool. Concentrate the arc more on the lower plate to prevent undercutting the upper plate. Watch the crater closely so it will form a bead with the correct contour.

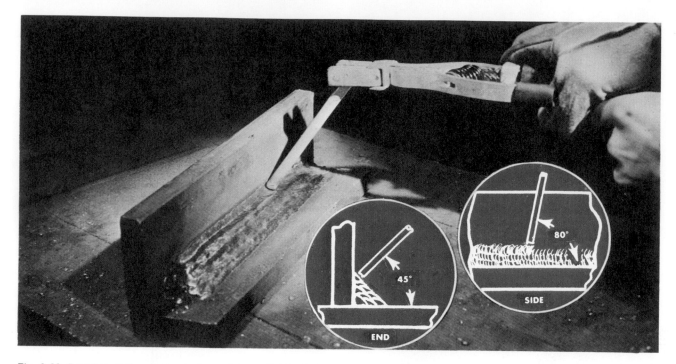

Fig. 9-20. Position of the electrode for welding a T-fillet joint. (The Lincoln Electric Co.)

Making a Multiple-Pass T-Fillet Joint

When a very strong T-joint is required, make a wider fillet along the seam. You can get a wider fillet by running several layers of beads as illustrated in Fig. 9-21. Deposit the first bead as described in making a single pass T-fillet joint. Remove the slag and lay the second bead over the first, weaving the electrode sufficiently to secure the desired width fillet. Deposit additional layers if necessary to get the right size fillet, but

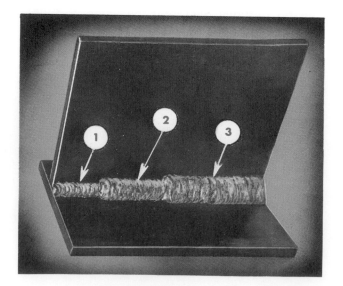

Fig. 9-21. A multiple-pass T-fillet joint.

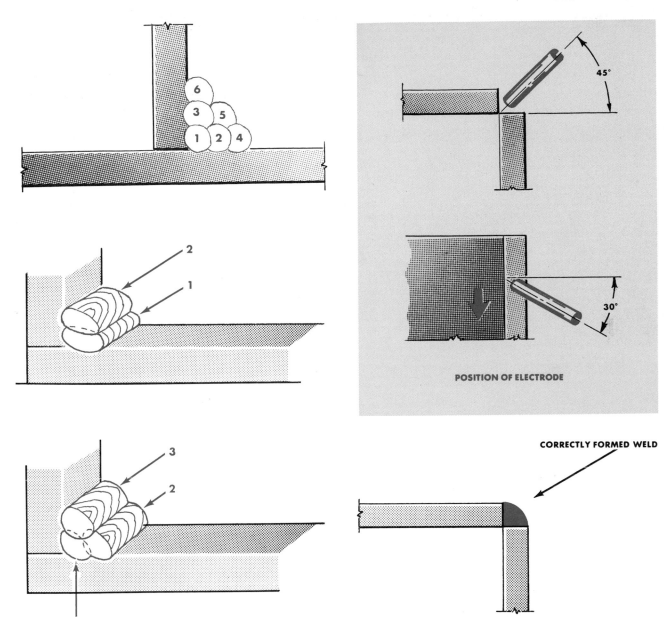

Fig. 9-22. Sequence of passes for a multiple T-fillet weld.

Fig. 9-23. An outside corner weld. The angle of the electrode is viewed from the side (top), and overhead (bottom).

be sure to clean off the slag after each pass. See Fig. 9-22.

How to Make an Outside Corner Weld

The outside corner weld, as shown in Fig. 9-23, is often used in constructing rectangular shaped objects such as tanks, metal furniture, and other machine sections where the outside corner must have a smooth radius. See Fig. 9-24.

To make an outside corner weld, tack the two plates and run a bead along the edge with the electrode held as indicated in Fig. 9-23. On light stock, one bead is usually enough. Heavy stock will probably require a series of passes to fill in the corner.

Fig. 9-24. An outside corner weld is used in fabricating this structure. (Hobart Brothers Co.)

To consider some of the problems that may be encountered look to the following *Check Chart.* Characteristic problems, their causes and the probable remedy may be found for arc welding procedures.

How to Weld Round Stock

A butt joint is generally used to weld round shafts or rods. First bevel both sides of the stock, leaving a shoulder in the center. Be sure to grind the edges so they have the same angle. Then

ARC WELDING PROBLEMS

CHARACTERISTICS	CAUSE	REMEDY
1. Unstable arc, arc goes out. Spatters over work.	Arc too long.	Shorten arc for proper penetration.
2. No penetration. Arc goes out often.	Not enough current for size of electrode.	Increase current. Use smaller electrode.
3. Loud crackling from arc. Flux melts too rapidly. Wide and thin bead, spatter in large drops.	Too much current for electrode. May be moisture in electrode cover.	Decrease current. Use larger electrode.
4. Weld remains in balls.	Wrong electrode.	Use proper electrode.
5. Difficulty in striking arc. Poor penetration.	Wrong polarity. Too little current.	Change polarity. Or, increase current.
6. Weak weld. Arc hard to start. Arc keeps breaking.	Dirty work.	Clean work. Remove slag from previous weld.
7. Arcing at ground clamp.	Poor ground.	Correct poor ground.

Fig. 9-25. The edges of the round stock are beveled on both sides and held in place with a piece of angle iron. (The Lincoln Electric Co.)

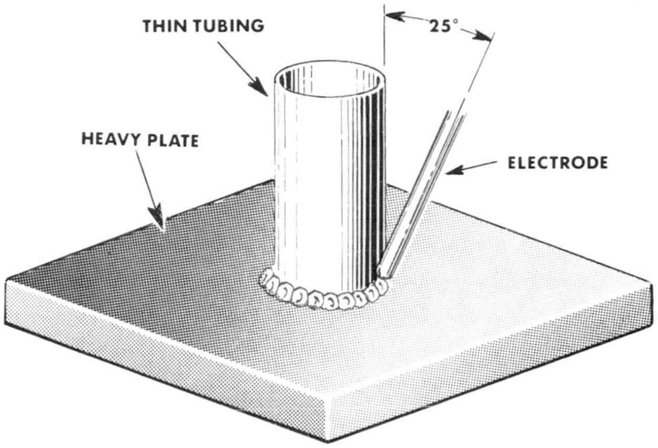

Fig. 9-26. Position of the electrode when welding pipe to a flat plate.

place the pieces in a vise or section of angle iron, as shown in Fig. 9-25, to hold them in position. To prevent the shaft from warping, deposit a small bead on one side and then lay a similar bead on the opposite side. Use a slight weaving motion on the final pass.

Sometimes you may have to weld a thin-walled pipe to a heavier flat plate. If you change the position of the electrode from the usual angle to approximately 25°, you can direct more heat on the flat plate and thus prevent the pipe from burning through. See Fig. 9-26.

Points to Remember

1. When making a root pass use smaller diameter electrodes and do not use any weaving motion while making the pass.

2. Use a slight but not heavy weaving motion in making a filler pass.

3. Place the edges of an open butt joint slightly apart to allow for joint expansion.

4. When the thickness of the metal exceeds $1/8''$ on a butt joint, always bevel the edges.

5. When welding a lap or T-joint, weld the edges on both sides if the structure is subjected to heavy stresses.

6. Tack the two pieces before starting to weld. This will keep them in position.

7. Hold the electrode at a 45° angle when welding a lap joint, keeping the arc more on the upper plate.

8. For an exceptionally strong lap or T-joint on heavy metal, use several layers of beads.

9. When a butt joint is used to weld a round shaft, bevel both sides to the same angle.

10. To weld a thin-walled pipe to a heavy flat plate, change the angle of the electrode so the arc is directed more on the plate.

QUESTIONS FOR STUDY AND DISCUSSION

1. What is the function of a root pass?

2. How many filler passes are used on a groove joint?

3. What is a cover pass and why is it used?

4. What is an advantage of welding in a flat position rather than in an overhead or vertical position?

5. What is meant by a single pass fillet weld?

6. When making a lap weld, how much should the edges overlap?

7. How is it possible to avoid undercutting when welding a lap joint?

8. What is the purpose of using a double fillet on a lap joint?

9. When should a multiple bead be used on a lap joint?

10. What are some of the factors that must be considered when using or making a T-joint?

11. When welding a T-joint, why should the arc favor the bottom plate?

12. How many passes should be made on an outside corner weld?

13. When is a butt joint used in welding?

14. What is the difference between an open and closed butt joint?

15. When should the edges of butt joints be beveled?

16. How should the edges of round stock be prepared for welding?

17. In what position should the electrode be held when welding a thin-walled pipe to a heavier flat plate?

CHAPTER 10 the horizontal position

On many jobs it is practically impossible to weld pieces in the flat position. Occasionally the welding operation must be done while the work is in a horizontal position. A weld is in a horizontal position when the joint is on a vertical plate and the line of weld runs on a line with the horizon as in Fig. 10-1.

To perform welds of this kind, you must use a slightly shorter arc at a slight reduction in amperage setting than you would for flat position welding. The shorter arc will minimize the tendency of the molten puddle to sag and cause overlaps. An overlap occurs when the puddle

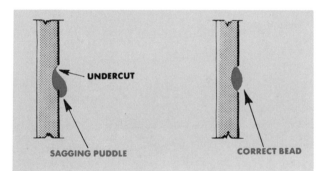
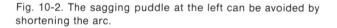

Fig. 10-2. The sagging puddle at the left can be avoided by shortening the arc.

runs down to the lower side of the bead and solidifies on the surface without actually penetrating the metal. See Fig. 10-2. A sagging puddle usually leaves an undercut on the top side of the seam as well as improperly shaped beads, all of which weaken a weld.

HOW TO HOLD THE ELECTRODE

For horizontal welding, hold the electrode so that it points upward 5° to 10°, and slants approximately 20° away from the deposited bead as illustrated in Fig. 10-3. In laying the bead, use a narrow weaving motion as shown in Fig. 10-4. By weaving the electrode, the heat will be distributed more evenly, thereby reducing still further any tendency for the puddle to sag. Keep the arc

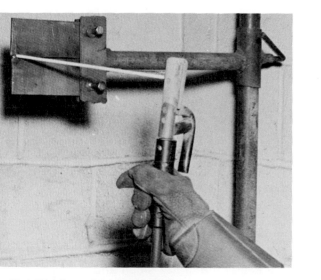

Fig. 10-1. Welding a horizontal seam.

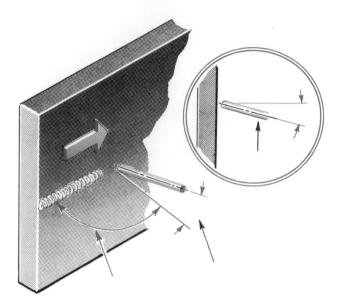

Fig. 10-3. Position of the electrode for horizontal welding.

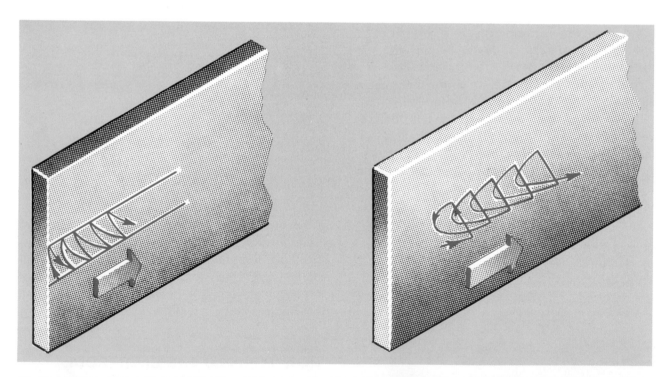

Fig. 10-4. The weaving pattern at left will result in a normal width bead, while the pattern at the right will result in a wider bead width.

as short as possible. If the force of the arc has a tendency to undercut the plate at the top of the bead, drop the electrode a little to increase the upward angle.

As the electrode is moved in and out of the crater, pause slightly each time it is returned. This keeps the crater small and the bead is less likely to drop.

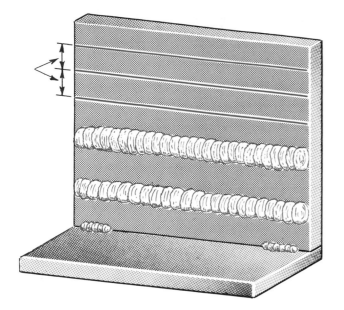

Fig. 10-5. Tack the plate so it is in a vertical position.

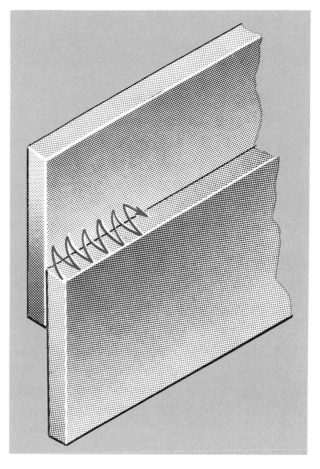

Fig. 10-6. This is how to make a single-pass lap weld.

Depositing Straight Beads in a Horizontal Position

1. Obtain a ¹/₄″ plate and draw a series of lines ¹/₂″ apart.

2. Fasten the plate on the bench in a vertical position as pictured in Fig. 10-5. To keep the piece in place, tack it to a flat piece.

3. Adjust the machine to the correct current and, with a weaving motion, deposit beads between the horizontal lines. Lay one bead by starting from the left side of the plate and working to the right. Then reverse the direction and run the bead from the right side to the left. Continue this operation until uniform beads can be made without overlapping and undercutting.

Making a Single-Pass Lap Joint in a Horizontal Position

1. Tack two ¹/₄″ plates to form a lap joint. Clamp the piece in a vertical position as shown in Fig. 10-6.

2. Run a single bead along the edge, using a slight weaving motion.

Watch the surface of the top plate closely to prevent any undercutting. Continue this operation on additional lap joints until a satisfactory weld is made.

Welding a Multiple-Pass Fillet T-joint in a Horizontal Position

1. Tack two plates to form a T-joint and fasten the base plate in a vertical position as illustrated in Fig. 10-7.

2. Run a bead along the root of the joint without any weaving motion. Clean the slag and deposit a second bead, using a slight weaving motion, penetrating the first bead and the plate. See Fig. 10-7.

3. Clean the slag off the second bead and deposit a third layer. Notice that the third bead penetrates into the first and second layers as well as into the upright plate. This penetration is important; otherwise a weak weld will result, and the layers may separate from each other.

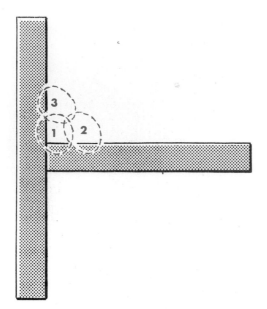

Fig. 10-7. Depositing a multiple bead on a T-joint.

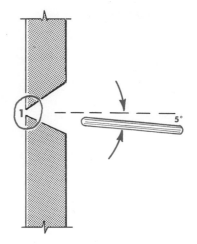

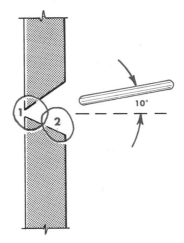

Welding a Multiple-Pass Butt Joint in a Horizontal Position

1. Obtain two pieces of ¹/₄″ steel plate and *bevel the edge of one plate.*

2. Tack the two plates together to form a butt joint, allowing ¹/₁₆″ space at the root opening. Fasten the plates in a vertical position with the beveled plate on top as shown in Fig. 10-8. The

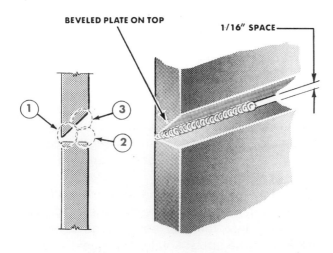

Fig. 10-8. Position of the butt joint for horizontal welding.

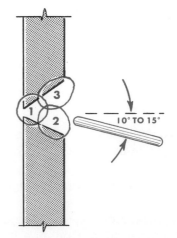

Fig. 10-9. Position of the electrode in horizontal butt welding.

plate that is not beveled should be on the bottom since its flat edge serves as a shelf, thus helping to prevent the molten metal from running out of the joint.

3. Deposit the first bead deep in the root of the joint. Remove the slag and lay the second bead. Then follow with a third bead.

4. On many welding jobs, the practice is to bevel both edges of the joint to form a 60° included angle. Since such a joint does not provide a retaining shelf for the bead as the one shown in Fig. 10-8, a little more skill is required to produce a satisfactory weld. In practicing this type of weld, notice in Fig. 10-9 the position of the electrode is changed.

The number of passes on the joint will depend on the thickness of the metal as well as the diameter of the electrode. The important thing is to secure sufficient penetration into each adjacent layer. It is common practice on a wide joint to finish the weld with a *wash bead,* commonly known as a *cover* or *cap pass,* as illustrated in Fig. 10-10, to produce a smooth finish. A wash bead is made by using a wide weaving motion that covers the entire area of the deposited beads.

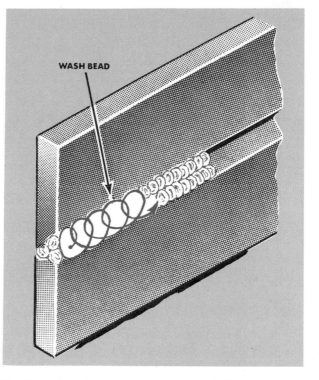

Fig. 10-10. A wash bead produces a smooth finish.

Points to Remember

1. Use a lower welding current and shorter arc for horizontal welding.

2. For horizontal welding, tilt the electrode upward 5° to 10° and slant it slightly away from the weld.

3. Use a slight weaving motion.

4. Do not allow the molten pool to sag and cause overlaps.

5. On a multiple pass weld, use a wash bead on the final pass.

QUESTIONS FOR STUDY AND DISCUSSION

1. Why is it essential to use a lower current and a shorter arc when welding in a horizontal position?

2. What can be done to avoid overlaps on horizontal welds?

3. In what position should the electrode be held for horizontal welding?

4. Why should a weaving motion be used when making horizontal welds?

5. What determines the number of passes that should be made on a weld?

6. What practice should be followed in beveling the edges for a multiple pass weld?

7. Where and why is a cover pass used?

CHAPTER 11 the vertical position

In the fabrication of many structures such as steel buildings, bridges, tanks, pipelines, ships, and machinery, the operator must frequently make vertical welds. A vertical weld is one with a seam or line of weld running up and down as shown in Fig. 11-1.

One of the problems of vertical welding is that gravity tends to pull down the molten metal from the electrode and plates being welded. To prevent this from happening, fast-freeze types of electrodes should be used. Puddle control can also be achieved by proper electrode manipulation and selecting electrodes specifically designed for vertical position welding.

POSITION AND MOVEMENT OF THE ELECTRODE

Vertical welding is done by depositing beads either in an upward or downward direction, (sometimes referred to as *uphill* and *downhill*) *Downward welding* is very practical for welding light gage metal because penetration is shallow, thereby forming an adequate weld without burning through the metal. Moreover, downward welding can be performed much more rapidly, which is important in production work. Although it is generally recommended for welding lighter materials it can be used for most metal thicknesses.

Fig. 11-1. After tacking the metal strips together, this operator lays vertical welds. (Hobart Brothers Co.)

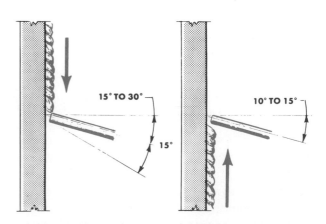

Fig. 11-2. Position of the electrode for downward (left) and upward (right) vertical welding.

On heavy plates of ¼″ or more in thickness, *upward welding* is often more practical, since deeper penetration can be obtained. Welding upward also makes it possible to create a shelf for successive layers of beads.

For downward welding, tip the electrode as in Fig. 11-2 left. Start at the top of the seam and move downward with little or no weaving motion. If a slight weave is necessary, swing the electrode so the crescent is at the top.

For upward welding, start with the electrode at right angles to the plates. Then, lower the rear of the electrode, keeping the tip in place, until the electrode forms an angle of 10°–15° with the horizontal as shown in Fig. 11-2 right.

Laying Straight Beads in Vertical-Downhill Method

Set up a practice piece in a vertical position with a series of straight lines drawn on it. Start at the top of the plate with the electrode pointed upward about 60° from the vertical plate. Keep the arc short and draw the electrode downward to form the bead. Travel just fast enough to keep the molten metal and slag from running ahead of the crater. Do not use any weaving motion to start with. Once this technique is mastered try weaving the electrode but very slightly with the crest at the top of the crater. See Fig. 11-3.

Laying Straight Beads in a Vertical Position—Uphill Method

1. Obtain a ¼″ plate and draw a series of straight lines. Then fasten the piece so the lines are in a vertical position.

2. Strike the arc on the bottom of the plate. As the metal is deposited, move the tip of the electrode upward in a rocking motion as shown in Fig. 11-4. This is often called a whipping motion. In rocking the electrode, do not break

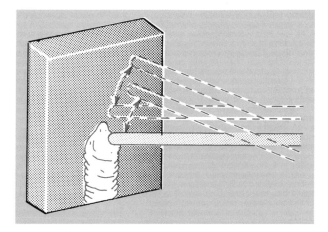

Fig. 11-4. A whipping motion helps to control the puddle in uphill welding.

the arc but simply pivot it with a wrist movement so the arc is moved up ahead of the weld long enough for the bead to solidify. Then return it to the crater and repeat the operation, working up along the line to the top of the plate. *Remember, do not break the arc while moving the electrode upward.* Withdraw it just long enough to permit the deposited metal to solidify and form a shelf so additional metal can be deposited. Continue to lay beads from bottom to top until each line is smooth and uniform in width.

Laying Vertical Beads with a Weaving Motion

On many vertical seams in uphill welding it is necessary to form beads of various widths. The width of the bead can be controlled by using one of the weaving patterns shown in Fig. 11-5. Each

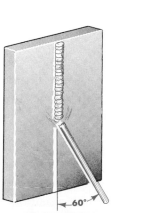

DOWNHILL WELDING WITH NO WEAVE MOTION **DOWNHILL WELDING WITH SLIGHT WEAVE MOTION**

Fig. 11-3. Downhill welding methods.

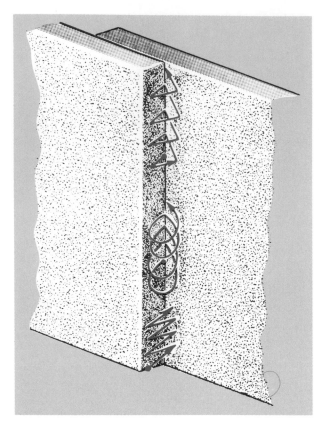

Fig. 11-5. Weaving patterns can be used to vary bead width.

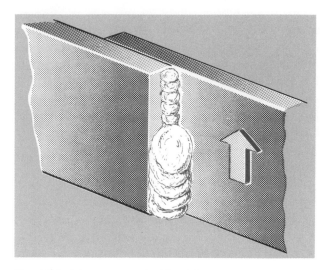

Fig. 11-6. A two-pass vertical lap joint.

pattern will produce a bead approximately twice the diameter of the electrode. Notice that each weave is shaped so the electrode can dig into the metal at the bottom of the stroke, and the upward motion momentarily removes the heat until the metal can solidify. When a smooth weld is required on the final pass of a wide joint, the *wash bead* should be used.

Before applying these weaving motions on actual joints, practice running them on a vertical plate. Continue this practice until a smooth bead of uniform width can be deposited.

Welding a Vertical Lap Joint

1. Obtain two ¼″ plates and tack them together to form a lap joint. Support the joint in a vertical position by tacking it to a flat scrap piece.

2. Deposit a small stringer bead in the root without any electrode motion.

3. Lay an additional layer as illustrated in Fig. 11-6. Use a weaving motion and work from the bottom to the top. Make certain that the second bead is thoroughly fused with the first.

Welding a Vertical Butt Joint

1. Obtain two ¼″ plates and bevel the edges to form a 60° V.

2. Tack the plates together with a ¹/₁₆″ root opening and fasten them in an upright position to provide a vertical butt joint.

3. Deposit a straight bead in the root opening and follow with additional layers as illustrated in Fig. 11-7. Be sure to clean off the slag after each bead and test to make certain that there is good fusion between the beads and sides of the V by bending it in a vise.

4. Try another butt weld, using either ³/₈″ or ¹/₂″ plate. Bevel the edges to form a 60° V with a ¹/₈″ root face. Tack the pieces together leaving a ¹/₈″ root opening. Support the joint in a vertical position with a backing strip. Deposit the necessary layer of beads and finish with a cover pass.

Welding a Vertical T-Joint

1. Obtain two ¼″ plates and tack them to form a T-joint. Support the joint in a vertical position.

2. Deposit a narrow, straight bead in the root.

3. Remove the slag and deposit at least one or two additional layers as shown in Fig. 11-8.

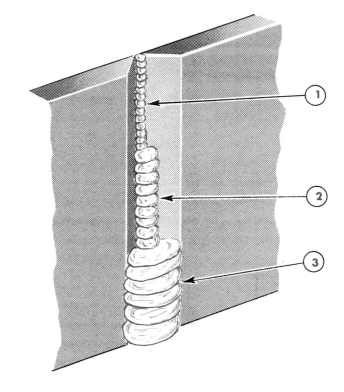

Fig. 11-7. Several layers of beads may be used on a vertical butt joint.

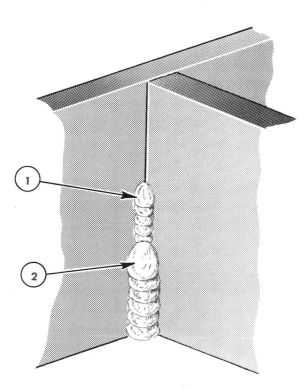

Fig. 11-8. A strong T-joint is attained by depositing 2 (left) and 3 beads.

Welding Technique for E-70XX Class Electrodes

Although the vertical techniques as previously described will generally prevail for all types of electrodes, sometimes a slight modification in procedure is advisable when using E-70xx class electrodes.

On downhill welding, drag the electrode lightly with a very short arc. Avoid a long arc since the weld depends on the molten slag for shielding. Stringer beads or small weaves are always preferred to wide weave passes. Use a lower current with DC than AC. Point the electrode directly into the joint and tip it forward only a few degrees in the direction of travel.

With uphill welding a triangular weave motion will often produce better results. Do not use a whipping motion or take the electrode out of the molten pool. Point the electrode directly into the joint and slightly upward to permit the arc force to assist in controlling the puddle. Use a current in the lower level of the recommended range.

Points to Remember

1. For welding light gage metal in a vertical position, the downward technique is more practical then upward.

2. For plates $1/4''$ or more in thickness, better results will be obtained by using the upward method of welding.

3. Rocking the electrode will often provide better control of the molten puddle in upward welding.

4. On grooved joints, always lay the first bead deep into the root opening.

QUESTIONS FOR STUDY AND DISCUSSION

1. In vertical welding, what can be done to prevent the molten puddle from sagging?

2. Why is welding downward on a vertical joint more applicable on light gage metal?

3. In what position should the electrode be held in downhill welding?

4. What motions should be used in downhill welding?

5. How should the electrode be held in making a vertical upward weld?

6. In laying a bead on a vertical seam, what is the advantage of applying a whipping action?

7. How can the width of a bead be controlled on an uphill weld?

8. How does welding with E-70xx electrodes differ from E-60xx electrodes?

CHAPTER 12 the overhead position

Welding in an overhead position is probably the most difficult operation to master. It is difficult because you must assume an awkward stance, and at the same time work against gravity, which exerts a downward force. See Fig. 12-1. In an overhead position the puddle has a tendency to drop, making it harder to secure uniform beads and correct penetration. Nevertheless, with a little practice it is possible to secure welds as good as those made in other positions.

POSITION FOR OVERHEAD WELDING

In learning to weld in an overhead position, some kind of fixture will be needed. The one shown in Fig. 12-2 is recommended, since the work can be adjusted to any height or position.

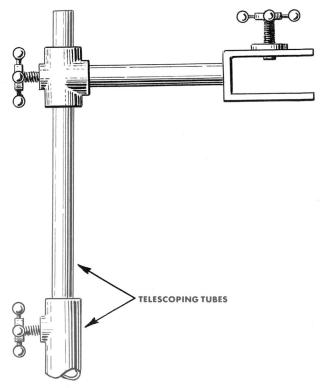

TELESCOPING TUBES

Fig. 12-1. Although it is a difficult position in which to work, overhead welding is often necessary.

Fig. 12-2. A jig is necessary to raise the work to the correct height and position.

119

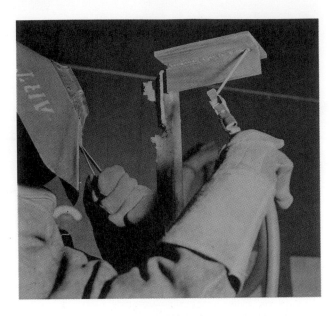

Fig. 12-3. Position of the holder and hand for overhead welding. (The Lincoln Electric Co.)

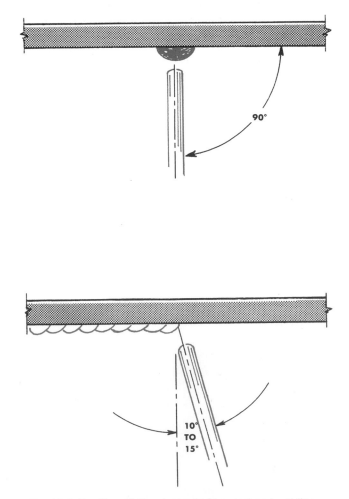

Fig. 12-4. Position of the electrode for overhead welding.

Caution: Since there is a possibility of some falling molten metal, one's clothing should be checked very carefully. Be sure the sleeves and pant cuffs are rolled down and a protective garment with a tight-fitting collar is zipped or bottoned up to the neck. Also it is a good idea to wear a cap and heavy duty shoes.

To start welding, place the electrode in the holder as illustrated in Fig. 12-3. Hold the electrode at right angles to the seam. Then tilt the rear of the electrode away from the crater until the electrode forms an angle of 10° to 15° with the horizontal as shown in Fig. 12-4. The line of weld may be in any direction—forward, backward, left, or right.

Grip the holder so the knuckles are up and the palm down. This prevents particles of molten metal from being caught in the hollow palm of the glove and allows the spatter to roll off the glove. Although the electrode can be held in one hand, sometimes it is better if it is held with both hands. See Fig. 12-5. To gain as much protection as possible from falling sparks and hot metal drippings, stand to the side rather than directly underneath the arc. The discomfort of the cable can be minimized by dropping it over the shoulder if you are welding in a standing position, or over the knees if in a sitting position. See Figs. 12-6 and 12-7.

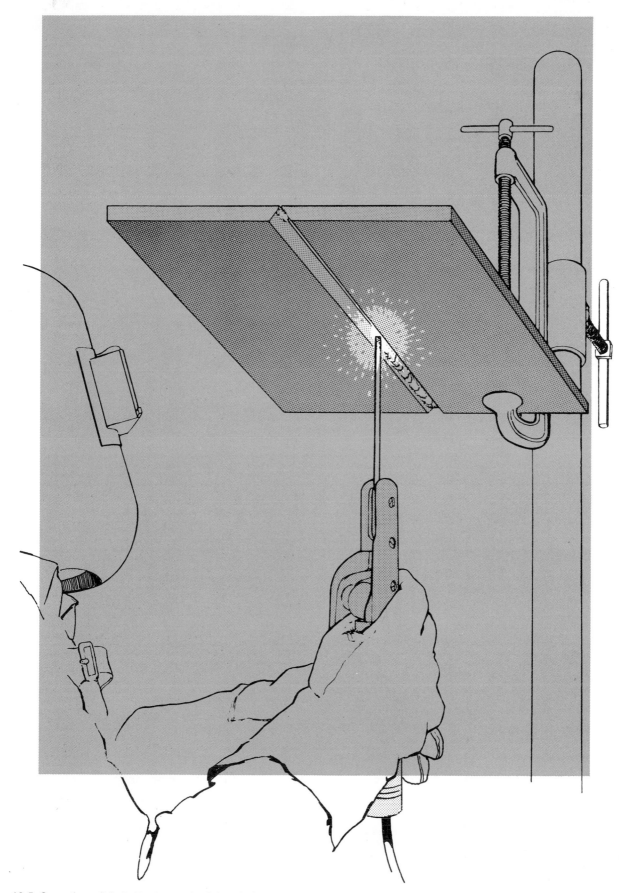

Fig. 12-5. Sometimes it is better to use both hands in overhead welding.

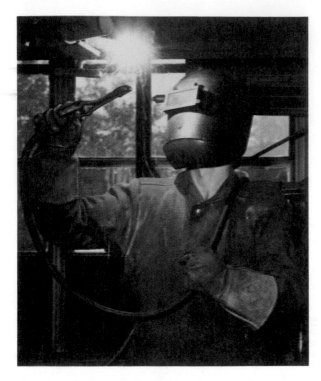

Fig. 12-6. If you are standing, drape the cable over your shoulder. (Hobart Brothers Co.)

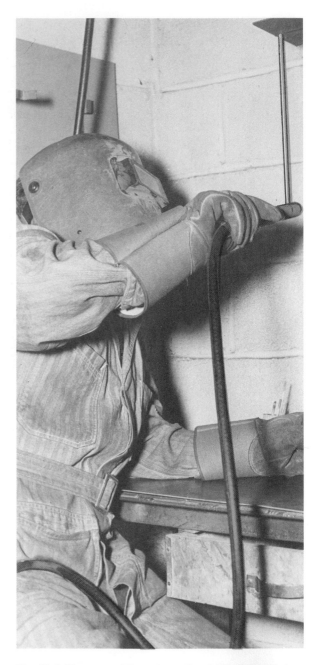

Fig. 12-7. If you are sitting, drape the cable over your knees.

Running Straight Beads in an Overhead Position

1. Clamp a ¼″ plate in the overhead jig.

2. Strike the arc and form a puddle as in flat position welding. Move the electrode forward keeping the arc as short as possible.

3. Run a series of straight beads without any weaving motion. To prevent the puddle from dropping, reduce the amperage slightly.

4. Continue to deposit straight beads until proper control of the puddle is mastered. Practice running the beads in one direction and then in another.

5. Obtain another ¼″ plate and practice weaving the arc as shown in Fig. 12-8.

Welding a Lap Joint in an Overhead Position

1. Tack two ¼″ plates to form a lap joint and clamp them in the overhead jig.

2. Hold the electrode so it bisects the angle between the plates and is inclined slightly away from the crater as illustrated in Fig. 12-9.

3. Lay the first bead deep in the root of the joint.

4. Remove the slag and deposit the second bead on the wide side of the plate. Clean off the slag and deposit the third bead. See Fig. 12-9.

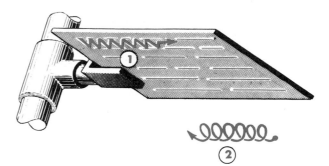

Fig. 12-8. Two kinds of weaving motions are used in overhead welding.

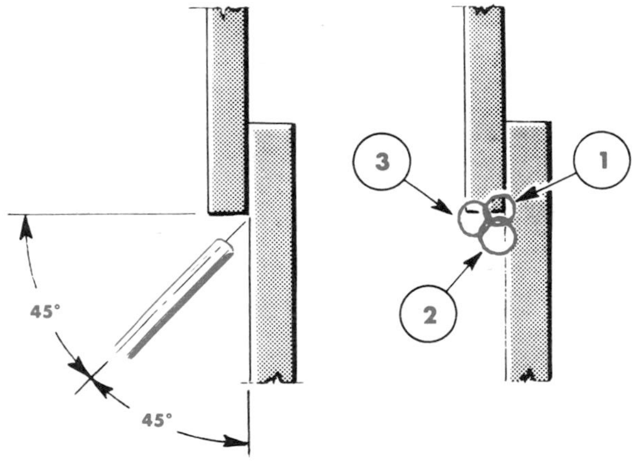

Fig. 12-9. Position the electrode at a 45° angle (left), and deposit three beads (right) to weld an overhead lap joint.

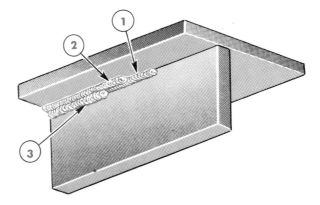

Fig. 12-10. Welding a T-joint in an overhead position.

Welding a T-Joint in an Overhead Position

1. Tack two ¼" plates to form a T-joint and clamp them in the overhead jig.

2. Deposit the first bead in the root of the V. Clean off the slag and deposit two additional beads as shown in Fig. 12-10.

Welding a Butt Joint in an Overhead Position

1. Bevel the edges of two ¼" plates and tack them together with a ¹⁄₁₆" root opening. Clamp the joint in the overhead jig.

2. Deposit three layers of beads as in Fig. 12-11. Clean off the slag after each layer.

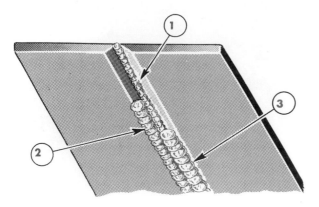

Fig. 12-11. Welding a butt joint in an overhead position.

Points to Remember

1. In overhead welding, incline the electrode 10° to 15° away from the crater.

2. Grip the electrode so that your knuckles face up and your palm down.

3. Stand to the side of the arc instead of directly underneath.

4. Drape the cable over the shoulder if welding in a standing position, or over your knees if in a sitting position.

5. Keep the arc length as short as possible.

6. Use a slight weaving motion to control the puddle.

7. Be sure proper clothing is worn.

QUESTIONS FOR STUDY AND DISCUSSION

1. Why is overhead welding more difficult?

2. How should the electrode be held for overhead welding?

3. Why should the holder be grasped so the palm of your hand is facing down?

4. What position should you assume for overhead welding?

5. What can be done to minimize the discomfort of the cable in the standing and in the sitting position?

6. What should be done to prevent the puddle from falling?

7. Why should sleeves and pant cuffs be turned down?

CHAPTER 13 cast iron

Most structures made of cast iron can be welded successfully. Because of the peculiar characteristics of cast iron, you will find that welding this metal requires a great deal more care than welding mild steel. However, if certain precautions are observed you should, with a little practice, be able to arc weld almost any cast iron piece. See Fig. 13-1.

TYPES OF CAST IRON

Cast iron is an iron-base material with a high percentage of carbon. The five types of cast iron are gray, white, malleable, alloy, and nodular.

Gray cast iron results when the silicon content is high and the iron is permitted to cool slowly. The combination of high silicon and slow cooling forces the carbon to separate in the form of graphite flakes, sometimes called free carbon. It is this separation of the carbon from the iron that makes gray cast iron so brittle.

Gray cast iron is used a great deal for machine castings. It can readily be identified by the dark gray, porous structure when the piece is fractured. If brought in contact with a revolving emery wheel, the metal gives off short streamers that follow a straight line and are brick red in color, with numerous fine, repeating yellow sparklers as shown in Fig. 13-2. Gray cast iron can be arc welded with comparative ease.

White cast iron possesses what is known as combined carbon. Combined carbon means that the carbon element has actually united with the iron instead of existing in a free state as in gray cast iron. This condition is brought about through the process of rapidly cooling the metal, leaving it very hard. In fact it is so hard that it is exceedingly difficult to machine, and special

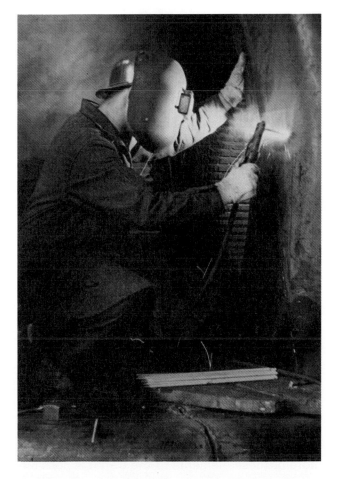

Fig. 13-1. This cast iron frame is being welded by the arc welding process. (Hobart Brothers Co.)

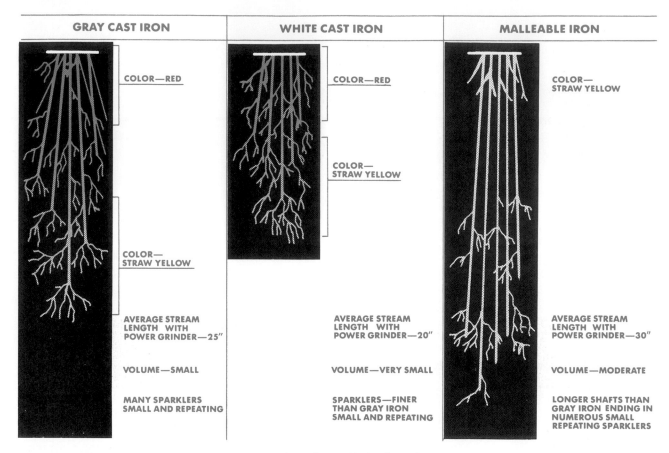

GRAY CAST IRON	WHITE CAST IRON	MALLEABLE IRON
COLOR—RED	COLOR—RED	COLOR—STRAW YELLOW
COLOR—STRAW YELLOW	COLOR—STRAW YELLOW	
AVERAGE STREAM LENGTH WITH POWER GRINDER—25″	AVERAGE STREAM LENGTH WITH POWER GRINDER—20″	AVERAGE STREAM LENGTH WITH POWER GRINDER—30″
VOLUME—SMALL	VOLUME—VERY SMALL	VOLUME—MODERATE
MANY SPARKLERS SMALL AND REPEATING	SPARKLERS—FINER THAN GRAY IRON SMALL AND REPEATING	LONGER SHAFTS THAN GRAY IRON ENDING IN NUMEROUS SMALL REPEATING SPARKLERS

Fig. 13-2. Notice how the spark characteristics vary for different kinds of cast iron.

cutting tools or grinders must be used to cut the metal. White cast iron is often used for castings with outer surfaces that must resist a great deal of wear. It is also used to make malleable iron castings.

The fracture of a piece of white cast iron will disclose a fine, silvery white, silky, crystalline formation. The spark test will show short streamers that are red in color. There are fewer sparklers than in gray cast iron and these are small and repeating as illustrated in Fig. 13-2.

Although white cast iron can be welded, generally speaking it has poor welding qualities.

Malleable cast iron is actually white cast iron which has been subjected to a long annealing process. The annealing treatment draws out the brittleness from the casting, leaving the metal soft but possessing considerable toughness and strength.

The fracture of a piece of malleable cast iron will indicate a white rim and a dark center. The spark test will show a moderate amount of short, straw-yellow streamers with numerous sparklers that are small and continue repeating as shown in Fig. 13-2.

Malleable cast iron can be welded; however, you must be sure that the metal is not heated above a critical temperature (approximately 1382°F or 750°C). If it is heated beyond this point, the metal reverts back to the original characteristics of white cast iron.

Alloy cast irons are those which contain certain alloying elements such as copper, aluminum, nickel, titanium, vanadium, chromium, molybdenum, and magnesium. By adding one or more of these elements to the iron it is possible to improve its tensile strength, machinability, fatigue resistance, and corrosion resistance. The alloy combinations cause the graphite to separate in a fine and evenly distributed structure, resulting in a cast iron possessing much higher mechanical properties.

Most alloy cast irons can be arc welded but greater precautions must be taken in the preheating and postheating stages to prevent the destruction of the alloying elements.

Nodular iron sometimes called ductile iron, has the ductility of malleable iron, the corrosion resistance of alloy cast iron and a tensile strength greater than gray cast iron. These special qualities are obtained by the addition of magnesium to the iron at the time of melting and then using special annealing techniques. The addition of magnesium and control of the cooling rate causes the graphite to change from a stringer structure to rounded masses in the form of spheroids or nodules. It is the formation of these nodular forms that gives the cast iron better mechanical properties.

Nodular iron can be arc welded providing adequate preheat and postheat treatments are used, otherwise some of the original properties are lost.

Preparing Cast Iron for Welding

Follow this procedure to prepare cast iron for welding:

1. Grind a narrow strip along each edge of the joint to remove the surface layer known as *casting skin.* See Fig. 13-3. Elimination of the surface layer is important because it is full of impurities. These impurities were embedded in the skin when the metal was poured into the sand mold. Unless they are removed, they will interfere with the fusion action of the weld metal.

2. V the edges as illustrated in Fig. 13-3. When the metal does not exceed $3/16''$ in thickness no V is necessary, but still remove the casting skin. On $3/16''$ to $3/8''$ metal, only a single V is required. The included angle of the V should be approximately 60°. Heavy cast iron pieces $3/8''$ or more in thickness should have a double V with a $1/16''$ to $3/32''$ root face. The included angle should be 60°.

3. If only a crack in a casting is to be welded, V the crack approximately $1/8''$ to $3/16''$ deep with a diamond point chisel, as shown in Fig. 13-4, or

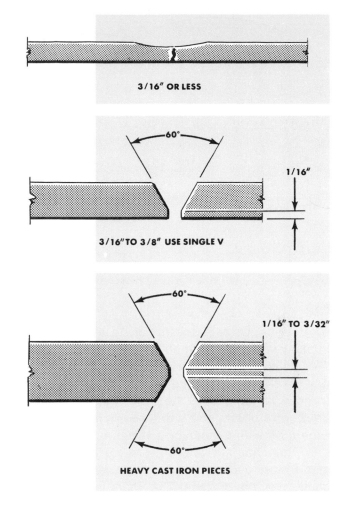

3/16″ OR LESS

60°

1/16″

3/16″ TO 3/8″ USE SINGLE V

60°

1/16″ TO 3/32″

60°

HEAVY CAST IRON PIECES

Fig. 13-3. This is how the edges should be prepared for welding cast iron.

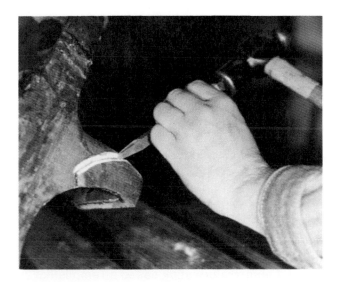

Fig. 13-4. V-ing a crack in a casting.

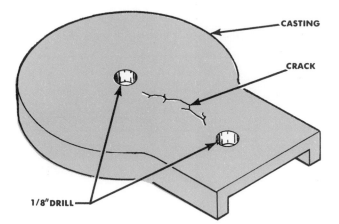

Fig. 13-5. Drill a ⅛" hole beyond each end of the crack.

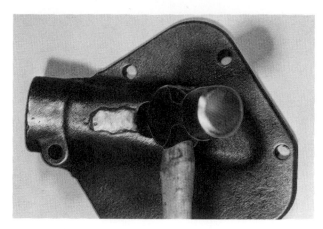

Fig. 13-6. Peen the bead while it is cooling.

by grinding. On sections that are less than ³/₁₆" in thickness, V only one-half the thickness.

4. Be sure the casting is entirely free from rust, scale, dirt, oil, and grease, as these substances, if trapped in the weld, will weaken it. Use a wire brush, and if grease or oil is present, wipe it off with a cleaning solvent.

5. Fine, hairline cracks in a casting can be made more visible by rubbing a piece of white chalk over the surface. The chalk leaves a wet line exactly where the crack is located.

6. During the welding operation, cracks have a tendency to extend. To prevent this from happening, drill a ⅛" hole a short distance beyond each end of the crack as illustrated in Fig. 13-5.

Keeping Cast Iron Cool

An important point in welding cast iron is to keep the piece as cool as possible. Otherwise, cracks will form. In the case of malleable cast iron, excessive heat will transform it into white cast iron. If possible, preheat the entire section with an oxy-acetylene torch.

CAUTION: Never heat it beyond a dull red color or a temperature exceeding 1200°F (approx. 650°C). Normally a preheat between 500° and 1200°F (260 to 650°C) is recommended. Correct preheat temperature for a welding can be determined in several ways: (1) Using a commercial temperature stick called a Tempilstik. (2) Placing a pointed wood stick on the heated surface, so that if it starts to burn the temperature is right for welding.

The preheat should be as uniform as possible over the entire casting and kept at this temperature until the welding is completed. Then the piece must be cooled slowly to room temperature. Very slow cooling is extremely essential with malleable and nodular iron.

When it is impossible to preheat the piece, it can be kept cool by running short beads 2" to 3" long. After a bead is deposited, allow it to cool until you can touch it with your hand before starting the next bead. *While cooling, peen the bead by striking it lightly with a hammer as shown in Fig. 13-6.* Peening helps to make the weld tight, and relieves stresses. However, peening can only be done on the machineable deposits, not the whole casting.

Electrodes for Welding Cast Iron

There are two main groups of electrodes for welding cast iron; machinable and non-machinable.

Machinable type electrodes are those whose deposits are soft and ductile enough so they can easily be machined after welding. See Fig. 13-7. They are used to repair all kinds of broken castings, correcting for machine errors, filling up defects, or to weld cast iron to steel.

There are two basic machinable electrodes. One has a 100 percent nickel core and the other

Fig. 13-7. An example of a weld made in a casting with machine type electrodes. (The Lincoln Electric Co.)

Fig. 13-8. A crack in a cast iron cylinder head is welded with non-machinable electrodes. (The Lincoln Electric Co.)

a nickel-iron base. Characteristics of these AWS classification electrodes are as follows: (Classification based on chemical composition of metal involved. The suffix CI designates an electrode for cast iron.)

ENi-CI	Nickel electrode	DCR or AC, general purpose welding especially for thin and medium sections, and castings with low phosphorus content, and where little or no preheat is used.
ENi-FeCI	Nickel iron electrode	DCR or AC for welding heavy sections, high-phosphorus castings, high- nickel alloy castings where high-strength welds are required, and welding nodular iron.

The non-machinable type electrode has a mild steel core with a heavy coating which melts at low temperatures, allowing the use of low welding current. It leaves a very hard deposit and is used only when the welded section is not to be machined afterwards. This electrode produces a tight and waterproof weld, making it ideal for repairing motor blocks, water jackets, transmission cases, compressor blocks, pulley wheels, pump parts, mower wheels, and other similar structures. See Fig. 13-8.

Cast iron electrodes are commonly known by the manufacturer's trade name such as:

Ferroweld —non-machinable —Lincoln
Softweld —machinable —Lincoln
Strongcast —non-machinable —Hobart
Softcast —machinable —Hobart
Airco 77 —non-machinable —Airco
Airco 375 —machinable —Airco

WELDING PROCEDURE

1. Set the machine for the correct amperage. Follow the recommendation of the electrode manufacturer. As a rule, the amperage setting for welding cast iron is lower than for welding mild steel.

2. Since it is important to keep the heat to a minimum, always use small diameter electrodes. Operators seldom weld with electrodes greater than 1/8″ in diameter.

3. Tip the electrode 5 to 10 degrees in the direction of travel, and deposit straight, progres-

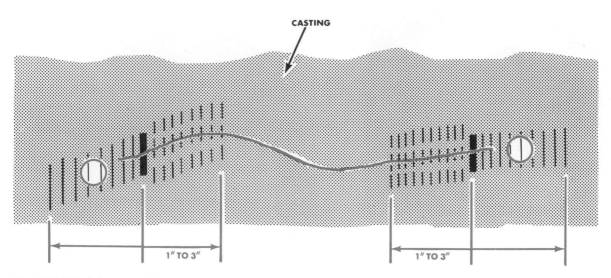

Fig. 13-9. This is how to weld a crack in a casting. Note the $1/8''$ holes beyond each end of the crack. These holes prevent the crack from extending. The starting points on each end are shown by the thick black lines.

sive stringer beads. Use a slightly longer arc than in welding mild steel. If more than one layer is to be deposited, use a slight weaving motion after the first bead is made. At no time should the electrode be manipulated so that the width of the deposit is greater than 3 times the electrode diameter.

4. Use a back-step welding technique when necessary so the crater of each bead lies on top of the previous bead.

5. When welding cracks in castings, start about $3/8''$ before the end of the crack and weld back to the hole, filling the hole. See Fig. 13-9. Then move slightly beyond the hole. Next move to the other end of the crack and do the same thing. Continue to alternate the weld on each end, limiting the length of each weld 1″ to 1$1/2$″ on thin material and 2″ to 3″ on heavier pieces. Always allow each section of weld to cool before starting the next, and peen each short bead.

Welding Sections with Pieces Knocked Out

To weld a section with one or more pieces broken out, fit the parts together. V the breaks and tack them before welding.

If the broken part cannot be made to fit in the casting, shape another piece, using mild steel. Place the new section in position and weld.

Studding Broken Castings

When a casting is 1$1/2$″ or more in thickness and is subjected to heavy stresses, use steel studs to strengthen the joint. Studding is not advisable on castings lighter than 1$1/2$″ because it tends to weaken rather than strengthen the joint. To apply studs proceed as follows:

1. V the crack.

2. Drill and tap $1/4''$ or $3/8''$ holes in the casting at right angles to the sides of the V. Space the holes so the center-to-center distance is equal to three to six times the diameter of the stud. See Fig. 13-10 for example.

3. Screw the studs into the tapped holes. The threaded end of the studs should be about $3/8''$ to $5/8''$ in length and should project approximately $1/4''$ to $3/8''$ above the casting.

4. Deposit beads around the base of the studs, welding them thoroughly to the casting. Remove the slag and deposit additional layers of beads to fill the V.

Brazing Cast Iron

The three common electrodes for brazing cast iron are ECuSn-A, ECuSn-C (both of which are of the copper-tin classification) and ECuAl-A2 (copper-aluminum classification). The main dif-

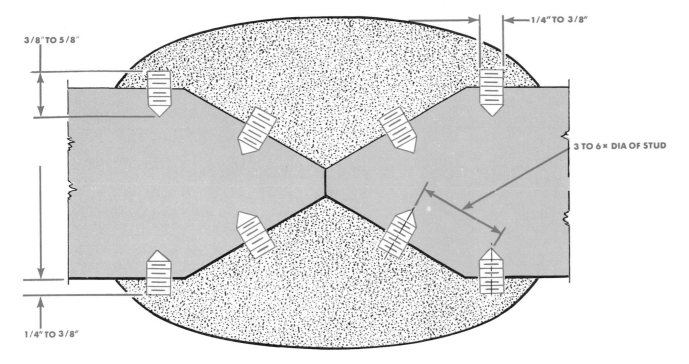

3/8″ TO 5/8″

1/4″ TO 3/8″

3 TO 6 × DIA OF STUD

1/4″ TO 3/8″

Fig. 13-10. Studs can be used to reinforce the weld in a broken casting.

ference between ECuSn-A and ECuSn-C is in the amount of tin they contain. The ECuSn-C has a higher percentage (8 percent) than the ECuSn-A (5 percent), thereby producing welds of greater hardness, tensile and yield strength. Both are used with DC reverse polarity and normally require that the area to be brazed be preheated to 400°F (204°C).

The ECuAl-A2 electrode has a relatively low melting point and high deposition rate at lower amperage which permits more rapid welding. This minimizes distortion and the formation of white cast iron in the fusion zone. The ultimate tensile and yield strength of the deposits are nearly double of the copper-tin deposits.

The ultimate success of any brazing operation with copper alloy electrodes depends considerably on the following:

1. Use of wide grooves.
2. Joints free of moisture, grease, oil, dirt.
3. A preheat between 300 to 400°F (about 150 to 205°C).
4. Use of the lowest possible amperage for good bonding.

5. Welding done at a fast rate to minimize dilution from the base metal.
6. Puddling should be avoided.
7. Parts cooled slowly with the use of powder lime or asbestos.

Points to Remember

1. Always remove the casting skin before welding cast iron pieces.
2. V the edges with a single or double V, depending upon the thickness of the metal.
3. Clean an old casting very carefully before welding.
4. To weld a crack in a casting, drill a 1/8″ hole a short distance beyond each end of the crack to prevent the crack from spreading.
5. Keep the casting as cool as possible.
6. Peen the bead to relieve stresses.
7. Use the correct type of electrode.
8. Use small diameter electrodes and maintain a lower amperage setting than for mild steel.
9. To weld cracks in castings, start the weld about 3/8″ away from the weld and travel to the

drilled hole. Alternate the weld on each end of the crack.

10. Run short beads of 1″ to 1 1/2″ on thin metal and 2″ to 3″ on heavy metal.

11. Reinforce heavy castings with studs for added strength.

12. Preheat the piece whenever possible, and cool slowly after the weld is completed.

QUESTIONS FOR STUDY AND DISCUSSION

1. What is the difference between gray, white, and malleable cast iron?

2. Why should the outer skin be removed before welding cast iron?

3. How should the joints be prepared for welding cast iron?

4. How should old castings be cleaned before welding?

5. What can be done to make fine cracks in castings more visible?

6. How can cracks in castings be prevented from spreading?

7. Why is it important to keep cast iron cool when welding? How should this be done?

8. If a casting is to be preheated, how can you determine when correct heat is reached?

9. Why should a bead be peened?

10. What type of electrode is used for welding the various types of cast iron?

11. Why is a lower amperage setting required for cast iron welding?

12. Why is it a good practice never to use electrodes greater than 1/8″ in diameter for welding cast iron?

13. What is the correct procedure for welding cracks in castings?

14. Why is it advisable to use studs in repairing some broken castings?

15. Why should studs be used only on heavy castings 1 1/2″ or more in thickness?

16. How far apart should studs be fastened before welding?

17. How deep should studs be driven in the castings? How much should they project above the surface?

18. What type of electrodes are used for brazing cast?

CHAPTER 14 *carbon steels*

Carbon steels can readily be arc welded but some require considerably more control of the welding process. For certain types, special electrodes must be used or some preheating and postheating treatment applied to secure sound welds with the required mechanical properties.

Welding Carbon Steels

An important point to remember in welding carbon steels is the effects of heat on the welded area. When steel is heated to a high temperature its structure undergoes a change. Depending on the amount of carbon present, the steel changes from a mixture of *ferrite* (pure iron) and *cementite* (iron and iron-carbide, sometimes called pearlite) to a solid solution known as *austenite,* where the carbon goes into a solution form and becomes evenly distributed. If an austenitic structure is cooled quickly, a *martensitic* condition develops which causes the carbon to precipitate and leave an extremely hard and brittle material. When this happens in a weld area, underbead cracking can be expected as well as cracks alongside the weld in the parent metal. Accordingly a welder must be aware, when welding certain types of carbon steels, of the effects of welding heat on the structure of the metal and take suitable action to prevent the formation of weld defects.

Carbon steels are those steels whose principal element is carbon. Steels in this group are referred to as low carbon, medium carbon, and high carbon.

Plain carbon steels are made in three grades: *killed, semikilled,* and *rimmed.* A killed steel is one that is deoxidized by adding silicon or aluminum (in the furnace ladle or mold) to cause it to solidify quietly without evolving gases. Killed steel is homogeneous, has a smooth surface and contains no blowholes. A semikilled steel is only partially deoxidized, while rimmed steel receives no deoxidizing treatment.

Welding low-carbon steels. Low-carbon steels are the easiest to weld. No particular control needs to be taken since the welding heat has no appreciable effect on the parent metal. Any mild steel coated electrodes in the E60XX or E70XX series will produce good welds. The choice of electrodes in these series is influenced by specific requirements such as depth of penetration, type of current, position of the weld, joint design, and deposition rate. See Chapter 6.

Welding medium-carbon steels. A medium-carbon steel is one whose carbon content ranges from 0.30 percent to 0.45 percent. Most medium-carbon steels are relatively easy to weld, especially with the availability of the E-70XX type electrodes. The E-7016, E-7018, and

E-7024 electrodes are frequently used because of their higher tensile strength and less tendency to produce underbead cracking particularly when no preheat can be applied. The E-6012 or E-6024 electrodes can also be used if simple precautions are taken and the cooling rate is sufficiently retarded to prevent excessive hardening of the weld.

Welding high-carbon steels. High-carbon steels are those whose carbon content is 0.45 percent or higher and are readily hardenable. These steels are considered more difficult to weld than other carbon steels. However, with proper care, high carbon steels can be arc welded successfully. Higher tensile strength electrodes in the E-80XX, E-90XX or E-100XX class are preferred because they minimize underbead cracking. For welding some high-carbon steels stainless steel electrodes, such as E-310-15 are often recommended. Generally, a preheat and post heat treatment is advisable for these steels.

Heat Control Processes

Preheating. Preheating involves heating the base metal to a relatively low temperature before welding is begun. Its main purpose is to lower the cooling rate of the weld, thereby reducing the thermal (heat) conductivity of the metal. The lower thermal conductivity allows a slower withdrawal of heat from the weld zone. This lessens the tendency for martensite to form and consequently, there is less likelihood for hard zones to develop in the surrounding weld area than if a weld joint is made without preheat. Preheating also burns grease, oil, and scale out of the joint and permits faster welding speeds. Preheating can be accomplished by moving an oxy-acetylene flame over the surface or placing the part in a heating furnace. Specifically, the advantages of a preheat treatment are:

1. Prevents cold cracks.
2. Reduces hardness in heat-affected zones.
3. Reduces residual stresses.
4. Reduces distortion.

Correct temperature is an important factor in preheating. Preheat temperatures for mild steel should be between 200° to 700°F (94 to 371°C), depending on the carbon content. The greater the carbon content, the higher the preheating temperatures. Preheating temperatures can be measured in various ways. Here are several:

1. Use of surface thermometers or thermocouples.
2. Marking the surface with a carpenter's blue chalk. A mark made with this chalk will turn to a whitish gray when the temperature reaches approximately 625°F (330°C).
3. Rubbing 50-50 solder on the surface. The solder starts to melt at 360°F (182°C).
4. Rubbing a pine stick on the heated surface. The pine stick chars at about 635°F (335°C).
5. Using tempilstiks which are crayons or liquids. When applied, they melt and change color at a specific temperature.

Process control. For welding most steels, the *control heat process* is often more economical than the use of preheating. The principle of this technique is to induce a large volume of heat into the base metal by welding with high-current at low speed or by making multiple-pass welds. The high-current and slow welding rate builds up considerable heat in the metal. This naturally slows up the rate of cooling and the subsequent prevention of hard crystalline zones near the weld area.

In multiple-pass welding the deposit of the first layer preheats the base metal. The heat of the next pass tempers the base metal adjacent to the first weld bead. Each successive pass then leaves just enough heat so there is no rapid cooling and thus no appreciable hardening.

Postheating. Postheating is intended primarily as a stress-relief treatment. For welding some of the higher range carbon steels, postheating is as important as preheating. Although preheating does control the cooling rate, nevertheless the possibility of stresses becoming locked in the welded area is always a factor. Unless these stresses are removed, cracks may develop when the piece cools completely, otherwise the part may become distorted, especially after a machining operation.

Postheating temperatures for stress relief should be in the range of 900° to 1250°F (472 to 677°C). Soaking period normally runs about one hour per inch of metal thickness.

Formation of Weld Cracks

Cracks in the weld metal may run in a longitudinal or transverse direction and are often invisible to the naked eye. The defects will usually show up under some type of ultrasonic, magnetic, or radiographic inspection.

Basically weld cracks occur when there is a shrinkage of the weld bead. Since high-carbon steels do not stretch very much because of their hardness, whenever a weld bead is deposited on the hard, rigid surface the generated contraction forces must be absorbed by the more ductile metal in the deposited weld bead.

Weld cracks can be avoided by keeping the weld penetration as low as possible thus minimizing any excessive stressing of the parent metal. Furthermore, if penetration is kept low there is less of a tendency for the weld bead to pick up an excessive amount of carbon from the base metal, thereby leaving the weld bead considerably more ductile and better able to absorb shrinkage stresses.

The use of low-hydrogen electrodes with iron-powder coatings produce minimum penetration and ductile welds. For welding high-carbon steels in the upper carbon range, special stainless steel electrodes of the E-310-15 type (25 percent chromium and 20 percent nickel) are frequently recommended because of their high ductility.

Crater cracks. These cracks occur across the weld bead crater and result from shrinkage. See Fig. 14-1. When a weld bead is deposited, solidification of the molten metal takes place from the sides and moves towards the center. The center of the crater cools rapidly because of the smaller amount of metal while the remaining bead cools more slowly, causing a concentration of stresses which eventually result in cracks. Crater cracks are particularly prevalent in thin concave fillet welds. See Fig. 14-2. Most crater cracks can be prevented by proper electrode manipulation. Care should be taken to completely fill up the craters and the weld bead rounded slightly by using a shorter arc.

Root cracks. In making a groove or fillet weld, the first pass is in the form of a narrow stringer bead along the weld seam. This is fol-

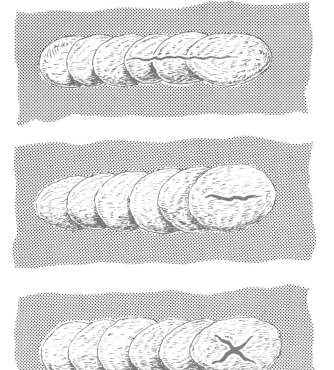

Fig. 14-1. Types of crater cracks.

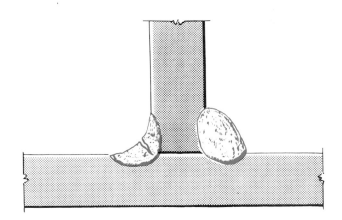

Fig. 14-2. Crater cracks will often occur on thin concave fillet welds.

lowed by one or more layers of weld beads. It is the first layer, or root bead, that is the most susceptible to cracking. See Fig. 14-3. The cracking is generally due to the excessive car-

Fig. 14-3. The root bead is often more susceptible to cracking than later beads.

bon which the bead picks up from the base metal, thereby making the weld metal hard and brittle. As the weld metal cools, it shrinks, and as additional layers of beads are deposited on the root weld, tensile stresses form which develop into cracks. Root cracks can be prevented by preheating the base metal, using a more ductile weld metal, and providing sufficient space between the plates so they can move as the weld cools.

Porosity. Porosity is a common problem in welding high-carbon steel. Molten high-carbon steel readily absorbs gases such as hydrogen and carbon monoxide which are released as the metal begins to cool. If this gas does not reach the surface before the metal begins to solidify it becomes entrapped in the metal, leaving small gas pockets or blowholes. The formation of blowholes is very common in welding steel with a high sulfur content. The presence of sulfur along with phosphorous and silicon when combined with other gases always generates a degree of porosity that seriously affects the strength of the weld.

Skillful welders usually can prevent porosity by proper manipulation of the electrode. The secret of gas-free welds is to keep the surface of the deposited metal pool fluid enough so the gas is rapidly released. Welding with low-hydrogen electrodes as a rule will prevent any extensive formation of porosity.

Excessive Hardening and Softening of Base Metal

As in the case of medium-carbon steel, the development of a hard layer in the weld zone is the result of too rapid cooling. The rapid cooling transforms the metal into a martensitic condition which leaves a very brittle area. The best procedure to avoid excessive hardening is to utilize a controlled system of preheating and postheating. Preheat treatment should be between 200° and 400°F (94 to 204°C) for steels with a carbon content of 0.45 to 0.65 percent and 400°-700°F (204 to 371°C) when the carbon content is over 0.60 percent. Postheating should be in the 1100° to 1200°F (592 to 650°C) range.

The strength of high-carbon steel is dependent upon its hardness which is obtained by a heat treating process. If the part is to be heat treated after welding, then excessive hardness or softness is not necessarily important. On the other hand, if the part is not to be heat treated then suitable precautions must be taken to prevent a loss of the hardness properties which are characteristic of high-carbon steel after heat treatment.

Welding will often have some softening effect on hardened high-carbon steel especially near the weld. To minimize both excessive hardening and softening, the use of a high-chromium, high-nickel stainless steel electrode is recommended. A small diameter electrode and an intermittent welding procedure will also reduce excessive softening and hardening of the base metal.

Points to Remember

1. Use E-60 or E-70 series electrodes to weld low-carbon steel.
2. Use low-hydrogen electrodes when welding medium-carbon steels.
3. Preheat steels having a high-carbon content to prevent cracking.
4. The control heat welding process will often eliminate the need for preheating carbon steels.
5. For many high-carbon steels a postheat treatment is often advisable for stress-relief.
6. Keep the cooling rate of the weld puddle as slow as possible in welding all kinds of high-carbon steel.
7. Low-hydrogen electrodes with iron powder coatings will usually minimize cracking in welding high-carbon steel.
8. Use correct electrode manipulation to allow gases to come to the surface in order to avoid gas pockets and blowholes.

QUESTIONS FOR STUDY AND DISCUSSION

1. What happens to high-carbon steel when it develops into a martensitic structure?

2. Why are low-carbon steels relatively easy to weld?

3. What type of electrodes are normally used for welding low-carbon steels?

4. When is a steel classified as a medium-carbon steel?

5. What type of electrodes are required for welding medium-carbon steel?

6. Why is a preheating treatment sometimes necessary in welding medium and high-carbon steels?

7. At what temperature should preheating be carried out?

8. What is meant by controlled heat welding?

9. What is the function of postheating?

10. At what temperature should postheating be done?

11. Why are high-carbon steels more difficult to weld?

12. How can weld cracks be avoided when welding high-carbon steel?

13. What are crater cracks and what causes them?

14. What are root cracks and why do they occur?

15. What causes porosity in a weld?

16. How can porosity in a weld be avoided?

17. Why must the condition of a heat treated part be considered before welding it?

18. How can excessive softening of a heat treated part be kept to a minimum in a welding operation?

19. What is the difference between a killed and a semikilled steel?

20. What is meant by a rimmed steel?

CHAPTER 15 alloy steels

An alloy steel is a steel which is mixed with one or more additional elements, other than carbon or iron, in large enough percentages to alter the characteristics and properties of the steel. The addition of such substances as manganese, nickel, chromium, tungsten, molybdenum, or vanadium produces a steel that is greater in strength and toughness. Alloy steels are known by their main alloying elements, such as manganese steel, chromium-nickel steel, molybdenum steel, etc.

Practically all alloy steels can be welded, but as a rule the welding operation is much more difficult to perform than in welding mild steel. This is due to the fact that the characteristics of the basic alloying ingredients are sometimes destroyed as the weld is made. Furthermore, unless precautions are taken, cracks appear near the welded areas, or slag inclusions and gas pockets form in the bead, all of which weaken the weld. Many of the difficulties, however, can be avoided or minimized by using special electrodes designed specifically for welding alloy steels.

Preheating and Postheating

Successful welding of many alloy steels requires a controlled rate of cooling because when heated to a high temperature and cooled rapidly, they readily harden. Welding these steels with-

out proper heat control produces an embrittlement in the heat-affected zone parallel to the weld joint. With proper preheating and postheating, the rate of cooling is delayed and consequently the metal near the weld zone does not harden appreciably.

The application of preheating and postheating is also an important factor in preventing weld cracks caused by shrinkage stresses. By slowing the rate of cooling, the stresses are more readily distributed throughout the weld and released while the metal is still hot.

The preheating and postheating temperatures will depend on the alloying content of the base metal. Some elements produce greater hardenability and therefore the cooling rate must be slower. In any event, the temperature should never exceed the original hardening and tempering temperatures unless the piece is to be heat treated again after the weld is completed.

When a preheat temperature is difficult to ascertain, the *clip test* is often used to make a rapid check. This test is not applicable for thin steels but will produce good results on heavy sections down to $3/8''$ in thickness.

The test involves the welding of a piece of low-carbon steel $1/2''$ in thickness and $2''$ or $3''$ square to the steel plate which is being checked for a preheat temperature. A convex contour fillet weld is made with an electrode and welding

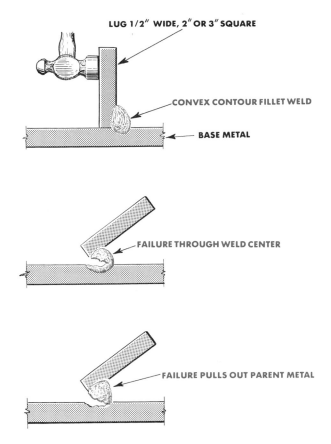

LUG 1/2" WIDE, 2" OR 3" SQUARE

CONVEX CONTOUR FILLET WELD

BASE METAL

FAILURE THROUGH WELD CENTER

FAILURE PULLS OUT PARENT METAL

Fig. 15-1. Clip test for finding out if preheating is needed.

current similar to the ones used for the welding job. The weld is allowed to cool for five minutes and then the clip is hammered until it breaks off. See Fig. 15-1A. If the lug breaks through the weld after a number of blows, the test indicates that no serious underbead cracking will result when the welding is carried out in the same manner at normal room temperature. If the lug breaks and pulls out some of the parent metal (Fig. 15-1B) the test shows that this particular steel must be preheated.

Welding Austenitic Manganese Steel

Austenitic manganese steel is a tough, non-magnetic alloy noted especially for its high strength, excellent ductility, and outstanding wear resistance. Welding this metal requires considerable attention since it is so sensitive to reheating. Any prolonged period of heating will embrittle the metal and result in loss of tensile strength and ductility. Consequently, low welding current and rapid welding rate without any extensive preheating is necessary.

There are two groups of manganese steel, known as low-manganese (2 percent manganese or less) and high-manganese (12–14 percent manganese minimum).

Low-manganese steels. These steels are used for fabricating parts or structures where they must withstand impact stresses and resist wear. To weld these metals a 0.5 percent molybdenum electrode of the E-7010 or E-7020 class is generally used. Since both electrodes are of the deep penetrating type, care must be taken to prevent an excessive amount of the parent metal from mixing into the weld. Slight preheating is advisable since it reduces underbead cracking. When preheating is impractical, the standard E-6012 electrode can be used. However, if this electrode produces a high number of cracks, the

TABLE 15-1. FILLER METALS RECOMMENDED FOR WELDING MANGANESE STEELS

FILLER METALS	TYPE OF FABRICATION
Austenitic manganese steel covered electrodes or tubular wire with less than 0.025% phosphorus.*	Recommended for joining austenitic manganese steel where the weldment is subjected to severe wear or structural stress or both. May be used also for multiple pass welding of manganese steel to carbon or low-alloy steel with the latter (the carbon and low-alloy steel only) slightly preheated.
Austenitic manganese steel with phosphorus over 0.025%, electrodes with or without covering.	Recommended for joining and filling only where structural stress and wear are light. For an extra factor of safety in fabrication welds, a root pass of stainless is suggested. Rods and wire in this class are not recommended for joining manganese steels to mild or low-alloy steels unless overlaid with stainless steel.
Nickel-chromium stainless steel electrodes.	Used for joining manganese steels to manganese, or to mild or low-alloy steels where weldments are subjected to little wear and moderate structural stress. NOTE: Welds cannot be oxy-acetylene cut.
Austenitic manganese-chromium (stainless) Electrodes (covered)	Recommended for joining manganese steel or manganese to mild or low-alloy steels where the weldment is subjected to wear and severe structural stress and where considerably higher yield strength is desired. NOTE: Welds cannot be oxy-acetylene cut.
Low phosphorus, composite austenitic manganese steel electrodes or tubular wire, or 309 or 310 stainless electrodes.†	Nonmagnetic applications for all types of stresses.

*Usually available only in composite or tubular form.
†18-8 stainless should be avoided in applications subjected to impact of the weld as some grades become magnetic after severe work hardening.
AWS

E-7016 or E-7018 should be substituted. See Table 15-1.

High-manganese steels. This steel is usually found in a cast form because of its tough core and hard abrasion resisting surface. This metal is used widely for stone crushing equipment parts, power shovel buckets, and other structures which are subjected to great wear. See Fig. 15-2.

Welding high-manganese steel involves joining two pieces, repairing cracks, or building up worn surfaces. Building up worn surfaces is a surfacing operation commonly referred to as *hardfacing*. A detailed description of hardfacing is presented in Chapter 31.

Welding high-manganese steel. To secure the best possible results in welding pieces of high-manganese steel, observe the following:

1. V the joint and clean the surfaces carefully.

2. Use the lowest possible current to prevent the formation of a brittle zone next to the weld.

3. The electrode most frequently recommended for high-manganese steel is a stainless steel 18-8 type. (E-308-15, E-308-16, E-309-15, E-309-16, E-310-15, E-310-16).

As a rule these electrodes are easier to apply and produce the most satisfactory welds. Other types of electrodes used for welding high-manganese steel are molybdenum-copper-manganese and nickel-manganese. However, more skill is needed to execute good welds with these electrodes.

4. Strike the arc ahead of the crater and continue the weld bead in the direction of travel. Keep the weld beads short (about 2") and allow them to cool between each run to avoid concentrated and prolonged heat in any localized area of the base metal. Do not continue to weld in an

Fig. 15-2. Because of its tough qualities, high-manganese steel is often used for power shovel buckets. (Hobart Brothers Co.)

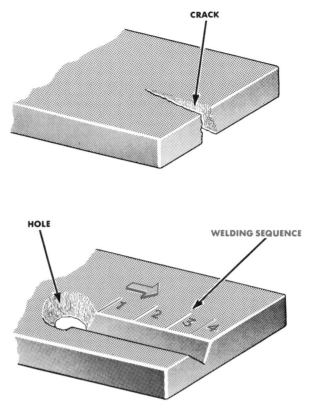

Fig. 15-3. Welding a crack in manganese steel.

area unless the temperature of the metal is below 750°F (397°C). Use tempilstiks to determine temperature by marking the base metal $3/8''$ to $1/2''$ from the weld. Actually you should be able to place your hand within 6″ to 8″ of the weld at any time. If you have trouble controlling the heat, place wet rags on areas adjacent to the weld.

5. Use a narrow weave with the electrode held at about a 45 degree angle in the direction of the weld bead travel.

6. Due to the high thermal expansion of manganese steel, many stresses form as the weld cools, causing cracks to develop during contraction. To reduce cracking, peen each run of weld bead when it is completed.

Repairs. To repair cracks in manganese steel follow this procedure:

1. Cut a hole at the end or ends of the crack to prevent stresses from spreading the crack further into the sound metal.

2. Gouge a U or V-groove in the crack and grind away all oxide from the weld area.

3. Deposit a root bead along section 1 as indicated in Fig. 15-3. Weld additional layers over the root bead until the groove is full. Then follow the same procedure for sections 2, 3, and 4 in order.

4. When the entire groove is filled, fill the hole or holes at the end of the original crack.

Welding Stainless Steel

The first stainless steel developed was the chromium-iron type, having as the main constituents chromium and iron. Later nickel was introduced into stainless steel, producing a metal that has been popularly referred to as *18-8.* (Approximately 18 percent chromium and 8 percent nickel.) Further developments have brought into use a whole series of stainless steels designed to meet more rigid fabricating demands.

How Stainless Steel is Classified

Stainless steel today is classified into two general AISI (American Iron and Steel Institute) series—200-300, and 400. Each series includes several different kinds of steel, all of course having some special characteristic. See Table 15-2.

The 400 series. The 400 series is further classified into two groups according to their crystalline structure. One group is known as ferritic and is non-hardenable and magnetic. The other group is referred to as martensitic, which is hardenable by heat treatment and is also magnetic.

Ferritic type stainless steels have better resistance to high temperature corrosion than the martensitic group. When subjected to the heat of welding they will develop some degree of brittleness. This brittleness may be reduced and ductility improved by cold working the weld area by peening and annealing.

Types 405 and 430 Ti (same as regular 430 but with titanium added) have been developed especially for welding.

TABLE 15-2.   TYPES AND WELDING CHARACTERISTICS OF STAINLESS STEELS.

CHROMIUM—NICKEL TYPES—NON-HARDENABLE

AISI	STRUCTURE	WELDING PROPERTIES
201	Austenitic	Very good, tough welds
202	Austenitic	Very good, tough welds
301	Austenitic	Very good, tough welds
302	Austenitic	Very good, tough welds
303	Austenitic	Fusion welding not recommended
304	Austenitic	Very good, tough welds
305	Austenitic	Very good, tough welds
308	Austenitic	Good, tough welds
309	Austenitic	Good, tough welds
310	Austenitic	Good, tough welds
316	Austenitic	Very good, tough welds
321	Austenitic	Very good, tough welds
347	Austenitic	Good, tough welds

CHROMIUM TYPES—NON-HARDENABLE

AISI	STRUCTURE	WELDING PROPERTIES
405	Ferritic	Good, fairly tough welds
409	Ferritic	Good
430	Ferritic	Fair, non-ductile welds
430Ti	Ferritic	Good
434	Ferritic	Fair, non-ductile welds
436	Ferritic	Fair, non-ductile welds
442	Ferritic	Fair, non-ductile welds
446	Ferritic	Fair, non-ductile welds

CHROMIUM TYPES—HARDENABLE

AISI	STRUCTURE	WELDING PROPERTIES
410	Martensitic	Fair, preheat 400–500°F, after welding anneal at 1250°F
414	Martensitic	Fair, preheat 400–500°F, after welding anneal at 1250°F
416	Martensitic	Poor, preheat 400–500°F, after welding anneal at 1250°F
420	Martensitic	Fair, preheat 400–500°F, after welding anneal at 1250°F
431	Martensitic	Fair, preheat 400–500°F, after welding anneal at 1250°F

The martensitic chromium steels will harden when cooled from welding temperatures. Since these steels are usually subjected to further heat-treating processes, the hardening effects present no problem, especially if the annealing or tempering treatment is begun immediately after the welding operation.

The 200–300 series. These steels have an austenitic structure which make them extremely tough and ductile in the as-welded condition. Hence, they are ideal for welding and require no annealing after welding if used in normal atmospheric conditions or when subjected to mildly corrosive actions. If they are to encounter severe corrosive conditions, it is advisable to anneal the welded structure.

Physical Properties of Stainless Steel

The coefficient of expansion of the 400 chromium types of steel is approximately the same as that of carbon steel. Consequently, the allowances for expansion are practically the same as those for carbon steel. The chromium-nickel 200–300 series have about a 50 to 60 percent greater coefficient of expansion than carbon steel and therefore require greater consideration in expansion control.

The heat conductivity of the 400 series is approximately 50 to 65 percent of carbon steel. With the 200–300 series the heat conductivity is almost 40 to 50 percent of carbon steel. Consequently in both series the heat is not conducted away as fast as in ordinary steel and as a result, stainless steels take longer to cool. This phenomenon is particularly important to remember in welding thin gages since there is greater danger of burning through the material.

Methods of Reducing Effects of Heat

Unfavorable effects of heat can be reduced substantially by means of chill plates. The use of chill plates such as copper pieces will help conduct the heat away. Rigid jigs and fixtures should be employed wherever possible, especially for the 200–300 series. When stainless steels are allowed to cool in a jig, warping and distortion is practically eliminated.

If jigs cannot be used, special welding procedures will be necessary to counteract expansion forces. The common practice is to resort to *skip* or *step-back* methods of welding. See Chapter 3.

Weldability of Stainless Steel

It is generally conceded that stainless steels in the 200–300 series have better welding qualities than those in the 400 series. However, this does not mean that the 400 series stainless steels are not weldable. Greater precautions simply have to be taken, especially in applying proper heat-treating methods after the weld is completed.

All methods of welding may be used in joining stainless steel. Oxy-acetylene welding is sometimes used for welding 20 gage and lighter stainless steel sheets, and the metallic arc for heavier plates. See Fig. 15-4.

Today the inert-gas-shielded arc is utilized a great deal more for welding stainless steel of all types because of the ease with which welds can be made and since with this process there is less danger of destroying the corrosion-resistant properties in the steel.

Choosing the Correct Joint Design

For thin metal, the flange-type joint is probably the most satisfactory design. Slightly heavier sheets, up to $1/8''$ in thickness, may be butted together. For plates heavier than $1/8''$, the edges should be beveled to provide a V so fusion can be obtained entirely to the bottom of the weld.

Since stainless steels have a much higher coefficient of expansion with lower thermal conductivity than mild steels, there are greater possibilities for distortion and warping. Therefore, whenever possible, clamps and jigs should be used to keep the pieces in line until they have cooled.

Stainless Steel Electrodes

In arc welding, flux coated electrodes should always be used. The flux shields the molten metal from air, preventing oxidation of the chromium and producing strong, corrosion-resistant welds. Also, the flux has a tendency to act as a stabilizing agent helping to maintain a steady arc with an even metal flow into the shielded area.

The slag formed by flux coated electrodes will flow to the surface where it should be brushed

Fig. 15-4. Large stainless steel vats for chemical processing are arc welded. (Hobart Brothers.)

off before subsequent beads are laid on the weld. To produce a good strong weld, the electrode should ordinarily be as low in carbon as possible. It is also desirable for the electrode coatings to be free from undesirable elements such as carbon.

The alloy content of electrodes normally should be higher than or the same as that of the base metal to compensate for expected alloy loss. Moreover, a columbium-bearing electrode must be used for both the columbium (Type 347) and the titanium (Type 321) stabilized grades. Chromium-nickel electrodes are often used in welding chromium grades due to the fact that they provide a ductile weld metal. See Table 15-3.

Stainless steel electrodes are identified in a different way than mild steel. For example, a standard 18-8 electrode for AC-DC current is designated as E-308-16. The prefix *E* indicates a metallic arc electrode. The next three digits are the AISI (American Iron and Steel Institute) sym-

TABLE 15-3.   ELECTRODES FOR WELDING STAINLESS STEEL.

AUSTENITIC	
201, 202	308-15, 16
301, 302, 304	
305, 308	308-15, 16
309	309-15, 16
310	310-15, 16
316	316-15, 16
321, 347	347-15, 16
FERRITIC	
405, 409	410-15, 16
430, 434	
436, 442	430-15, 16
446	446-15, 16
MARTENSITIC	
410, 414	
416, 420	410-15, 16
431	430-15, 16

bols for a particular type of metal. Thus 308 represents a metal containing 18 percent chromium and 8 percent nickel. The last two digits following the dash may be either 15 or 16; the 1 indicates an all-position welding, and the 5 or 6 specifies the type of covering and applicable welding current. The 5 designates a lime-coated electrode for DC reverse polarity. The 6 is an electrode designed for AC and DC reverse polarity and has a titanium type covering.

Electrode Selection[1]

The selection of the proper electrode for stainless steel application is in most cases a more critical choice than with mild steel because of the number of types and grades of stainless steel and the varying degrees of severity of heat, corrosion media, etc., to which the weldment will be subjected. Selecting the right electrode for most satisfactory results is a matter of analyzing all the conditions applying to the particular job. To determine the right type and size of electrode best suited to a given set of conditions the following factors must be considered:

1. The analysis of the base metal to be welded.
2. Dimensions of the section to be welded.
3. Type of welding current available.
4. Welding position or positions to be used.
5. The fit-up of the section to be welded.
6. Specific properties of the weld deposit.
7. Requirements of a specific code, standard or specification.
8. Selection of the electrode must be exercised carefully because of the high cost of the material to be welded.
9. Not only must the stainless steel weld metal have sufficient tensile strength and ductility but it must also have corrosion resistance equivalent to the parent metal. Hence, an electrode having an analysis comparable to the base metal should be used.

Welding Current

Both direct and alternating current are used in the arc welding of stainless steel. Reversed po-

larity will produce deeper weld penetration and more consistent fusion when welding stainless steel sheets and light plates with direct current.

Since stainless steel has a lower melting point than mild steel, at least 20 percent less current is recommended than would ordinarily be used for mild steel. Also, the low thermal conductivity of stainless localizes the heat from the arc along the weld, again lowering current requirements.

Welding Procedure

To produce good welds, all butting edges should be squared. Stainless steel sheets 18 gage and lighter are fitted with no gap. Heavier gage sheets and plates are set up with gaps.

It is necessary to have the material to be welded free from scale, grease, and dirt in order to prevent weld contamination.

To start arc welding, the arc is struck by touching the metal electrode on the work and quickly withdrawing it a short distance (enough to maintain the proper arc). The tendency for the electrode to stick, or freeze, to the work may be overcome by using a striking motion similar to that used when striking a match. Any coating on the electrode tip must be removed before an arc can be struck.

To maintain the arc, the electrode should be fed continuously into the arc, to compensate for metal deposited, and also moved rapidly in a continuous movement in the direction of the welding.

To finish the weld or break the arc, the electrode should be held close to the work, thus shortening the arc, then moved quickly back over the finished bead.

In order to reduce weld oxidation and porosity, the arc should be as short as possible. Too long an arc is inefficient and increases spattering. Vertical and overhead welding require a short arc together with smaller diameter electrodes than are used in horizontal welding.

After welding, all slag, scale, and discoloration should be completely removed from the weld bead and adjacent base metal. See Fig. 15-5. Light weld discoloration may be removed electrolytically. Scale or oxide is best removed by

---

1. Airco

Fig. 15-5. Appearance of a butt and fillet weld made on stainless steel plate.

grinding, pickling, or sandblasting. When grinding, refinish with progressively finer grits. The smoother and cleaner the surface of any stainless part, the better the corrosion resistance.

Flat position welding[1].   In welding butt joints in the flat position, the current selected should be high enough to insure ample penetration with good wash-up on the sides. When several beads are required it is advisable to use a number of small beads rather than to try and fill up the groove with one or two passes. A fairly short arc should be maintained, and any weaving should be limited to $2\frac{1}{2}$ times the electrode diameter. In general, it is good practice to hold the electrode vertical or very slightly tilted in the direction of travel. The latter case should only be used with small diameter electrodes. For lack of a hard and fast rule it may be said that the correct position is one that gives a clean pool of metal and which solidifies uniformly as the work progresses. The

movement of the electrode across the pool controls the flow of the metal and slag. Any weave technique employed should be in the form of a U for best results.

Vertical welding[1].   The welding of butt joints in the vertical position progressing upward can be accomplished with a reduced current from that used in the flat positions for a given electrode diameter. Oscillation or whipping is not recommended but instead a motion in the form of a V may be used for the first pass. The point of the V is the root of the joint, hesitating momentarily at this point to assure adequate penetration and to bring the slag to the surface. The arc is then brought out on one side of the point about $\frac{1}{8}''$ and immediately returned to the root of the joint.

After the momentary pause at the root the procedure is repeated on the other side of the weld. Electrodes of $\frac{3}{16}''$ in diameter may be used on sections heavy enough to give rapid dissipation of the heat, but $\frac{5}{32}''$ diameter electrodes are the generally accepted maximum size for less massive sections. Welding should progress from the bottom upward except for single pass corner welds with $\frac{3}{32}''$ and smaller diameter electrodes, which may be used from the top downward. In the usual vertical fillet weld the electrode is inclined slightly below the horizontal position (holder end lower than arc end) and the weave motion should be rapid across the center of the bead.

Overhead welding[1].   Stringer beads are recommended when welding in the overhead position, since attempts to carry a large puddle of molten metal will result in an irregular convex bead. To assure best results, a short arc should be maintained and the machine should be set properly thereby providing good penetration of the base metal.

Horizontal fillet welding[1].   Horizontal fillet welds and lap welds require a machine setting high enough to give good penetration into the root of the joint and a well-shaped bead. Too low a current is easily recognized, since difficulty will be experienced in controlling or concentrating

1. Airco

the arc in the joint, and a convex bead with poor fusion will result.

When two legs of equal thickness are being welded, the electrode should be held at equal distance from each face and tilted slightly forward in the direction of travel. If one leg is lighter than the other, the electrode should be pointed towards the heavier leg. Undercutting on the vertical is caused by dwelling too long on that leg or by too high a current.

Welding Low Alloy Molybdenum Steels

Two of the more common molybdenum alloy steels are known as carbon-moly and chrome-moly. Carbon-moly steels have high strength especially at high temperatures. For this reason they are used extensively in piping systems where high pressures at high temperatures are encountered.

Chrome-moly steels are widely used for highly stressed parts. Many aircraft components are fabricated from this metal, such as landing gear supports, tubular frames and engine mounts. See Fig. 15-6.

Low-carbon Moly steels with carbon content below 0.15 percent are readily weldable in much the same way as mild steel. Usually E-7010, E-7011, E-7015, E-7016, E-7018, E-7020, E-70270

Fig. 15-6. Using the Tig process to weld stainless steel pipe. (American Welding Society.)

electrodes will provide approximately the same tensile strength as the base plate strength. Where the tensile strength of the weld need not be as high as that of the base metal, ordinary E-6010 electrodes can be used.

Some preheating is advisable at temperatures of 400° to 650°F (204° to 339°C). When the thickness of the metal is over ³/₈″, stress relief is often necessary after the welding is completed. Stress relief is accomplished at temperatures of 1200° to 1250°F (650 to 677°C) for periods of one hour for each inch of thickness followed by slow cooling in the furnace at a rate of 200° to 250°F (94 to 121°C) per hour. When the heat reaches about 150°F (65°C) the rest of the cooling may be done in still air.

Chrome-moly steels having a low carbon content may be welded with E-6012, E-6013, or E-6024 electrodes. These electrodes will pick up enough alloy from the parent metal to produce the necessary tensile strength in the weld. For higher carbon chrome-moly steels special low hydrogen electrodes of the E-7015 or 7016 type are used with adequate preheat and postheat treatment to prevent extreme brittleness of the metal at the fusion zone.

Welding Low Alloy Nickel Steels

The addition of nickel (3 to 5 percent) to steel greatly increases the elastic properties as well as the strength, toughness, and corrosion resistance.

To weld low alloy nickel steels where the tensile strength must be equal to the base plate, an E-7010 or E-7020 electrode is used. As a rule, standard E-6012 electrodes are best for welding thin sheets because they penetrate less than the E-70 series.

If there are persistent signs of cracking during welding, it is best to switch to a low hydrogen electrode of the E-7010, E-7015, or E-7016 series. On heavy sections, preheating to a dull red is generally advisable.

Welding Clad Steels

Clad steels are those which have a mild steel core and one or both sides covered with a thin

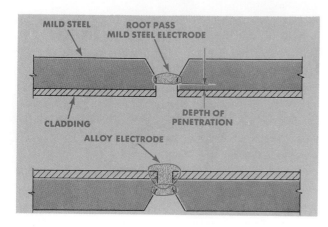

Fig. 15-7. Welding clad steel.

layer of some other metal. The covering may consist of zinc, aluminum, chromium, or various forms of nickel-copper alloys or chromium-nickel alloys.

Light-gage materials are usually welded with the Tig process. Shielded metal arc or Mig welding is preferred for thicker materials. In making any clad weld there should be a good joint fitting. With a poor fit-up the molten clad material is likely to penetrate the steel base, thereby forming hard, brittle welds with low ductility. Care also must be taken to use correct type electrodes. Remember the purpose of cladding is to impart a corrosive resistance coating to the base metal. Unless the electrodes correspond as much as possible in composition to the cladding material, the benefit of the cladding is lost.

Groove welds. A 1/16″ root face should be provided above the cladding in a groove weld. See Fig. 15-7. The lip acts as a barrier to prevent the cladding material from flowing into the mild steel. The weld should be made on the steel side with the root bead fused to the base of the steel lip but not penetrating into the cladding. Normally a larger bevel than used for regular steel welding and smaller diameter electrodes are necessary to control the depth of penetration of the root bead.

Welding electrodes for the mild steel side should be E-6020 or E-6030 when welding in the flat position. In vertical or overhead positions E-6010 or E-6011 electrodes are recommended. After sufficient passes are made with mild steel electrodes the final pass on the clad side should be with a correct alloy electrode.

Points to Remember

1. Some preheating is required when welding most alloy steels.

2. Use E-7016 or E-7018 to weld low-manganese steels.

3. Use the lowest possible current when welding high-manganese steel.

4. Best results are obtained with E-308, E-309, or E-310 electrodes for high-manganese steel welding.

5. Peen each bead to minimize cracking when welding manganese steel.

6. Arc welding is more satisfactory when welding stainless steel material over 20 gage in thickness.

7. When welding stainless steel, use a short arc with only a slight weaving motion.

8. Be sure to use the right kind of electrodes for the type of stainless steel to be welded.

9. In flat position welding of stainless steel hold the electrode vertical or tilted only slightly in the direction of travel.

10. When making vertical welds on stainless steel, avoid any whipping action of the electrode, instead use a motion in the form of a V.

11. Use E-70XX type electrodes to weld low-carbon moly steels.

12. Keep preheating temperatures between 400° to 650°F (204° to 339°C) for low-carbon moly steels.

13. Chrome moly steels can be welded with E-6012, E-6013, or E-6024 electrodes.

14. Use E-7010 or E-7020 electrodes to weld low-alloy nickel steel.

15. If cracks appear in the process of welding low-alloy nickel steel, switch to E-7010, or to E-7016 electrodes.

QUESTIONS FOR STUDY AND DISCUSSION

1. Why is some form of preheating recommended when arc welding alloy steels?

2. What is the purpose of a clip test and how is it conducted?

3. What are some of the basic characteristics of austenitic manganese steel?

4. Why must the lowest possible current be used when welding manganese steel?

5. What type of electrode produces the most satisfactory results on manganese steel?

6. What should be done to prevent cracks forming when welding high-manganese steel?

7. How are stainless steels classified?

8. What are the particular qualities of stainless steel that makes this metal so valuable?

9. Why are chill plates frequently used when one is welding stainless steel?

10. How does the symbol classification of stainless steel electrodes differ from mild steel electrodes?

11. Why is less current required in welding stainless steel?

12. Why are martensitic types of stainless steels more difficult to weld?

13. What is the difference between low-carbon moly steel and chrome-moly steel?

14. What kind of electrodes are usually recommended for welding low-carbon moly steel?

15. Why is some form of stress relief often used after welding moly steels?

16. What are clad steels?

17. What precautions should be taken to weld clad steels?

CHAPTER 16 *non-ferrous metals*

Non-ferrous metals are those which do not contain iron—such as aluminum, copper, brass, bronze, Monel, and Inconel. Continued experiment and development have now made possible successful welding of these metals. Actually, with a little practice, you can master the art of welding non-ferrous metals as easily as you can weld most steels.

Welding Aluminum

Aluminum plays an important role in the manufacture of every conceivable type of equipment ranging from railroad cars, trucks, and buildings, to cooking utensils. See Fig. 16-1. In most cases, welding is used in fabricating these aluminum products.

Aluminum weighs only about one-third as much as other commonly used metals; it has a high strength-to-weight ratio, is highly resistant to corrosion, possesses great electrical conductivity, and permits unusual ease in fabrication.

Most types of aluminum used in making commercial products are readily weldable either by oxy-acetylene, metallic arc, or gas-shielded arc process.

The welding technique employed depends on such factors as experience of the welder and kind of work to be done. In some instances it is more satisfactory to weld with the oxy-acetylene

flame, while for other jobs the task is simplified if the welding is performed with the metallic arc or gas-shielded arc.

Classification of Aluminum

Aluminum is classified into three main groups: commercially pure aluminum, wrought alloys, and casting alloys. *Commercially pure aluminum* has a purity of at least 99 percent, with the remaining 1 percent consisting of iron and silicon. Since it lacks alloying ingredients, this aluminum does not possess a very high tensile strength. One of its chief qualities is its ductility, which makes the metal especially adaptable for pressing and forming operations.

Wrought alloys are those which contain one or more alloying elements and possess a much higher tensile strength. The main alloying elements are copper, manganese, magnesium, silicon, chromium, zinc, and nickel. The wrought aluminums are either non-heat-treatable or heat-treatable. The non-heat-treatable types are those which are not hardenable by any forms of heat-treatment; their varying degrees of hardness are controlled only by cold working. The heat-treatable alloys are those in which hardness and strength are further improved by subjecting them to heat-treating processes.

Casting alloys are used to produce aluminum

Fig. 16-1. Welding is used in fabricating many aluminum products, from quarter million pound jet airplanes, to kitchen equipment. (Aluminum Company of America)

castings, the metal being poured into a sand or permanent metal mold. A great many of the castings are weldable, but extreme care must be exercised when welding the heat-treated types to prevent any loss of the characteristics achieved by the heat-treating process.

Method of Designating Aluminum Alloys

Wrought aluminum and wrought aluminum alloys are designated by a four digit system. The first digit shows the alloy groups. See Table 16-1. Thus 1xxx indicates aluminum of at least 99.00 percent purity, 2xxx indicates an aluminum alloy in which copper is the major alloying element, and 3xxx an aluminum alloy with manganese as the major alloying element.

Pure aluminum. In the 1xxx groups for aluminum of at least 99.00 percent purity, the last two of the four digits in the designation indicate the minimum aluminum percentage. These dig-

TABLE 16-1. DESIGNATIONS FOR ALUMINUM.

Aluminum, 99.00 percent minimum and greater 1xxx

	MAJOR ALLOY	AA NUMBER
	Copper	2xxx
Aluminum	Manganese	3xxx
alloys	Silicon	4xxx
grouped	Magnesium	5xxx
by major	Magnesium and Silicon	6xxx
alloying	Zinc	7xxx
elements	Other element	8xxx

its are the same as the two digits to the right of the decimal point in the minimum aluminum percentage when it is expressed to the nearest 0.01 percent.

The second digit in the designation indicates modifications in impurity limits. If the second digit in the designation is zero, it indicates that there is no special control on individual impurities; while integers 1 through 9, which are assigned consecutively as needed, indicate special control of one or more individual impurities. Thus 1030 indicates 99.30 percent minimum aluminum without special control on individual impurities and 1130, 1230, 1330, etc. indicate the same purity with special control on one or more impurities. Likewise 1075, 1175, etc. indicate 99.75 percent minimum aluminum; and 1097, 1197, etc., indicate 99.97 percent.

Aluminum alloys. In the 2xxx through 7xxx alloy groups the last two of the four digits in the designation have no special significance but serve only to identify the different alloys in the group.

The second digit in the alloy designation indicates alloy modifications. If the second digit in the designation is zero, it indicates the original alloy; while integers 1 through 9, which are assigned consecutively, indicate the various alloy modifications.

Temper designations. Temper designation is based on the sequence of basic treatments used to produce the various tempers. The temper designation follows the alloy designation and is separated from it by a dash. Basic temper designations consist of letters. Subdivisions of the basic tempers, where required, are indicated by digits following the letter, and they specify sequences of basic treatments.

Basic temper designations:

F = as fabricated, no effort is made to control the mechanical properties

O = fully annealed—applies to wrought products only

H = strength improved by strain-hardening—applies to wrought products only. Digits which follow the letter indicate the manner and extent to which strain-hardening is achieved

W = Solution heat-treated—applies to alloys which age at room temperature after solution heat-treatment

T = Thermally treated to produce stable tempers. The T is always followed by one or more digits

Subdivisions of H temper. The first digit which follows the H-temper aluminum indicates specific operations, as follows:

H1 = Strain-hardened only
H2 = Strain-hardened and partially annealed
H3 = Strain-hardened and stabilized

The second digit following the H1, H2 and H3 classifications indicates the degree of strain-hardening. These numbers run from 1 to 8, with 8 representing a temper having the highest ultimate tensile strength.

Examples:

1100-0—fully annealed
3003-H18—cold worked, hardest temper
3003-H24—cold worked and partially annealed
5052-H32—cold worked and stabilized

Subdivisions of T temper. The thermally treated aluminums are designated by temper numbers of 1 to 10 and have the following meanings. (The temper designation follows the alloy designation and is separated from it by a dash.)

T1 = Strength is increased by room temperature aging

T2 = Annealed to improve ductility and dimensional stability, cast products only

T3 = Solution heat-treated and then cold worked

T4 = Solution heat-treated and naturally aged to a stable condition

T5 = Artificially aged

T6 = Solution heat-treated and then artificially aged

T7 = Solution heat-treated and then stabilized

T8 = Solution heat-treated, cold-worked, and then artificially aged

T9 = Solution heat-treated, artificially aged, and then cold-worked

T10 = Artificially aged and then cold worked

Examples:

2017-T4
6070-T2
7075-T8

Cast aluminum designation system. Cast aluminum alloys have the same four digit numerical designation as used to identify aluminum and aluminum alloys. The first digit indicates the alloy group as shown in Table 16-2.

TABLE 16-2.   DESIGNATIONS FOR CAST ALUMINUM.

Aluminum, 99, percent or greater purity	1xx.x
Aluminum alloys	
Copper	2xx.x
Silicon, with added Copper and/or Magnesium	3xx.x
Silicon	4xx.x
Magnesium	5xx.x
Zinc	7xx.x
Tin	8xx.x
Other elements	9xx.x

The second two digits identify the aluminum alloy or indicate the aluminum purity. The last digit, which is separated from the others by a decimal point, indicates the product form: that is, casting or ingot. A zero is used to designate a casting and 1 or 2 as an ingot. Numbers 1 and 2 simply specify the chemical composition limits of the ingot.

Metallurgical Aspects[1]

In high-purity form aluminum is soft and ductile. Most commercial uses, however, require greater strength than pure aluminum affords. This is achieved in aluminum first by the addition of other elements to produce various alloys which singly or in combination impart strength to the metal. Further strengthening is possible by means which classify the alloys roughly into two categories, non-heat-treatable and heat-treatable.

Non-heat-treatable alloys. The beginning or initial strength of alloys in this group depends upon the hardening effect of elements such as manganese, silicon, iron and magnesium, singly or in various combinations. The non-heat-treatable alloys are usually designated, therefore, in the 1000, 3000, 4000, or 5000 series. Since these alloys are work-hardenable, further strengthening is made possible by various degrees of cold working, denoted by the H series of tempers. Alloys containing appreciable amounts of magnesium when supplied in strain-hardened tempers are usually given a final elevated-temperature treatment called stabilizing to insure stability of properties.

Heat-treatable alloys. The initial strength of alloys in this group is enhanced by the addition of alloying elements such as copper, magnesium, zinc, and silicon. Since these elements singly or in various combinations show increasing solid solubility in aluminum with increasing temperature, it is possible to subject them to thermal treatments which will impart pronounced strengthening.

The first step, called heat treatment or solution heat treatment, is an elevated-temperature pro-

---

1. Aluminum Company of America, Aluminum Association.

cess designed to put the soluble element or elements in solid solution. This is followed by rapid quenching, usually in water, which momentarily freezes the structure and for a short time renders the alloy very workable. It is at this stage that some fabricators retain this more workable structure by storing the alloys at below freezing temperatures until they are ready to form them. At room or elevated temperatures the alloys are not stable after quenching, and precipitation of the constituents from the supersaturated solution begins.

After a period of several days at room temperature, a process termed aging or room-temperature precipitation begins and the alloy becomes considerably stronger. Many alloys approach a stable condition at room temperature, but some alloys, particularly those containing magnesium and silicon or magnesium and zinc, continue to age-harden for long periods of time at ambient or ordinary room temperature.

By heating for a controlled time at slightly elevated temperatures, even further strengthening is possible and properties are stabilized. This process is called artificial aging or precipitation hardening. By the proper combination of solution heat treatment, quenching, cold working and artificial aging, the highest strengths are obtained.

Clad alloys. The heat-treatable alloys in which copper or zinc are major alloying constituents, are less resistant to corrosive attack than the majority of non-heat-treatable alloys. To increase the corrosion resistance of these alloys in sheet and plate form they are often clad with high-purity aluminum, a low magnesium-silicon alloy, or an alloy containing 1 percent zinc. The cladding, usually from 2½ to 5 percent of the total thickness on each side, not only protects the composite due to its own inherently excellent corrosion resistance, but also exerts a galvanic effect which further protects the core material.

Special composite may be obtained such as clad non-heat-treatable alloys for extra corrosion protection, for brazing purposes, or for special surface finishes. Some alloys in wire and tubular form are clad for similar reasons and experimental extrusions have also been clad.

WELDING CHARACTERISTICS OF ALUMINUM

Non-heat-treatable wrought aluminum alloys in the 1000, 3000, 4000 and 5000 series are weldable. The heat of the welding may remove some of the material's strength developed by cold working but the strength will never be below that of the material in its fully annealed condition. Nevertheless, the welding procedure should confine the concentration of heat in the narrowest zone possible.

The heat-treatable alloys in the 2000, 6000, and 7000 series can be welded except that the 2000 and 7000 are not generally recommended for oxy-acetylene welding, nor should the 7000 types be welded with the metallic arc. See Table 16-3.

Higher welding temperatures and speeds are needed to penetrate these alloys which is impossible with an oxy-acetylene flame. In most in-

TABLE 16-3. WELDABILITY OF ALUMINUM.

	WELDING PROCESS		
TYPE	GAS	METAL-ARC	GAS SHIELDED-ARC
1060	A	A	A
1100	A	A	A
3003	A	A	A
3004	A	A	A
5005	A	A	A
5050	A	A	A
5052	A	A	A
2014	X	C	C
2017	X	C	C
2024	X	C	C
6061	A	A	A
6063	A	A	A
6070	C	B	A
6071	A	A	A
7070	X	X	A
7072	X	X	A
7075	X	X	C

A—Readily weldable
B—Weldable in most applications—may require special technique
C—Limited weldability
X—Not recommended

stances resistance welding is preferred for high strength alloys.

Welding Procedure

The following paragraphs describe the general procedure for arc welding aluminum alloys.

Type of electrode. A heavy shielded electrode containing 95 percent aluminum and 5 percent silicon is recommended for welding most alloys. See Table 16-4 for correct diameter of electrode.

A common aluminum electrode is the E-Al-43 (AWS classification). This electrode has a coating that readily dissolves the aluminum oxide and provides a smooth operating arc. It is an all-position electrode using DC current with reverse polarity. The electrode can be used to weld aluminum sheets, plates or castings. It produces homogeneous, dense welds, free from cracks.

Position of the weld. While welding can be performed in any position, the task is simplified and the quality of the completed joint is more satisfactory if the welding is done in a flat position as shown in Fig. 16-2. Welding aluminum in an overhead position is particularly difficult and should be avoided whenever possible.

Fig. 16-2. Welding brackets to the head of a tank is simplified when performed in a flat position. (Aluminum Company of America)

Current setting. For the best results, use reversed polarity. See Table 16-4 for the correct setting for various metal thicknesses.

Type of joint. Due to the difficulty of controlling the arc at low currents, it is not very practical to arc weld material less than 1/16″ in thickness.

TABLE 16-4. ELECTRODE SIZE, CURRENT, AND NUMBER OF PASSES FOR ARC WELDING ALUMINUM.

METAL THICKNESS (INCHES)	ELECTRODE (INCH) DIA	APPROXIMATE CURRENT (AMPERAGE)	NUMBER OF PASSES (BUTT, LAP, FILLET)
0.081	1/8	60	1
.102	1/8	70	1
.125	1/8	80	1
.156	1/8	100	1
.188	5/32	125	1
.250	3/16	160	1
.375	3/16 for laps and fillets 1/4 for butts	200	2
.500	3/16 for laps and fillets 1/4 for butts	300	3
1.000	5/16	450	3

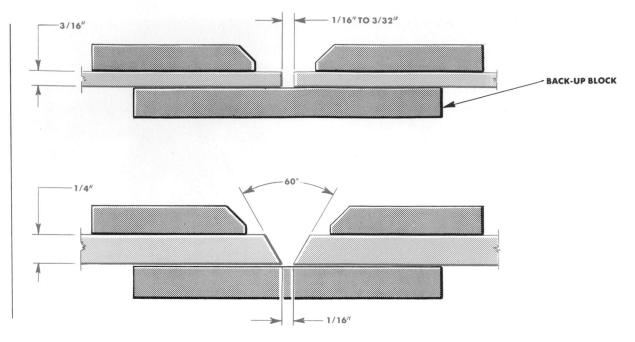

Fig. 16-3. Bevel the edges for welding heavier aluminum.

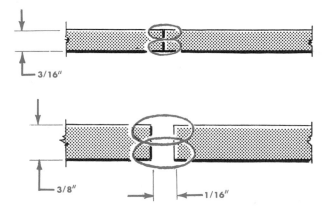

Fig. 16-4. For some aluminum welds run a bead on both sides.

Since it is easy to secure good penetration in aluminum, no edge preparation is necessary on material ¼″ or less in thickness. On heavier metal, it is better to bevel the edges as shown in Fig. 16-3. For some butt welds, beveling can be dispensed with on stock ³/₁₆″ to ³/₈″ in thickness by running a bead on both sides as in Fig. 16-4. Copper back-up blocks should be used when-ever possible, especially on plates ⅛″ or less in thickness.

Striking the arc. Brush the end of the electrode across the surface as in striking a match. During arc starting, the electrode melts very fast, and the metal solidifies rapidly when the arc is extinguished. If you try to just touch the plate, the electrode will freeze.

Manipulating the electrode. Keep the arc as short as possible, with the electrode coating almost touching the molten pool of metal. Hold the electrode about perpendicular to the work at all times. Direct the arc so both edges of the joint are uniformly heated. Move the electrode either forward or backward along the seam and advance it at such a rate as to produce a uniform bead as illustrated in Fig. 16-5. Refrain from weaving the electrode as in welding mild steel. Just keep it moving in a straight line. Before starting a new electrode, remove the slag from the crater at approximately one inch back of the crater. Then quickly move the electrode back about ½″ over the finished weld and proceed forward as soon as the crater is completely remelted.

Fig. 16-5. This is how an aluminum butt weld should appear.

When making a lap weld or fillet weld, hold the electrode so that the angle between the electrode and the plate is approximately 45°. Manipulate the electrode with a small rotary motion, playing the arc first on the vertical member and then on the horizontal plate.

Cleaning the finished weld. After a weld is completed, it is important that you remove the slag over the bead. This slag is solidified flux from the coating on the electrode. If it is not removed, it will attack the aluminum, especially when moisture is present. To remove the flux, crack it off with a chipping tool and then give it a vigorous brushing with a wire brush. For some welded aluminum products the slag is removed by means of a diluted acid solution followed by a warm water rinse.

Welding Aluminum Castings

Most aluminum alloy castings can be welded by the metallic arc process as shown in Fig. 16-6. However, considerable precautions must be taken to prevent the formation of cracks during contraction on cooling and the loss of mechanical properties through excessive heat. To weld a casting successfully proceed as follows:

1. Thoroughly remove oil, grease, and dirt from the weld area with a suitable solvent.

2. Chip out a 45° bevel on sections ³/₁₆″ thick or heavier.

3. Clamp the pieces to be welded to hold them in correct alignment.

4. Preheat the entire casting between 500° and 800°F (260° to 427°C) in a furnace or with a gas torch. Be careful not to overheat the casting.

5. Use an electrode having approximately the same composition as the base metal. The E-Al-43 electrodes usually will produce a good weld.

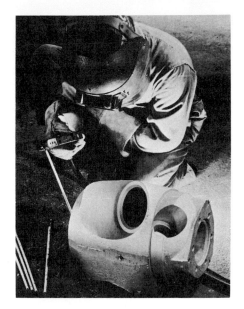

Fig. 16-6. Arc welding an aluminum alloy casting. (Hobart Brothers Co.)

Welding Copper

Copper is a soft, tough, and ductile metal. It cannot be heat-treated but will harden when cold-worked. Commercially available coppers are divided into two groups: oxygen-bearing copper and oxygen-free copper.

Oxygen-bearing copper. This electrolytic tough pitch copper is practically 99.9 percent pure and is considered to be the best conductor of heat and electricity. A small amount of oxygen in the form of copper oxide is uniformly distributed throughout the metal but it is insufficient to affect the ductility of the copper. However, if heated above 1680°F (965°C) for prolonged periods the copper oxide tends to migrate to the grain boundaries, causing a reduction in strength and ductility. Also when exposed to this temperature the copper will absorb carbon monoxide and hydrogen which react with the copper oxide and release carbon dioxide or water vapor. Since these gases are not soluble in copper they exert pressure between the grains and produce internal cracking and embrittlement.

Oxygen-free copper. This group of copper contains a small percentage of phosphorus or some other deoxidizer, thereby leaving the metal free of oxygen and consequently no copper oxide. The absence of copper oxide gives the metal superior fatigue resistance qualities and better cold working properties over the oxygen-bearing copper.

Weldability. Oxygen-bearing copper is not recommended for gas welding because it causes embrittlement. Some welds can be made with the shielded metal-arc process in situations where the tensile strength requirements are extremely low (19,000 psi or less), providing a high welding current and high travel speeds are used. The high current and speed will not allow embrittlement to develop.

Deoxidized copper is the most widely used type for fabrication by welding. A properly made weld will have a tensile strength of about 30,000 psi. This copper can be welded with all standard welding processes including oxy-acetylene, shielded metal-arc, gas tungsten-arc, and gas metal-arc. Since copper has a very high coefficient of expansion, contraction precautions must be taken. Provide suitable jigging and clamping to prevent movement while cooling. Contraction forces will often cause cracking during the cooling temperature range.

Due to the rapid heat transfer of copper, the metallic arc is not too practical for welding copper sheets over 1/4″ in thickness. Operators prefer the carbon arc (Chapter 29) or some gas shielded arc process.

Special metallic arc electrodes have been developed to weld sheet copper, the most common are phosphor bronze, silicon bronze, and aluminum bronze.

For copper sheets 1/4″ or less in thickness, the metallic arc actually is the simplest process to use. With this method it is possible to concentrate the heat and bring about instant fusion of the parent metal and electrode.

The welding procedure for welding copper is much the same as welding other metals with coated electrodes.

Joint design must include relatively large root openings and groove angles. Tight joints should be avoided to prevent buckling, poor penetration, slag inclusions, undercutting, and porosity. Copper backing strips are often advisable.

Welding Brass

Brass is a copper-zinc alloy. Since application of heat tends to vaporize the zinc, arc welding this metal is somewhat difficult. When the zinc volatilizes, the zinc fumes and oxides often obscure vision and make welding hard to perform. Furthermore, the formation of oxides produces a dirty surface which ruins the wetting properties of the molten metal.

To arc weld brass, use heavily coated phosphor-bronze electrodes. Be sure there is an abundance of fresh air circulating around the area to remove the harmful zinc oxide fumes. Best results will be obtained in welding brass if small deposits of metal are made at a time.

Welding Bronze

Bronze is a copper-tin alloy. It possesses higher mechanical properties than either brass or copper. Actually it has about the same mechanical properties as mild steel, but with the corrosion resistance of copper. Hence, bronze is often used in fabricating products requiring superior fatigue-resisting qualities.

Since the thermal conductivity of bronze is near that of steel, this metal can easily be welded. Either the carbon arc or metallic arc may be used. With the metallic arc process, a heavily coated phosphor-bronze electrode is necessary and the current set in reversed polarity. It is very important that the metal be absolutely clean to get sound welds.

Welding Monel and Inconel

Monel metal is an alloy containing approximately 67 percent nickel, 30 percent copper, and small quantities of other ingredients such as iron, aluminum, and manganese.

Inconel is an alloy having about 80 percent nickel, 15 percent chromium, and 5 percent iron.

Both Monel and Inconel are not hardened by regular heat treating processes. Their high strength is obtained by cold working such as rolling or drawing. These metals are extensively used as corrosion resisting linings in tanks and liquid carrying vessels.

Monel and Inconel can be welded with good results by the metallic arc process. The operation is performed almost with as much ease as in welding mild steel. Although these metals may be welded in any position, better results are obtained if welded in a flat position. In general, arc welding should not be attempted on sheets lighter than 0.050″ (18 gage).

The procedure for arc welding Monel or Inconel is as follows:

1. Remove the thin, darkly colored oxide film around the area to be welded. The oxide can be removed by grinding, sandblasting, rubbing with emery cloth, or pickling.

2. No preheating is necessary to arc weld these metals.

3. Use a heavy coated electrode especially designed for welding Monel and Inconel. Reverse polarity will produce the best results.

4. For welds in a flat position, hold the electrode at an angle of about 20° from the vertical, ahead of the puddle. In this position it is easier to control the molten flux and to eliminate slag trappings. To make welds in other positions, hold the electrode approximately at right angles to the plate.

5. Whenever it is necessary to withdraw the electrode, draw the arc slowly from the crater. Such a procedure permits a blanket of flame to cover the crater, protecting it from oxidation while the metal solidifies.

6. Avoid depositing wide beads. Hold the arc weaving motion to a minimum.

Points to Remember

Welding Aluminum

1. Use a heavy-coated electrode containing about 95 percent aluminum and 5 percent silicon such as E-Al-43.

2. Try to place the weld in a flat position.

3. Use a low current in reverse polarity.

4. Bevel the edges of metal ¼″ or more in thickness.

5. Use a brushing motion to start the electrode arc.

6. Keep the arc as short as possible and avoid any weaving motion.

7. Clean the bead after completing the weld.

8. To weld aluminum casting, remove all oil, grease, and dirt from the weld area, plus any oxide coating of aluminum.

9. Chip out a groove in the area to be welded.

10. Preheat casting between 500° and 800°F (260 to 427°C).

Welding Copper

1. Use metallic arc process on sheets ¼″ or less in thickness.

2. Use phosphor-bronze heavily coated electrodes.

3. Whenever possible use the carbon arc to weld heavy copper sections.

4. Carry out metallic arc process on copper the same as in welding mild steel.

Welding Brass

1. Use heavily coated phosphor-bronze electrodes.

2. Make small deposits of beads at a time.

3. Have plenty of fresh air circulating to remove harmful fumes.

Welding Bronze

1. Use heavily coated phosphor-bronze electrodes.

2. Set current in reversed polarity.

3. Clean edges to be welded.

4. Follow same technique as in welding mild steel.

Welding Monel and Inconel

1. Place metal in flat position if at all possible.

2. Do not weld sheets lighter than 18 gage in thickness.

3. Remove the oxide film from the surface to be welded.

4. Do not preheat.

5. Use heavily coated electrodes especially designed for these metals.

6. Use reverse polarity current.

7. Withdraw the arc slowly from the crater to prevent oxidation.

8. Deposit narrow beads.

QUESTIONS FOR STUDY AND DISCUSSION

1. What is meant by a non-ferrous metal?

2. What are some of the outstanding properties of aluminum?

3. What is the difference between non-heat-treatable and heat-treatable aluminum?

4. In the four digit system used to identify aluminum, what does each digit represent?

5. What temper designations are used for non-heat-treatable aluminum?

6. Why are some heat-treatable aluminum alloys difficult to weld?

7. What kind of an electrode should be used to weld non-heat-treatable aluminum?

8. When should the edges of aluminum be beveled for welding?

9. Why should a brushing motion be used in starting an aluminum electrode?

10. Why should the slag be removed from a weld made on aluminum?

11. What type of copper is the easiest to weld?

12. Why is the metallic arc process better for welding copper sheet which is less than $1/4''$ in thickness?

13. Why is it difficult to arc weld brass?

14. Brass is an alloy consisting of what elements?

15. What are the principal alloying ingredients in bronze?

16. What is the advantage of bronze over brass or copper?

17. What are the chief ingredients in Monel and Inconel?

18. What must be done to the surface of Monel and Inconel before welding?

19. When welding Monel or Inconel what type of current should be used?

20. In what position should the electrode be held when welding Monel and Inconel?

CHAPTER 17 gas tungsten arc—Tig

The primary consideration in any welding operation is to produce a weld that has the same properties as the base metal. Such a weld can only be made if the molten puddle is completely protected from the atmosphere during the welding process. Otherwise atmospheric oxygen and nitrogen will be absorbed in the molten puddle, and the weld will be porous and weak. In gas-shielded arc welding, a gas is used as a covering shield around the arc to prevent the atmosphere from contaminating the weld.

Originally gas-shielded arc welding was developed to weld corrosion resistant and other difficult-to-weld metals. Today the various gas-shielded arc processes are being applied to all types of metals. Gas-shielded arc welding will eventually displace much of the Shielded Metal-Arc and Oxy-Acetylene production welding due to the superiority of the weld, greater ease of operation, and increased welding speed. In addition to manual welding, the process can be automated, and in either case can be used for both light and heavy gage ferrous and non-ferrous metals. See Fig. 17-1.

SPECIFIC ADVANTAGES OF GAS-SHIELDED ARC

Since the shielding gas excludes the atmosphere from the molten puddle, welded joints are stronger, more ductile, and more corrosion-resistant than welds made by most other welding processes. The gas-shielded arc particularly simplifies the welding of non-ferrous metals since no flux is required. Whenever a flux is needed, there is the problem of removing traces of the flux after welding. Furthermore, with the use of flux there is always the possibility that slag inclusions and gas pockets will develop.

Another advantage of the gas-shielded arc is that a neater and sounder weld can be made because there is very little smoke, fumes, or sparks to contend with. Since the shielding gas around the arc is transparent, the welder can clearly observe the weld as it is being made. More important, the completed weld is clean and free of complications often encountered in other forms of metallic-arc welding.

Welding can be done in all positions with a minimum of weld spatter. Inasmuch as the weld surface is smooth, there is a substantial saving in production cost because little or no metal finishing is required. Also, there is less distortion of the metal near the weld.

Types of Gas-Shielded Arc Processes

There are two general types of gas-shielded arc welding. Gas Tungsten-Arc (Tig), and Gas Metal-Arc (Mig). Each has certain distinct advantages, however both produce welds that are deep penetrating and relatively free from atmospheric contamination. Most industrial metals can be welded easily with either the Tig or Mig

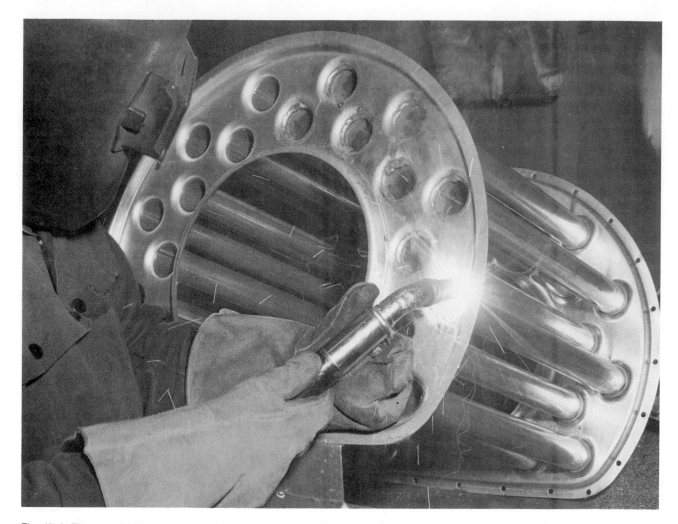

Fig. 17-1. The gas-shielded arc process is used extensively in industrial applications. (Hobart Brothers Co.)

process. These include such metals as aluminum, magnesium, low-alloy steel, carbon steel, stainless steel, copper, nickel, Monel, Inconel, titanium, and others.

Both welding processes can be semi-automatic or fully automatic. In semi-automatic the operator controls the speed of travel and direction. With the automatic process the weld size, weld length, rate of travel, and starting and stopping are all controlled by the equipment.

TIG Welding (GTAW)

In the gas-tungsten arc process, a virtually non-consumable tungsten electrode is used to provide the arc for welding. During the welding

cycle a shield of inert gas expels the air from the welding area and prevents oxidation of the electrode, weld puddle, and surrounding heat-affected zone. See Fig. 17-2.

In Tig welding, the electrode is used only to create the arc. It is not consumed in the weld. In this way it differs from the regular shielded metal-arc process, where the stick electrode is consumed in the weld. For joints where additional weld metal is needed, a filler rod is fed into the puddle in a manner similar to welding with the oxy-acetylene flame process. See Fig. 17-3.

This type of welding is often referred to as *Heliarc* (Linde) or *Heliwelding* (Airco) which are manufacturers' tradenames.

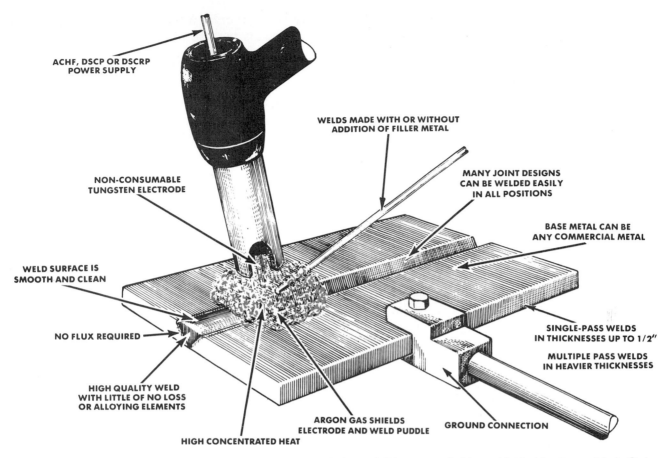

ACHF, DSCP OR DSCRP
POWER SUPPLY

WELDS MADE WITH OR WITHOUT
ADDITION OF FILLER METAL

MANY JOINT DESIGNS
CAN BE WELDED EASILY
IN ALL POSITIONS

NON-CONSUMABLE
TUNGSTEN ELECTRODE

BASE METAL CAN BE
ANY COMMERCIAL METAL

WELD SURFACE IS
SMOOTH AND CLEAN

NO FLUX REQUIRED

SINGLE-PASS WELDS
IN THICKNESSES UP TO 1/2"

MULTIPLE PASS WELDS
IN HEAVIER THICKNESSES

HIGH QUALITY WELD
WITH LITTLE OF NO LOSS
OR ALLOYING ELEMENTS

ARGON GAS SHIELDS
ELECTRODE AND WELD PUDDLE

GROUND CONNECTION

HIGH CONCENTRATED HEAT

Fig. 17-2. In Tig welding, a non-consumable tungsten electrode is used. It is surrounded by a shield of inert gas. (Linde Co.)

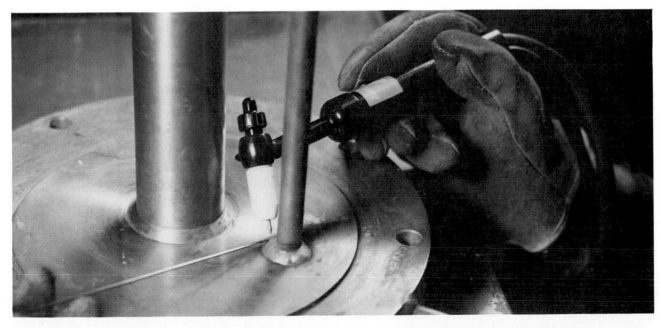

Fig. 17-3. In Tig welding, when additional weld metal is needed, a filler rod is fed into the weld puddle. (Airco)

WELDING MACHINES

Any standard DC or AC arc welding machine can be used to supply the current for Tig welding. However, it is important that the generator or transformer have good current control in the low range. This is necessary in order to maintain a stable arc. It is especially necessary when welding thin gage materials. If an old DC machine which has poor low-range current control is to be used, then it is advisable to install a resistor in the ground line between the generator and the work bench. Such a resistor will enable the electrical system to provide a very low, stable arc current.

An AC machine must be equipped with a high-frequency generator to supply an even current. Remember that in an AC welder, the current is constantly reversing its direction. Every time the current flow changes direction, there is a very short interval when no current is flowing. This causes the arc to be unstable, and sometimes go out. With a high-frequency generator in the system a more even current flow is possible. This stabilizes the arc.

Both the resistor for the DC machine, and the high frequency generator for the AC welder can be obtained from most welding equipment dealers. Automatic or semi-automatic controls are also manufactured. These can be installed on the machines to regulate the flow of gas and the water supply for cooling.

Specially designed machines for tungsten inert-gas welding with all of the necessary controls are available. Many power supply units are made to produce both AC and DC current. See Figs. 17-4 and 17-5.

The choice of an AC or DC machine depends on what weld characteristics may be required. Some metals are joined more easily with AC

Fig. 17-4. A complete facility for tungsten inert gas (Tig) welding. (Miller Electric Manufacturing Co.)

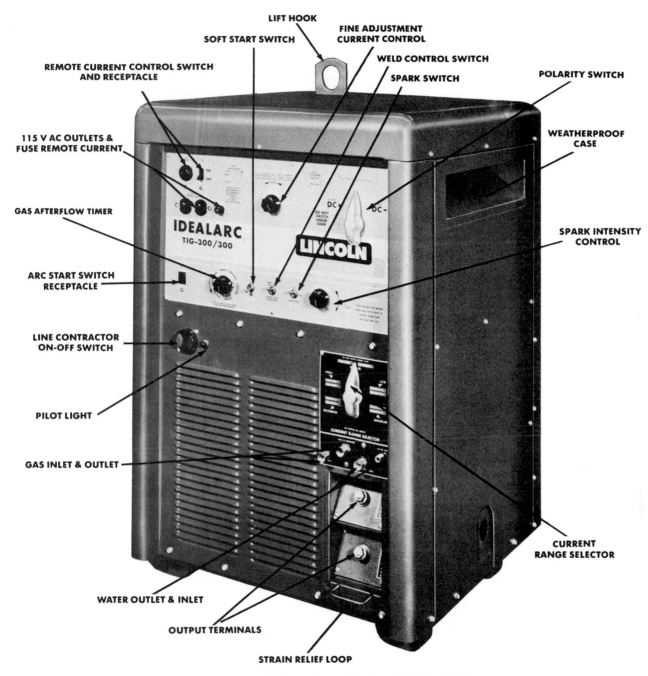

LIFT HOOK

SOFT START SWITCH

FINE ADJUSTMENT
CURRENT CONTROL

WELD CONTROL SWITCH

REMOTE CURRENT CONTROL SWITCH
AND RECEPTACLE

SPARK SWITCH

POLARITY SWITCH

115 V AC OUTLETS &
FUSE REMOTE CURRENT

WEATHERPROOF
CASE

GAS AFTERFLOW TIMER

SPARK INTENSITY
CONTROL

ARC START SWITCH
RECEPTACLE

LINE CONTRACTOR
ON-OFF SWITCH

PILOT LIGHT

GAS INLET & OUTLET

CURRENT
RANGE SELECTOR

WATER OUTLET & INLET

OUTPUT TERMINALS

STRAIN RELIEF LOOP

Fig. 17-5. Special machines are manufactured to do Tig welding. (The Lincoln Electric Co.)

current while with others better results are obtained when DC current is used. See Table 17-1. To understand the effects of the two different currents an explanation of their behavior in a welding process is necessary.

Direct Current Reverse Polarity (DCRP)

With direct current the welding circuit may be either straight or reverse polarity. When the machine is set for straight polarity the flow of

TABLE 17-1.   CURRENT SELECTION FOR TIG WELDING.

METAL	AC CURRENT with High Frequency Stabilization	DC CURRENT	
		Straight Polarity	Reverse Polarity
Magnesium up to 1/8″ thick	1	NR	2
Magnesium above 3/16″ thick	1	NR	NR
Magnesium castings	1	NR	2
Aluminum	1	NR	2
Aluminum castings	1	NR	NR
Stainless steel up to 0.050″	1	2	NR
Stainless steel 0.050″ and up	2	1	NR
Brass alloys	2	1	NR
Silver	2	1	NR
Hastelloy alloys	2	1	NR
Silver cladding	1	NR	NR
Hard-facing	1	2	NR
Cast iron	2	1	NR
Low carbon steel 0.015″ to 0.030″	2	1	NR
Low carbon steel 0.030″ to 0.125″	NR	1	NR
High carbon steel 0.015″ to 0.030″	2	1	NR
High carbon steel 0.030″ and up	2	1	NR
Deoxidized copper up to 0.090″	NR	1	NR

Key: 1. Excellent operation—best recommendation
       2. Good operation—second recommendation
       NR = Not recommended

electrons from the electrode to the plate creates considerable heat in the plate. In reverse polarity, the flow of electrons is from the plate to the electrode, thus causing a greater concentration of heat at the electrode. See Fig. 17-6. The intense heat at the electrode tends to melt off the end of the electrode and may contaminate the weld. Hence, for any given current, DCRP requires a larger diameter electrode than DCSP. For example, a 1/16″ diameter tungsten electrode normally can handle about 125 amperes in a straight-polarity circuit. However, if reverse polarity is used with this amount of current the tip of the electrode will melt off. Consequently, a 1/4″ diameter electrode will be required to handle 125 amperes of welding current.

Polarity also affects the shape of the weld. DCSP produces a narrow deep weld whereas DCRP with its larger diameter electrode and lower current forms a wide and shallow weld. See Fig. 17-7. For this reason *DCRP is never used in gas tungsten-arc welding except occasionally for welding aluminum and magnesium.* These metals have a heavy oxide coating which

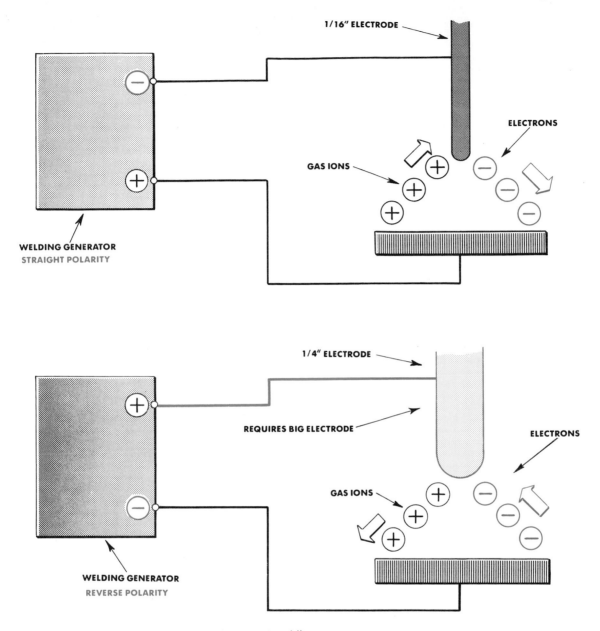

Fig. 17-6. Straight and reverse polarity in direct current welding.

is more readily removed by the greater current cleaning action of DCRP.

The same cleaning action is present in the reverse-polarity half of the AC welding cycle. No other metals require the kind of cleaning action that is normally needed on aluminum and magnesium. The cleaning action develops because of a bombardment of positive charged gas ions that are attracted to the negative charged work-piece. These gas ions when striking the metal have sufficient power to break the oxide and dislodge it from the surface. Generally speaking, better results are obtained in welding aluminum and magnesium with alternating current.

Direct Current Straight Polarity (DCSP)

Direct current straight polarity is used for welding most metals because better welds are

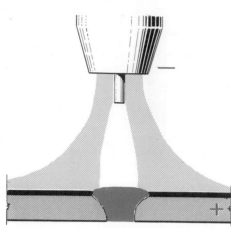

DC STRAIGHT POLARITY
DEEP PENETRATION—NARROW WELD

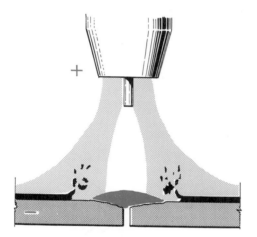

WELDING WITH DCSP PRODUCES DEEP PENETRATION BECAUSE IT CONCENTRATES HEAT AT THE JOINT

DC REVERSE POLARITY
SHALLOW PENETRATION — WIDE WELD

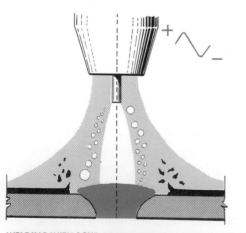

WELDING WITH DCRP PRODUCES GOOD CLEANING ACTION BUT WELD PENETRATION IS SHALLOW

A   H   F
DEEP PENETRATION—WIDE WELD

WELDING WITH ACHF COMBINES THE DESIRED CLEANING ACTION, ON THE POSITIVE HALF OF EACH CYCLE, WITH THE HEATING REQUIRED FOR GOOD WELD PENETRATION, ON THE NEGATIVE SWING

Fig. 17-7. The different types of operating current directly affect weld penetration, contour, and metal transfer. (Linde Co.)

achieved. With the heat concentrated at the plate the welding process is more rapid, there is less distortion of the base metal, and the weld puddle is deeper and narrower than with DCRP. Since more heat is liberated at the puddle, smaller diameter electrodes can be used.

Alternating Current (ACHF)

AC welding is actually a combination of DCSP and DCRP. Notice in Fig. 17-8A that half of each complete AC cycle is DCSP and the other half is DCRP.

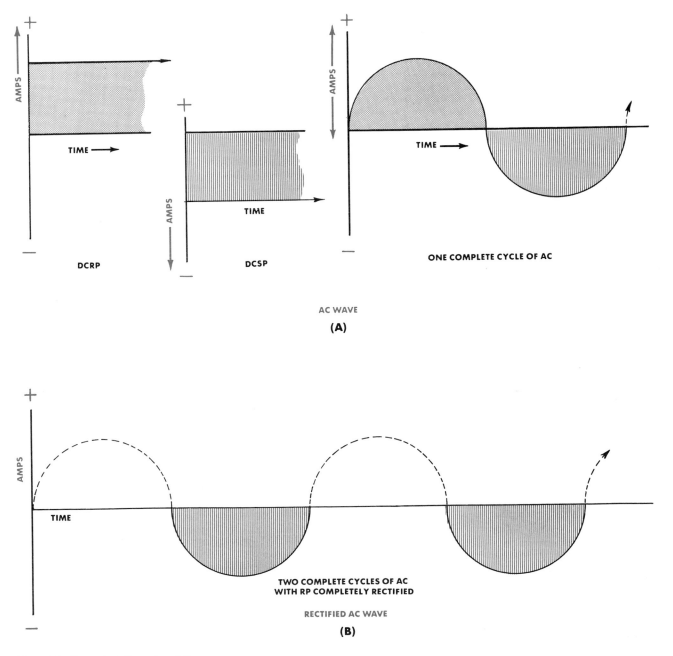

Fig. 17-8. Characteristics of an AC wave.

Unfortunately oxides, scale and moisture on the workpiece often tend to prevent the full flow of current in the reverse-polarity direction. If no current whatsoever flowed in the reverse-polarity direction, the current wave would resemble something as shown in Fig. 17-8B. During a welding operation the partial or complete stoppage of current flow (rectification) would cause the arc to be unstable and sometimes even go out. To prevent such rectification, AC welding machines incorporate a high-frequency current flow unit. The high-frequency current is able to jump the gap between the electrode and the workpiece, piercing the oxide film and forming a path for the welding current to follow.

WELDING EQUIPMENT

Torches

Manually operated welding torches are constructed to conduct both the welding current and the inert gas to the weld zone. These torches are either air or water-cooled. See Figs. 17-9 and 17-10. Air-cooled torches are designed for welding light gage materials where low current values are used. Water-cooled torches are recommended when the welding requires amperages over 200A. A circulating stream of water flows around the torch to keep it from overheating.

The tungsten electrode which supplies the welding current is held rigidly in the torch by means of a collet that screws into the body of the torch. A variety of collet sizes are available so different diameter electrodes can be used. Gas is fed to the weld zone through a nozzle which consists of a ceramic cup. Gas cups are threaded into the torch head to provide directional and distributional control of the shielding gas. The cups are interchangeable to accommodate a variety of gas flow rates.

Some torches are equipped with a gas lens to eliminate turbulence of the gas stream which tends to pull in air and cause weld contamination. Gas lenses have a permeable barrier of concentric fine-mesh stainless steel screens that fit into the nozzle. See Fig. 17-11.

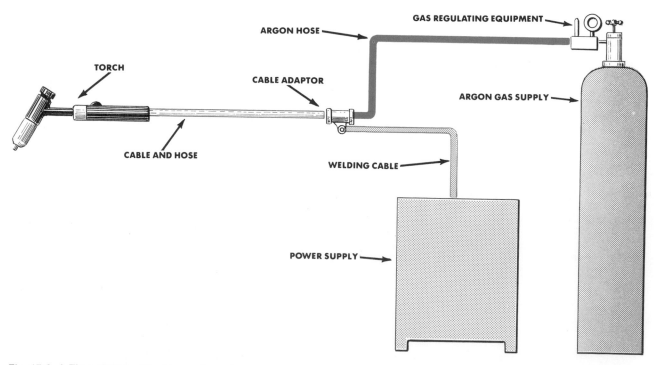

Fig. 17-9. A Tig welding unit with an air-cooled torch.

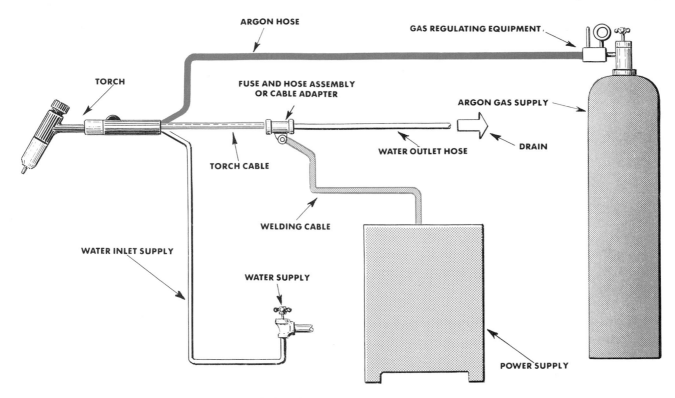

ARGON HOSE

GAS REGULATING EQUIPMENT

TORCH

FUSE AND HOSE ASSEMBLY
OR CABLE ADAPTER

ARGON GAS SUPPLY

WATER OUTLET HOSE

DRAIN

TORCH CABLE

WELDING CABLE

WATER INLET SUPPLY

WATER SUPPLY

POWER SUPPLY

Fig. 17-10. A Tig welding unit with a water-cooled torch.

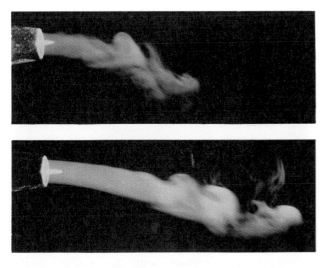

Fig. 17-11. Gas lens eliminates turbulence of the gas stream.

TABLE 17-2. APPROXIMATE CUP ORIFICE FOR TIG WELDING.

TUNGSTEN ELECTRODE diameter (inches)	CUP ORIFICE diameter (inches)
1/16	1/4 –3/8
3/32	3/8 –7/16
1/8	7/16–1/2
3/16	1/2 –3/4

Pressing a control switch on the torch starts the flow of both current and gas. On some equipment the flow of current and gas is energized by a foot control. The advantage of the foot control is that the current flow can be better controlled as the end of the weld is reached. By gradually decreasing the current it is less likely for a cavity to remain in the end of the weld puddle and less danger of cutting short the shielding gas.

Gas cups vary in size. The size to be used depends upon the type and size of torch and the diameter of the electrode. See Table 17-2. (It is a good practice to follow the manufacturer's recommendations.)

Electrodes

Basic diameters of non-consumable electrodes are $1/16''$, $3/32''$, and $1/8''$. They are either pure tungsten, or alloyed tungsten. The alloyed tungsten electrodes usually have one to two percent thorium or zirconium. The addition of thorium increases the current capacity and electron emission, keeps the tip cooler at a given level of current, minimizes movement of the arc around the electrode tip, permits easier arc starting, and the electrode is not as easily contaminated by accidental contact with the workpiece. The two percent thoria electrodes normally maintain their formed point for a greater period than the one percent type. The higher thoria electrodes are used primarily for critical sheet metal weldments in aircraft and missile industries. They have little advantage over the lower thoria electrode for most steel welds. The introduction of the striped electrode combines the advantage of the pure, low, and high thoriated tungsten electrodes. This electrode has a solid stripe of two percent thoria inserted in a wedge the full length of the electrode.

The diameter of the electrode selected for a welding operation is governed by the welding current to be used. Larger diameter tungsten electrodes are required with reversed polarity than with straight polarity. (See Tables 17-4 thru 17-9, pages 184 through 187 for recommended sizes of electrodes, current, and material thickness for Tig welding.)

Electrode shapes. To produce good welds the tungsten electrode must be shaped correctly. The general practice is to use a pointed electrode with DC welding, and a spherical end with AC welding. See Fig. 17-12.

It is also important that the electrode be straight, otherwise the gas flow will be off-center from the arc.

Shielding Gas

Shielding gas for gas tungsten-arc welding can be argon, helium, or a mixture of argon and helium. Argon is used more extensively because it is less expensive than helium. Argon is 1.4 times as heavy as air and 10 times as heavy as helium. There is very little difference between the viscosity of these two gases. Since argon is heavier than air it provides a better blanket over the weld. Moreover, there is less clouding during the welding process with argon and consequently it permits better control of the weld puddle and arc.

Argon normally produces a better cleaning action, especially in welding aluminum and magnesium with alternating current. With argon there is a smoother and quieter arc action. The lower arc voltage characteristic of argon is particularly advantageous in welding thin material because there is less tendency for burning through the metal. Consequently, argon is used most generally for shielding purposes when welding materials up to $1/8''$ in thickness by manual welding or when operating with low-speed welding machine.

The use of argon also permits better control of the arc in vertical and overhead welding. As a rule, the arc is easier to start in argon than in helium and for a given welding speed the weld produced is narrower with a smaller heat-affected zone.

Where welding speed is important, especially in machine welding or in welding heavy materials and metals having high heat-conductivity, helium is sometimes used. Higher welding speeds are possible with helium because higher-arc voltage can be obtained at the same current. Since the arc voltage in helium is higher, a lower current is possible to get the same arc

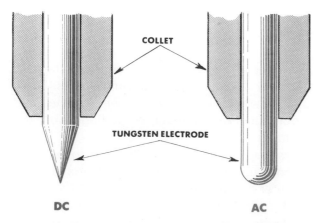

Fig. 17-12. Be sure the tungsten electrode is shaped with proper angle.

TABLE 17-3. SELECTION OF GASES.

METAL	TYPE	GAS	RESULT
Al	Manual welding	Argon	Better arc starting, cleaning action and weld quality; lower gas consumption
		Helium	High welding speeds possible
	Machine welding	Argon-Helium	Better weld quality, lower gas flow than required with straight helium
Mg	0–1/16″	Helium	Controlled penetration
	0–1/16″ +	Argon	Excellent cleaning, ease of puddle manipulation, low gas flows
Mild Steel	0–1/8″	Argon	Ease of manipulation, freedom from overheating
	0–1/8″ +		(Mig process preferred)
	Spot welding	Argon	Generally preferred for longer electrode life
			Better weld nugget contour
			Ease of starting, lower gas flow
		Argon-Helium	Helium addition improves penetration on heavy gage metal
	Manual welding	Argon	Better puddle control, especially for position welding
SS	Machine welding	Argon	Permits controlled penetration on thin gage material (up to 14 gage)
		Argon-Helium	Higher heat input, higher welding speeds possible on heavier gages
		Argon-Hydrogen (65%–35%)	Prevents undercutting, produces desirable weld contour at low current levels, requires lower gas flows
		Helium	Provides highest heat input and deepest penetration
Cu & Ni Cu–Ni Alloys (Monel & Inconel)		Argon	Ease of obtaining puddle control, penetration, and bead contour on thin gage metal
		Argon-Helium	Higher heat input to offset high heat conductivity of heavier gages
		Helium	Highest heat input for high welding speed on heavy metal sections
Ti		Argon	Low gas flow rate minimize turbulence and air contamination of weld; improved metal transfer; improved heat affected zone
		Helium	Better penetration for manual welding of thick sections (inert gas backing required to shield back of weld against contamination)
Si Bronze		Argon	Reduces cracking of this 'hot short' metal
Al Bronze		Argon	Less penetration of base metal

power. Hence, welds can be made at higher speeds inasmuch as the increase in power comes from the increase in voltage rather than in current. A mixture of argon and helium is often used in welding metals that require a higher heat input. See Table 17-3 for the recommended selection of gases.

Argon and helium are supplied in steel cylinders containing approximately 330 cubit feet at a pressure of 2000 psi. Either a single or two stage regulator may be used to control the gas flow or a specially designed regulator containing a flowmeter, as shown in Fig. 17-13, is used. The advantage of the flowmeter is that it provides better gas flow control. The flowmeter is calibrated either to show the flow of gas in cubic feet per hour (cfh) or liters per minute (lpm).

The correct flow of argon to the torch is controlled by turning the adjusting screw on the regulator. The rate of flow required depends on the thickness of the metal to be welded.

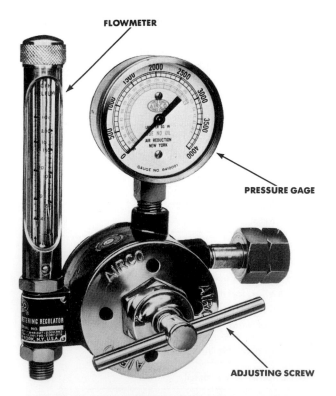

Fig. 17-13. An argon regulator with flowmeter. (Air Reduction Sales Co.)

Filler Rod

When Tig welding heavy gage metals a filler rod is required. Normally filler metal is not necessary on light gage materials since they can be made to readily flow together. Occasionally filler metal is added on thin pieces when it is essential to reinforce the joint.

Filler metal must be of the same composition as that of the base metal. Thus mild steel rods are used to weld low-carbon steel, aluminum rods for welding aluminum, copper rods for joining copper, and so on. Sometimes strips of the parent metal will serve as satisfactory filler metal.

Special filler rods are available for Tig welding. These rods are similar in classification to the filler wires used for Mig welding. See Chapter 19. The copper-coated mild steel rods used for oxyacetylene welding are not suitable for Tig welding since they tend to contaminate the tungsten electrode. The special Tig filler rods contain a greater amount of deoxidizers, thereby producing less spattering in the weld and sounder weld joints.

In general, the diameter of the filler rod should be about the same as the thickness of the metal to be welded.

Protective Equipment

A helmet like the one used in metallic arc welding is required to protect the welder from arc radiation. The shade of the lens to be used depends upon the intensity of the arc.

Besides the regulation helmet, protective clothing such as an apron and gloves must be worn whenever welding with gas tungsten arc.

Joint Preparation

Regardless of the type of joint used, proper cleaning of the metal is essential. All oxidation, scale, oil, grease, dirt, and other foreign matter must be removed.

The following are the most common joints designed for Tig welding:

Butt joint. For light materials the *square-edge butt joint,* as shown in Fig. 17-14, is the easiest to prepare and can be welded with or

Fig. 17-14. Square-edge butt joint.

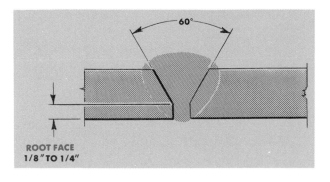

Fig. 17-15. Single-V butt joint.

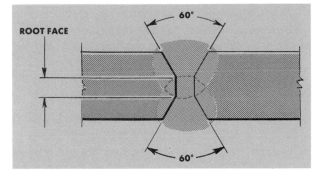

Fig. 17-16. Double-V butt joint.

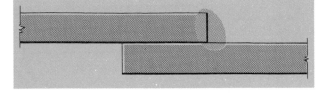

Fig. 17-17. Lap joint.

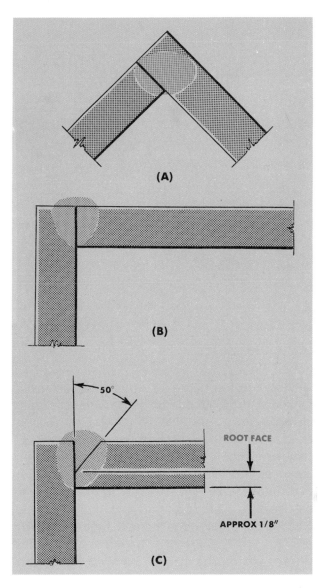

Fig. 17-18. Corner joints.

without filler rod. If the weld is to be made without filler rod, extreme care must be taken to avoid burning through the metal.

The *single-V butt joint* is preferable on material ranging in thickness from ³/₈″ to ¹/₂″ in order to secure complete penetration. The included angle of the V should be approximately 60° with a root face of about ¹/₈″ to ¹/₄″ as illustrated in Fig. 17-15.

Double-V butt joint is needed when the metal exceeds ¹/₂″ in thickness and the design is such that the weld can be made on both sides. With a double V there is greater assurance that penetration will be complete. See Fig. 17-16.

Lap joint. The only special requirement for making a good lap weld is to have the pieces in close contact along the entire length of the joint

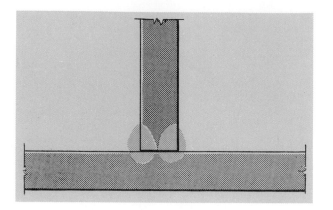

Fig. 17-19. T-joint.

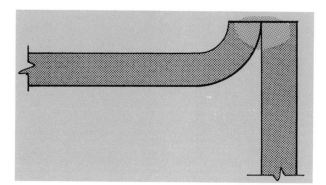

Fig. 17-20. Edge joint.

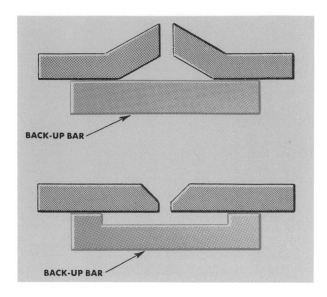

Fig. 17-21. A back-up bar should not touch the weld zone.

as shown in Fig. 17-17. On metal $^1/_4''$ or less in thickness, the weld can be made with or without filler rod. As a rule, the lap joint is not recommended for material exceeding $^1/_4''$ in thickness.

Corner joint. On light material up to $^1/_8''$ in thickness, no filler rod is required for a corner joint. See A, B of Fig. 17-18. With heavier metal the use of a filler rod is advisable. If the metal exceeds $^1/_4''$, one edge of the joint should be beveled. See Fig. 17-18C. The number of passes will depend on the size of the V and the thickness of the metal.

T-joint. Filler rod is necessary to weld T-joints regardless of the thickness of the metal. As a rule, a weld should be made on both sides of the joint as illustrated in Fig. 17-19. The number of passes over the seam will depend on the thickness of the material and the size of the weld to be made.

Edge joint. The edge joint is suitable only on very light material. No filler rod is needed to make this weld. See Fig. 17-20.

Backing the Weld

For many welding jobs, some suitable backing is necessary. On light gage metals, backing is used to protect the underside of the weld from atmospheric contamination and burning through. On heavier stock, back-up bars draw some of the heat generated by the intense arc.

The type of material used for back-up bars depends on the metal to be welded. Copper bars are suitable for stainless steel. When welding aluminum or magnesium, steel or stainless steel back-up bars are needed.

The back-up bar should be designed so it does not actually touch the weld zone. See Fig. 17-21.

WELDING PROCEDURE

Preliminary Steps

Before starting to weld, follow these steps:

1. Check all electrical circuit connections to make sure they are tight.

2. Check for the proper diameter electrode

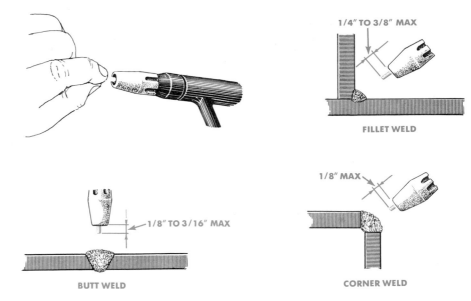

Fig. 17-22. Adjust the electrode so it extends beyond the edge of the gas cup.

1/4″ TO 3/8″ MAX

FILLET WELD

1/8″ TO 3/16″ MAX

BUTT WELD

1/8″ MAX

CORNER WELD

and cup size. (Follow manufacturer's recommendations.)

3. Adjust the electrode so it extends about $1/8''$ to $3/16''$ beyond the end of the gas cup for butt welding and approximately $1/4''$ to $3/8''$ for fillet welding. See Fig. 17-22.

4. Check the electrode to be certain that it is firmly held in the collet. Test it by placing the end against a solid surface and pushing the torch down gently but firmly. If the electrode moves into the nozzle, tighten the collet holder or gas cup. Be careful not to overtighten the gas cup because this will strip the threads very easily.

5. Set the machine for the correct welding amperage. (See Tables 17-4 to 17-9.)

6. If a water-cooled torch is to be used, turn on the water.

7. Turn on the inert gas and set to the correct flow.

Starting the arc. If you are using an AC machine, the electrode does not have to touch the metal to start the arc. To get the arc going, first turn on the welding current and hold the torch in a horizontal position about 2″ above the work. See Fig. 17-23. Swing the end of the torch toward the workpiece so the end of the electrode is $1/8''$ above the plate, as in Fig. 17-24. The high-frequency current will jump the gap between the electrode and the plate, establishing

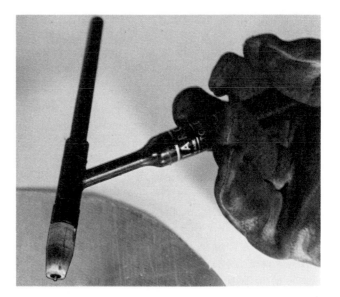

Fig. 17-23. To start the AC arc, first hold the torch in this manner.

the arc. *Be sure the downward motion is made rapidly* to provide the maximum amount of gas protection to the weld zone.

If a DC machine is used, hold the torch in the same position, but in this case the electrode must touch the plate to start the arc. When the arc is struck, withdraw the electrode so it is about $1/8''$ above the workpiece.

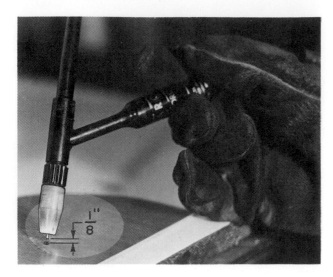

Fig. 17-24. Establish the AC arc by moving the tip of the electrode to within $\frac{1}{8}''$ of the plate.

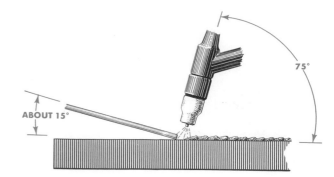

Fig. 17-25. Hold the torch at a 75° angle.

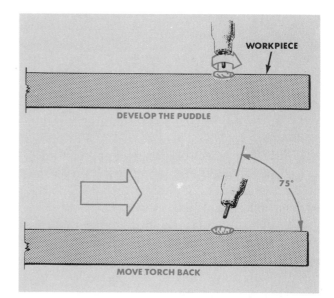

Fig. 17-26. Start the puddle by a circular movement of the torch. Once the puddle is formed, advance the puddle without circular motion.

To stop the arc on the AC or DC machine, snap the electrode back to the horizontal position. Make this movement rapidly to avoid marring or damaging the weld surface.

Some machines are equipped with a foot pedal to permit a gradual decrease of current. With such a control it is easier to fill the crater completely and prevent crater cracks.

If you are using a water-cooled cup, do not allow the cup to come in contact with the work when the current is on. The hot gases may cause the arc to jump from the electrode to the cup instead of to the plate, thereby damaging the cup. Be sure that the water flow is set according to the manufacturer's recommendations.

Welding a butt joint. Hold the torch at a 75° angle to the surface of the work as illustrated in Fig. 17-25. Preheat the starting point of the weld by moving the torch in small circles as shown in Fig. 17-26. As soon as the puddle becomes bright and fluid, move the torch slowly and steadily along the joint to form a uniform bead. No circular motion of the torch is necessary.

If the filler rod is to be added, hold the rod about 15° from the work as shown in Figs. 17-25 and 17-27. As the puddle becomes fluid, move the arc to the rear of the puddle and add the rod by touching the leading edge of the puddle. Remove the rod and bring the arc back to the leading edge of the puddle. Repeat this sequence for the entire length of the seam.

Welding a lap and T-joint. To weld a lap or T-joint without filler rod, first form a puddle on the bottom piece. After the puddle is formed, shorten the arc to about $\frac{1}{16}''$. Then rotate the torch directly over the joint until the pieces are joined. After the welding is started, no further torch rotation is necessary. Move the torch along the seam with the end of the electrode just above the edge of the top sheet.

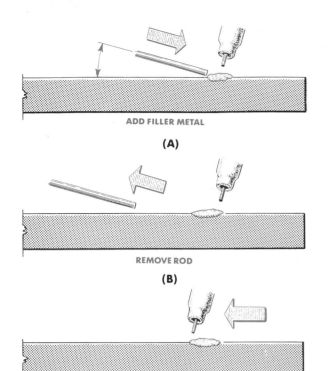

ADD FILLER METAL

(A)

REMOVE ROD

(B)

MOVE TORCH TO LEADING EDGE OF PUDDLE

(C)

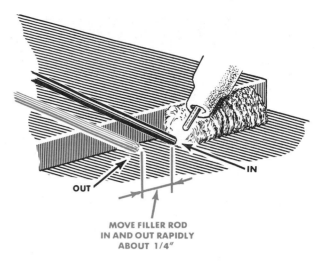

MOVE FILLER ROD
IN AND OUT RAPIDLY
ABOUT 1/4"

OUT   IN

Fig. 17-28. Welding a lap joint with a filler rod.

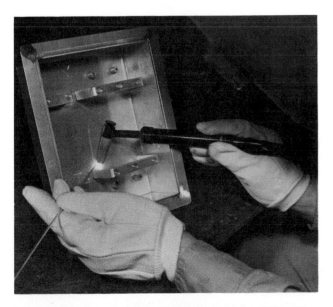

Fig. 17-27. Follow the steps in the top view in adding filler rod. The bottom view shows the weld in progress.

In welding a lap joint, you will find that the puddle forms a V shape. The center of the V is called a *notch,* and the speed at which this notch travels determines how fast the torch should be moved. Do not get ahead of it. See Fig. 17-28.

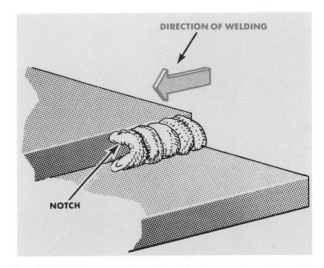

DIRECTION OF WELDING

NOTCH

Fig. 17-29. Advance the torch just fast enough so the notch continues to form and move forward.

Make certain that this notch is completely filled for the entire length of the seam. Otherwise there will be insufficient fusion and penetration.

If a filler rod is to be used on the lap joint, dip the end of the rod in and out of the puddle about every 1/4" travel of the puddle as shown in Fig. 17-29. Watch carefully to avoid laying bits of filler rod on the cold, unfused base metal. If you add just the right amount of rod at the correct moment, you will get a uniform bead of the proper proportions.

Welding a corner joint. A corner joint does not need any filler rod. Start the puddle at the

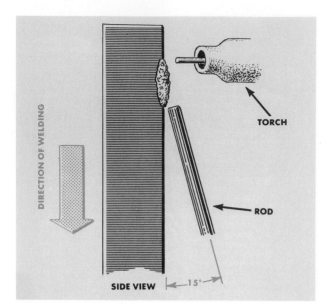

Fig. 17-30. Positioning the torch and rod for vertical downward welding.

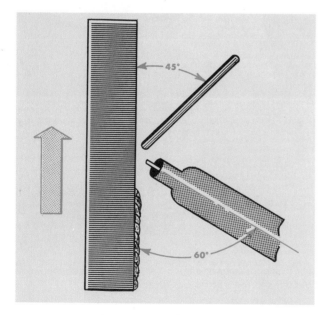

Fig. 17-31. Position of the torch and rod for vertical upward welding.

beginning edge and move the torch straight along the seam. If you find that the molten metal has a tendency to roll off the edge, your speed is too slow. On the other hand, if the completed portion of the weld is rough and uneven, then your speed is too fast.

Vertical welding. Vertical Tig welding on thin material is usually done in a downward position to achieve an adequate weld without burning through the metal. When filler rod is to be used,

add it from the bottom or leading edge of the puddle as shown in Fig. 17-30.

On heavier materials, an upward welding technique is preferred since deeper penetration can be achieved. Upward welding generally requires a filler rod. Notice in Fig. 17-31 the position of the torch and rod.

Horizontal welding. Start the arc about 1/2" from the right of the joint. Once the arc is started, move it to the beginning of the joint.

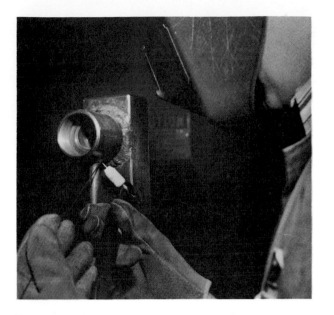

Fig. 17-32. Position of torch and rod for Tig horizontal welding.

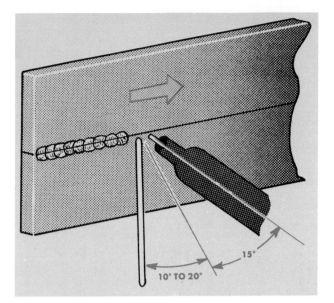

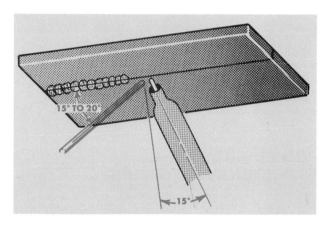

Fig. 17-33. Position of the torch and rod for Tig overhead welding.

Hold the torch and rod as shown in Fig. 17-32. Dip the filler rod into the front of the puddle and preferably on the high side as the torch is moved along the joint. While the rod is dipped into the puddle, withdraw the torch slightly. This allows the molten metal to solidify and prevents it from sagging. Keep the arc length as close as possible to the electrode diameter. Good arc length and correct speed eliminates undercutting and permits complete penetration.

Overhead welding. When Tig welding in an overhead position the current should be reduced between 5 to 10 percent to what is normally used for flat position welding. Slightly reduced current will give better control of the weld puddle. Both the torch and rod should be held somewhat as in flat welding. See Fig. 17-33. A smaller bead is advisable since it is less affected by the pull of gravity. Dip the filler rod in and out as in other welding positions. By pulling the arc back a little further the filler metal will solidify faster.

MECHANIZED TIG WELDING

Although mechanized Tig welding is not as common as mechanized Mig welding, nevertheless it does serve as a rapid and efficient welding process in many fabrication situations. In a Tig mechanized system the filler metal is automatically fed from a wire feeder that runs to a wire torch mounted behind the Tig welding torch. See Fig. 17-34.

Linde has developed what is known as a Tig Hot-Wire Welding process which produces qual-

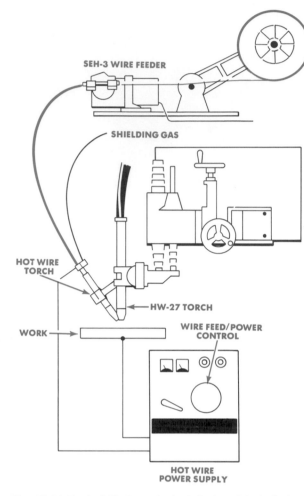

Fig. 17-34. Typical Tig hot-wire installation. (Linde Co.)

Fig. 17-35. The hot-wire welding torch is mounted behind the Tig welding torch to supply molten metal to the weld puddle. (Linde Co.)

ity welds at the speed of Mig welding. The filler wire is pre-heated to a molten state as it enters the weld puddle, leaving the Tig arc completely free to concentrate on the weld and not the wire. Wire melting is accomplished by passing an AC current through the wire by means of an AC power unit. The power unit is regulated so the wire reaches the melting point as it enters the weld puddle.

By attaching the hot wire torch behind the welding torch, Fig. 17-35, the operator is given an unobstructed view of the actual weld. By pre-heating the filler wire, weld porosity is eliminated and welds are made with greater quality and speed.

PULSED-CURRENT GAS TUNGSTEN-ARC WELDING

A pulsed-current process used with a gas tungsten-arc has been developed by Hobart Brothers Company whereby two levels of welding current, high and low, are used instead of the ordinary single level current. This high-and-low level generates a pulsating current or arc which produces overlapping spot welds. See Fig. 17-36. The arc spot welds are actually formed by the high level current. Then when the current switches to a low level the welds are allowed to cool and partially solidify between each spot. The welding controls are set so the arc spot overlaps will produce a continuous seam.

Pulsed current Tig welding can be manual or automatic and with or without filler rod. The process is particularly adapted for welding very

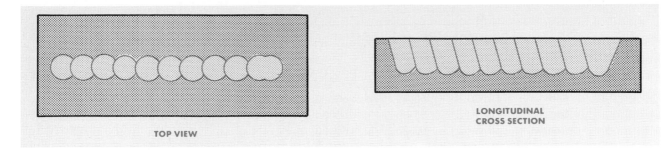

TOP VIEW

LONGITUDINAL
CROSS SECTION

Fig. 17-36. Weld produced by the pulsed-current—GTAW process. (Hobart Brothers Co.)

Fig. 17-37. Welding machine for Tig pulsed-current welding. (Hobart Brothers Co.)

thin materials where critical control of metallurgical factors is necessary. A pulsed current permits more tolerance of edge misalignment, greater variations in back-up bar and fixturing, better root penetration and less distortion.

Another distinct advantage of the pulsed arc is in welding curved seams or pipes. Normally with other welding techniques a change in current or travel speed must be made when changing posi-

tions around curved edges to insure uniform weld appearance. The pulsed current is more tolerant of welding position and allows continuous welding without having to vary travel speed, voltage or current. Fig. 17-37 illustrates a rectifier-type Tig pulsed-current welding machine.

TIG WELDING COMMON METALS

The actual technique of Tig welding common metals such as aluminum, magnesium, copper, stainless steel, carbon steels and low-alloy steels is virtually the same. In general these metals can be Tig welded more easily and with better results than by the oxy-acetylene or shielded arc process. Several specific welding characteristics for each of these metals are described in the paragraphs that follow.

Aluminum. Non-heat-treatable wrought aluminum alloys in the 1000, 3000 and 5000 series are readily weldable. The heat-treatable alloys in the 2000, 6000, and 7000 series can be welded but higher welding temperatures and speeds are needed. Elimination of weld cracking in these alloys can often be achieved with a rod having a higher alloy content than that of the base metal.

While welding can be performed in any position the task is simplified and the quality of the completed joint is more satisfactory if the weld is done in a flat position. Copper back-up blocks should be used wherever possible especially on plates $1/8''$ or less in thickness. In most cases, the torch should be moved in a straight line without a weaving motion.

TABLE 17-4.   TIG WELDING—ALUMINUM.

STOCK THICKNESS (inches)	TYPE OF JOINT	AMPERES, AC CURRENT			ELECTRODE (inches) dia	ARGON FLOW 20 psi		FILLER ROD (inches) dia
		FLAT	HORIZONTAL & VERTICAL	OVERHEAD		lpm	cfh	
1/16	Butt	60–80	60–80	60–80	1/16	7	15	1/16
	Lap	70–90	55–75	60–80	1/16	7	15	1/16
	Corner	60–80	60–80	60–80	1/16	7	15	1/16
	Fillet	70–90	70–90	70–90	1/16	7	15	1/16
1/8	Butt	125–145	115–135	120–140	3/32	8	17	1/8
	Lap	140–160	125–145	130–160	3/32	8	17	1/8
	Corner	125–145	115–135	130–150	3/32	8	17	1/8
	Fillet	140–160	115–135	140–160	3/32	8	17	1/8
3/16	Butt	190–220	190–220	180–210	1/8	10	21	5/32
	Lap	210–240	190–220	180–210	1/8	10	21	5/32
	Corner	190–220	180–210	180–210	1/8	10	21	5/32
	Fillet	210–240	190–220	180–210	1/8	10	21	5/32
1/4	Butt	260–300	220–260	210–250	3/16	12	25	3/16
	Lap	290–340	220–260	210–250	3/16	12	25	3/16
	Corner	280–320	220–260	210–250	3/16	12	25	3/16
	Fillet	280–320	220–260	210–250	3/16	12	25	3/16

psi—pounds per square inch
lpm—liters per minute
cfh—cubic feet per hour

Best results are obtained by using ACHF current with argon as a shielding gas. See Table 17-4 for recommended welding requirements, of gas flow, current, etc.

Magnesium. The welding characteristics of magnesium are somewhat comparable to those of aluminum. Both, for example, have high heat conductivity, low melting point, high thermal expansion, and oxidize rapidly.

With gas tungsten-arc several current variations are possible. DC reverse polarity with helium gas produces wider weld deposits, higher heat, larger heat-affected zone, and shallower penetration. AC current with superimposed high frequency; and helium, argon, or a mixture of these gases, will join material ranging in thickness from 0.20″ to over 0.25″. Both DCRP and AC current provide excellent cleaning action of the base metal surface. Direct current straight polarity with helium as a shielding gas produces a deep penetrating arc but no surface cleaning. This technique is used for mechanized butt welding of sheets up to 1/4″ in thickness without beveling. See Table 17-5 for recommended welding requirements.

Copper and copper alloys. Deoxidized copper is the most widely used type for Tig welding.

Copper alloys such as brass, bronze, and copper alloys of nickel, aluminum, silicon and beryllium are also readily Tig welded. DCSP is generally used for welding these metals. However ACHF or DCRP is often recommended for beryllium copper or for copper alloys less than 0.040″ thick. Workpieces thicker than 1/4″ should be preheated to approximately 300-500°F (145° to 260°C) prior to welding. A forehand welding technique will usually produce the best results. See Tables 17-6 and 17-7 for specific welding requirements.

Always be sure that there is good ventilation when welding copper or copper alloys. The fumes of these metals are highly toxic, therefore, a high-velocity ventilating system is absolutely necessary.

TABLE 17-5. TIG WELDING—MAGNESIUM

STOCK THICKNESS (inch)	TYPE OF JOINT	AMPERES AC CURRENT FLAT POSITION	WELDING ROD (inches) dia	ARGON FLOW 15 psi lpm	cfh	REMARKS
0.040	Butt	45	3/32, 1/8	6	13	Back-up
.040	Butt	25	3/32, 1/8	6	13	No backing
.040	Fillet	45	3/32, 1/8	6	13	
.064	Butt	60	3/32, 1/8	6	13	Back-up
.064	Butt and corner	35	3/32, 1/8	6	13	No backing
.064	Fillet	60	3/32, 1/8	6	13	
.081	Butt	80	1/8	6	13	Back-up
.081	Butt, corner and edge	50	1/8	6	13	No backing
.081	Fillet	80	1/8	6	13	
.102	Butt	100	1/8	9	19	Back-up
.102	Butt, corner and edge	70	1/8	9	19	No backing
.102	Fillet	100	1/8	9	19	
.128	Butt	115	1/8, 5/32	9	19	Back-up
.128	Butt, corner and edge	85	1/8, 5/32	9	19	No backing
.128	Fillet	115	1/8, 5/32	9	19	
3/16	Butt	120	1/8, 5/32	9	19	1 pass
3/16	Butt	75	1/8, 5/32	9	19	2 passes
1/4	Butt	130	5/32, 3/16	9	19	1 pass
1/4	Butt	85	5/32	9	19	2 passes

TABLE 17-6. TIG WELDING—COPPER ALLOYS.

STOCK THICKNESS (inches)	TYPE OF JOINT	DC CURRENT—STRAIGHT POLARITY AMPERES FLAT	HORIZONTAL & VERTICAL	OVERHEAD	ELECTRODE (inches) dia	ARGON FLOW (20 psi) lpm	cfh	FILLER ROD (inches) dia
1/16	Butt	100–120	90–110	90–100	1/16	6	13	1/16
	Lap	110–130	100–120	100–120	1/16	6	13	1/16
	Corner	100–130	90–110	90–110	1/16	6	13	1/16
	Fillet	110–130	100–120	100–120	1/16	6	13	1/16
1/8	Butt	130–150	120–140	120–140	1/16	7	15	3/32
	Lap	140–160	130–150	130–150	1/16, 3/32	7	15	3/32
	Corner	130–150	120–140	120–140	1/16	7	15	3/32
	Fillet	140–160	130–150	130–150	1/16, 3/32	7	15	3/32
3/16	Butt	150–200	—	—	3/32	8	17	1/8
	Lap	175–225	—	—	3/32	8	17	1/8
	Corner	150–200	—	—	3/32	8	17	1/8
	Fillet	175–225	—	—	3/32	8	17	1/8
1/4	Butt	150–200	—	—	3/32	9	19	1/8, 3/16
	Lap	250–300	—	—	1/8	9	19	1/8, 3/16
	Corner	175–225	—	—	3/32	9	19	1/8, 3/16
	Fillet	175–225	—	—	3/32	9	19	1/8, 3/16

TABLE 17-7.   TIG WELDING—DEOXIDIZED COPPER.

STOCK THICKNESS (inches)	TYPE OF JOINT	DC CURRENT STRAIGHT POLARITY AMPERES FLAT POSITION	ELECTRODE (inches) dia	ARGON FLOW 20 psi lpm	cfh	FILLER ROD (inches) dia
1/16	Butt	110–140	1/16	7	15	1/16
	Lap	130–150	1/16	7	15	1/16
	Corner	110–140	1/16	7	15	1/16
	Fillet	130–150	1/16	7	15	1/16
1/8	Butt	175–225	3/32	7	15	3/32
	Lap	200–250	3/32	7	15	3/32
	Corner	175–225	3/32	7	15	3/32
	Fillet	200–250	3/32	7	15	3/32
3/16	Butt	250–300	1/8	7	15	1/8
	Lap	275–325	1/8	7	15	1/8
	Corner	250–300	1/8	7	15	1/8
	Fillet	275–325	1/8	7	15	1/8
1/4	Butt	300–350	1/8	7	15	1/8
	Lap	325–375	1/8	7	15	1/8
	Corner	300–350	1/8	7	15	1/8
	Fillet	325–375	1/8	7	15	1/8

Linde Co.

Stainless steel. Stainless steels, especially those in the 300 series, are very easy to Tig weld. Either direct-current straight-polarity or alternating current with high-frequency stabilization can be used. The gas tungsten-arc is particularly adaptable for welding light gage stainless steel.

The procedure for welding martensitic stainless steels is the same as for welding austenitic stainless steels. If a filler rod is used it should have a slightly higher chromium content. Danger of cracking is greatly reduced if the metal is preheated to a temperature of 300° to 500°F (145° to 260°C). See Table 17-8 for specific welding requirements.

Carbon steels. Gas tungsten-arc is being used more extensively in welding low and medium-carbon and low-alloy steels because of the ease with which the welding can be accomplished and the greater protection from atmospheric contamination. For economic reasons gas tungsten-arc welding is limited to materials under 1/4″ in thickness. When the gas tungsten-arc process is used without filler rod there may

be evidence of some pitting in the weld. This porosity can be eliminated by lightly brushing the seam with a mixture of aluminum powder and methyl alcohol. Ordinarily when filler rods are used they should contain sufficient deoxidizers to prevent porosity. Remember oxyacetylene rods are not suitable for Tig welding.

High-carbon steels are weldable, but preheating, special welding techniques, and post-welding stress relief are required. Steels with a high amount of carbon are subject to rapid grain growth and unless precautions are taken, the welded area loses its toughness, strength, and ductility. Quite often after-welding heat treating brings about a grain refinement of the steel but this process adds to the problems and economics of fabrication.

Very high-carbon steels are rarely recommended for Tig welding as the required welding temperature tends to destroy their mechanical properties. The usual practice in repairing broken parts made with such steels is to use a brazing process where the heat is not sufficient

TABLE 17-8. TIG WELDING—STAINLESS STEEL.

STOCK THICKNESS (inches)	TYPE OF JOINT	DC CURRENT STRAIGHT POLARITY amperes			ELECTRODE (inches) dia	ARGON FLOW 20 psi		FILLER ROD (inches) dia
		flat	horizontal & vertical	overhead		lpm	cfh	
1/16	Butt	80–100	70–90	70–90	1/16	5	11	1/16
	Lap	100–120	80–100	80–100	1/16	5	11	1/16
	Corner	80–100	70–90	70–90	1/16	5	11	1/16
	Fillet	90–110	80–100	80–100	1/16	5	11	1/16
3/32	Butt	100–120	90–110	90–110	1/16	5	11	1/16
	Lap	110–130	100–120	100–120	1/16	5	11	1/16
	Corner	100–120	90–110	90–110	1/16	5	11	1/16
	Fillet	110–130	100–120	100–120	1/16	5	11	1/16
1/8	Butt	120–140	110–130	105–125	1/16	5	11	3/32
	Lap	130–150	120–140	120–140	1/16	5	11	3/32
	Corner	120–140	110–130	115–135	1/16	5	11	3/32
	Fillet	130–150	115–135	120–140	1/16	5	11	3/32
3/16	Butt	200–250	150–200	150–200	3/32	6	13	1/8
	Lap	225–275	175–225	175–225	3/32	6	13	1/8
	Corner	200–250	150–200	150–200	3/32	6	13	1/8
	Fillet	225–275	175–225	175–225	3/32	6	13	1/8
1/4	Butt	275–350	200–250	200–250	1/8	6	13	3/16
	Lap	300–375	225–275	225–275	1/8	6	13	3/16
	Corner	275–350	200–250	200–250	1/8	6	13	3/16
	Fillet	300–375	225–275	225–275	1/8	6	13	3/16

TABLE 17-9. TIG WELDING—PLAIN CARBON AND LOW-ALLOY STEELS.

STOCK THICKNESS	DC CURRENT STRAIGHT POLARITY (amperes)	FILLER ROD dia (inches)	ARGON FLOW (psi)	
			lpm	chf
0.035	100	1/16	4–5	8–10
.049	100–125	1/16	4–5	8–10
.060	125–140	1/16	4–5	8–10
.089	140–170	1/16	4–5	8–10

to affect their metallurgical structure. Table 17-9 includes specific welding requirements for Tig welding plain carbon and low-alloy steels.

Points to Remember

1. Tig welding can be used for joining practically all metals and alloys in various thicknesses and types of joints.

2. Be sure to use a cup of the correct size. Nozzles having too small an orifice tend to overheat and either crack or deteriorate very rapidly.

3. A water-cooled torch is recommended for welding currents that are above 200 or 250 amperes.

4. Argon is generally the inert gas recommended for Tig welding.

5. The power source can be either a DC or AC machine. With a DC machine, better penetration is usually obtained with straight polarity. For some metals better welds are made with an AC machine having a high-frequency voltage, than with a DC machine.

6. The diameter of the electrode depends on the kind and thickness of the metal to be welded. Make certain the tip is properly shaped for the type of current used.

7. When welding light gage metals, it is often necessary to use back-up bars.

8. Before starting to weld, always check to make sure the electrode extends the correct distance.

9. Follow the recommendations for the correct gas flow, otherwise the shielding gas will not be effective.

10. If filler rod is to be used, be sure it is of the right diameter.

11. Always make sure water is flowing before attempting to operate the shielded-gas spot welding gun.

12. Never attempt to adjust the tungsten electrode without first shutting down the power supply machine.

QUESTIONS FOR STUDY AND DISCUSSION

1. What are some of the advantages of gas-shielded arc welding compared to other welding processes?

2. What is meant by Tig welding?

3. What kind of metal can be welded with the Tig process?

4. When is a filler rod used in Tig welding?

5. In Tig welding, what type of power supply unit may be used?

6. Why should an AC machine be of the high-frequency type?

7. What polarity is used in Tig welding?

8. What determines whether an air-cooled or water-cooled torch is used?

9. Why is it important to use a cup of the correct size?

10. What is meant by an inert gas?

11. Why is argon generally considered a better shielding gas for Tig welding than helium?

12. What is the function of a flowmeter in a gas regulator assembly?

13. In Tig welding, what results can be expected when DCRP or DCSP current is used with respect to heat distribution?

14. What determines the size of the tungsten electrode to be used for welding?

15. What should be the shape of the tungsten electrode for DC and AC welding?

16. How is the arc started and stopped in Tig welding?

17. How far should the electrode extend beyond the end of the gas cup on the torch?

18. What is the proper torch angle for welding a butt joint?

19. What precaution should be observed when using a water-cooled torch?

20. When using a filler rod, how should it be manipulated?

21. What is Tig hot-wire welding?

22. How are welds produced by the Tig pulsed current process?

CHAPTER 18 Tig spot welding

For many years spot welding was confined to the conventional process commonly referred to as resistance spot welding. With this method, the material to be joined is placed between two copper electrodes which are brought together, and when current is turned on, sufficient heat is generated to fuse the pieces together. The limitation of this process is that pressure must be applied on both sides of the pieces and the work has to be of size to conveniently feed into the spot welding machine.

The development of Tig spot-welding now makes it possible to produce localized fusion similar to resistance spot-welding without requiring accessibility to both sides of the joint. A special tungsten arc gun is applied to one side of the joint only. Heat is generated from resistance of the work to the flow of electrical current in a circuit of which the work is a part. A comparison of weld cross-sections made by resistance spot-welding and Tig spot-welding is shown in Fig. 18-1.

The Tig spot-welding process has a wide range of applications in fabricating sheet metal products involving joints which are impractical to resistance spot-welding because of the location of the weld or the size of the parts or where welding can be made only from one side.

Equipment

Any DC power supply providing up to 250 amperes with a minimum open circuit voltage of 55 volts can be adapted for spot welding. The gun has a nozzle with a tungsten electrode. See

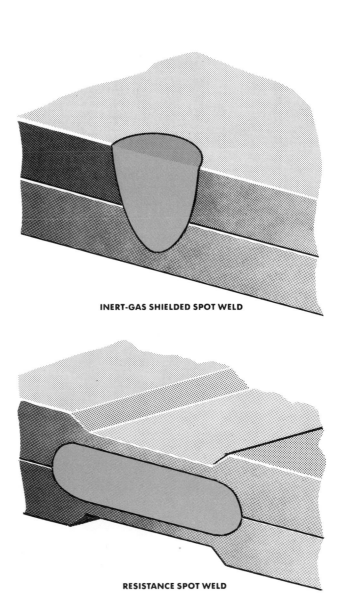

INERT-GAS SHIELDED SPOT WELD

RESISTANCE SPOT WELD

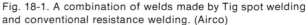

Fig. 18-1. A combination of welds made by Tig spot welding and conventional resistance welding. (Airco)

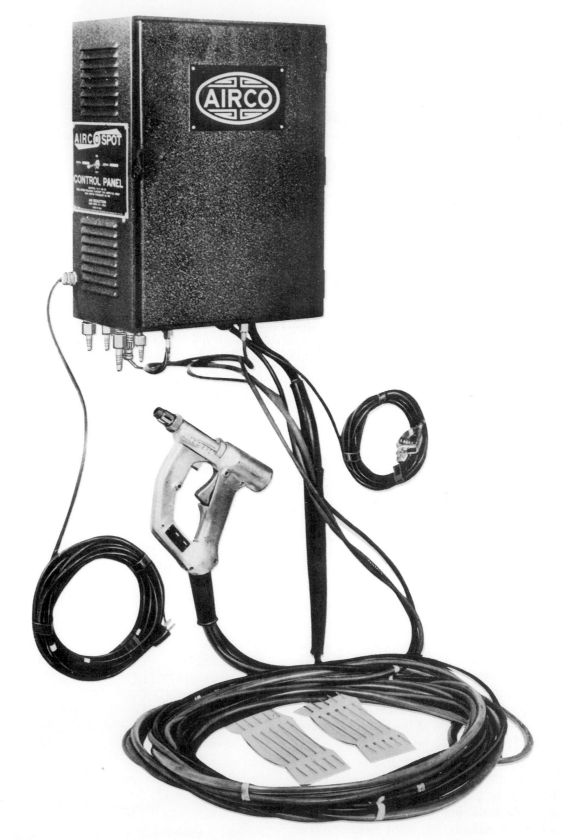

Fig. 18-2. Tig spot-welding gun. (Airco)

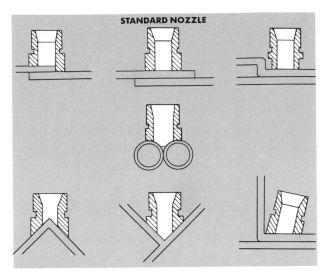

Fig. 18-3. Tig spot-welding gun nozzle can be shaped for a variety of welding jobs (Airco)

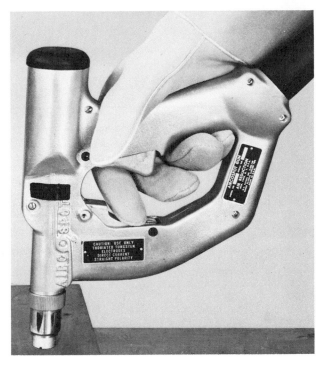

Fig. 18-4. To make a spot weld, place the gun against the work and pull the trigger. (Airco)

Fig. 18-2. Various shape nozzles are available to meet particular job requirements. The standard nozzle can also be machined to permit access in tight corners or its diameter reduced to weld on items such as small holding clips. As a matter of fact the nozzle can be shaped for a variety of welding functions as indicated in Fig. 18-3.

For most operations a $1/8''$ diameter electrode is used. The end of the electrode should normally be flat and of the same diameter as the electrode. However, when working at low amperage settings (100 amperes or less) better results will be obtained if the end of the electrode is tapered slightly to provide a blunt point approximately one half the diameter of the electrode. This will prevent the arc from wandering.

Whenever the end of the electrode balls excessively after only a few welds have been made, it is usually an indication of excessive amperage, dirty material, or insufficient shielding gas.

Making a Weld

To make a spot weld, the end of the gun is placed against the work and the trigger is pulled. See Fig. 18-4. Squeezing the trigger starts the flow of cooling water and shielding gas and also advances the electrode to touch the work. At the same time, the electrode automatically retracts, establishing an arc which is extinguished at the end of a preset length of time. The electrode is usually set at the factory to provide an arc length of $1/16''$ which has been found to be generally satisfactory for practically all types of welding applications.

Amperage. The amperage required for a weld will naturally be governed by the thickness of the metal to be welded. The major effect of increasing the amperage, when both pieces are approximately the same thickness, is to increase the penetration. However, it also tends to increase the weld diameter somewhat as shown by the dotted line in Fig. 18-5 top. Increasing the amperage, when the bottom part is considerably heavier than the top part, will result in an increase in weld diameter with little or no increase in penetration as shown in Fig. 18-5 bottom.

Weld time. Weld time is set on the dial in the control cabinet. The dial is calibrated in 60ths of a second and is adjustable from 0 to 6 seconds. The effect of increasing the weld time is to

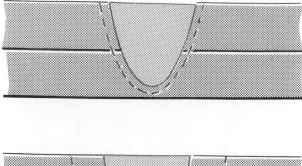

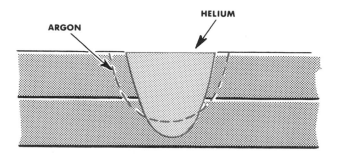

Fig. 18-5. Increasing the amperage will increase the weld diameter. (Airco)

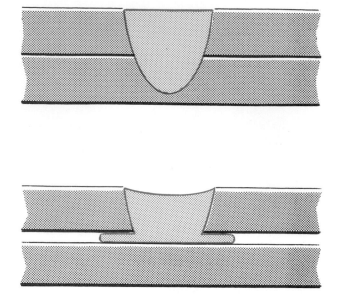

Fig. 18-7. Good surface contact is important in making a sound spot weld. (Airco)

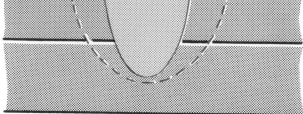

Fig. 18-6. The effects of shielding gas in making a spot weld. (Airco)

increase the weld diameter. But in so doing it also increases the penetration.

Shielding gas. Helium will produce greater penetration than argon, although argon will produce a larger weld diameter. See Fig. 18-6. Gas flow should be set at approximately 6 cfh.

Surface condition and surface contact. Mill scale, oil, grease, dirt, paint, and other foreign materials on or between the contacting surfaces will prevent good contact and reduce the weld strength. The space between the two contacting surfaces resulting from these surface conditions or poor fit-up acts as a barrier to heat transfer and prevents the weld from breaking through

into the bottom piece. Consequently, good surface contact is important for sound welds. See Fig. 18-7.

Backing. Although Tig spot welding can be done from one side only, it is obvious that the bottom piece must have sufficient rigidity to permit the two parts to be brought into contact with pressure applied by the gun. If the thickness, size or shape of the bottom part is such that it does not provide this rigidity, then some form of backing support or jigging will be required. Backing may be either of steel or copper.

Points to Remember

1. Use the correct shape nozzle for the required job.

2. Shape the end of the electrode to meet specific job requirements.

3. Set the amperage to suit the thickness of the metal to be spot welded.

4. Be sure to remove all scale, oil, grease, dirt and paint from the outer and contacting surfaces of the joint.

5. Provide sufficient backing so the two parts can be brought into contact when the gun is applied.

QUESTIONS FOR STUDY AND DISCUSSION

1. What is the advantage of Tig spot-welding over the conventional resistance spot welding?

2. What type of equipment is required for Tig spot-welding?

3. What diameter tungsten electrode is normally used for Tig spot-welding?

4. How is the penetration affected by increasing or decreasing the amperage in Tig spot-welding?

5. What is meant by weld-time in Tig spot-welding?

6. What inert gases are used for spot-welding?

7. Why must there be good surface contact for proper Tig spot-welding?

8. When is backing or jigging necessary in Tig spot-welding?

CHAPTER 19 *gas metal arc—Mig*

The Gas Metal Arc welding process (Mig) sometimes referred to as GMAW, uses a continuous consumable wire electrode. The molten weld puddle is completely covered with a shield of gas. The wire electrode is fed through the torch at pre-set controlled speeds. The shielding gas is also fed through the torch. See Figs. 19-1 and 19-2.

The welding can be completely automatic or semi-automatic. When completely automatic, the wire feed, power setting, gas flow and travel over the workpiece are pre-set and function automatically. When semi-automatic, the wire feed, power setting and gas flow are pre-set, but the torch is manually operated. The welder directs the torch over the weld seam, holding the correct arc-to-work distance and speed. See Figs. 19-3 and 19-4.

Mig welding is sometimes referred to by the tradename of the manufacturer such as *Micro-wire Welding* (Hobart), *Aircomatic Welding* (Airco), *Sigma Welding* (Linde), and *Millermatic Welding* (Miller).

SPECIFIC ADVANTAGES OF MIG WELDING

The following are considered to be some of the more important advantages of Mig welding:

1. Since there is no flux or slag and very little spatter to remove, there is a considerable saving in total welding cost. Generally, weld clean-up is often more costly than welding, thus eliminating clean-up time saves money.

2. Less time is required to train an operator. As a matter of fact welding operators who are proficient in other welding processes can usual-

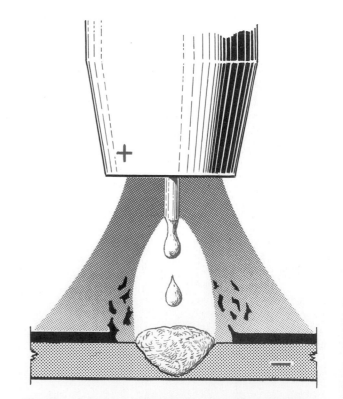

Fig. 19-1. High temperature electric arc melts advancing wire electrode into a globule of liquid metal. Wire is fed mechanically through the torch. Arc heat is regulated by conditions pre-set on the power supply. (Linde Co.)

194

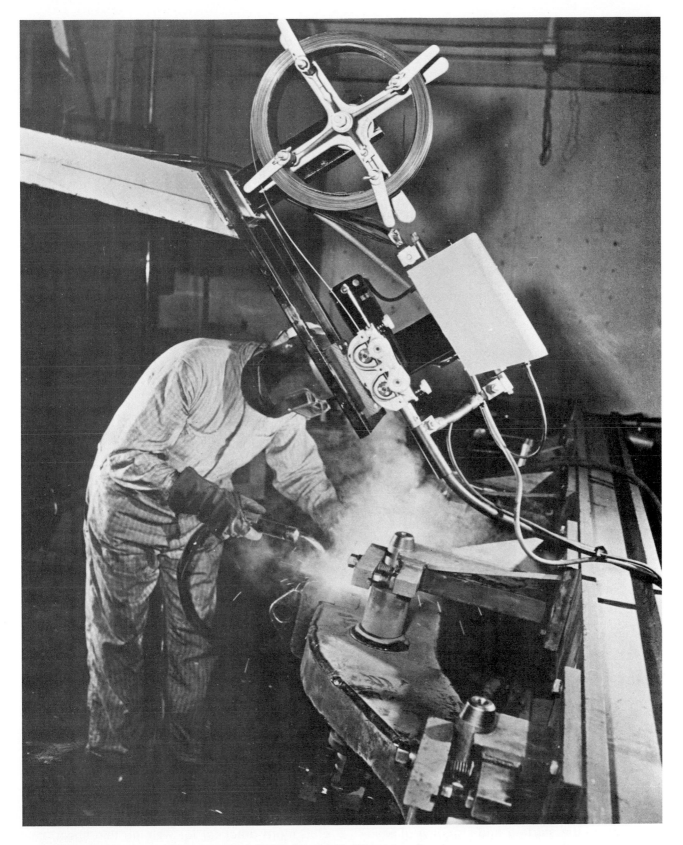

Fig. 19-2. In Mig welding a continuous wire is fed to the puddle. (Chemetron Corp.)

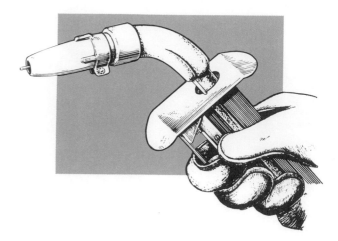

Fig. 19-3. A semi-automatic welder. (Hobart Brothers Co.)

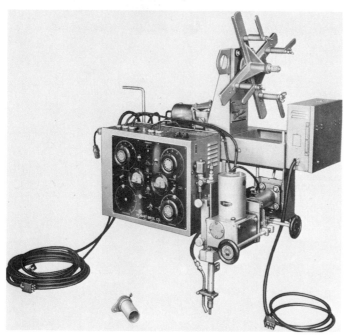

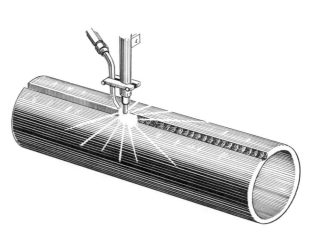

Fig. 19-4. One of the many types of automatic welders. (Hobart Brothers Co.)

ly master the technique of Mig welding in a matter of hours. All the operator has to do is pull the gun trigger and weld. His main concern is to watch the angle of the welding gun and speed of travel, and wire stick-out.

3. The welding process is faster, especially when compared with metallic arc stick welding. There is no need to start and stop in order to change electrodes. As a rule weld failures are often due to the starting and stopping of welding. This usually causes slag inclusions, cold

lapping and other problems like crater cracking.

4. Because of the high speed of the Mig process, better metallurgical benefits are imparted to the weld area. With faster travel there is a narrower heat-affected zone and consequently less molecular disarrangement, less grain growth, less heat transfer in the parent metal, and greatly reduced distortion.

5. Although originally Tig welding was considered more practical for welding thin sheet, because of its lower current, the development of

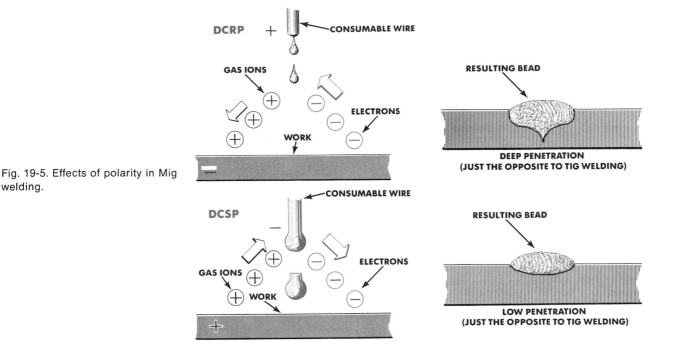

Fig. 19-5. Effects of polarity in Mig welding.

the short circuiting transfer technique now makes it possible to weld thin stock equally as effectively with the Mig process.

6. Since Mig welding has deep penetrating characteristics, narrower beveled joint design can be used. Furthermore, the size of fillet welds is reduced by comparison to other welding methods.

Welding current. Different welding currents have a large effect on the results obtained in gas metal-arc welding. Optimum efficiency is achieved with direct current reverse polarity (DCRP). See Fig. 19-5. The heat in this instance is concentrated at the weld puddle and therefore provides deeper penetration at the weld. Furthermore, with DCRP there is greater surface cleaning action which is important in welding metals having heavy surface oxides such as aluminum and magnesium.

Straight polarity (DCSP) is impractical with Mig welding because weld penetration is wide and shallow, spatter is excessive, and there is no surface cleaning action. The ineffectiveness of straight polarity largely results from the pattern of metal transfer from the electrode to the weld puddle. Whereas in reverse polarity the transfer is in the form of a fine spray, with straight polarity the transfer is largely of the erratic globular type. The use of AC current is never recommended since the burn-offs are unequal on each half-cycle.

TYPES OF METAL TRANSFER

When welding with consumable wire electrodes, the transfer of metal is achieved by three methods: spray transfer, globular transfer, and short circuiting transfer. The type of metal transfer that occurs will depend on electrode wire size, shielding gas, arc voltage, and welding current.

Spray transfer. In spray transfer very fine droplets or particles of the electrode wire are rapidly projected through the arc plasma from the end of the electrode to the workpiece in the direction in which the electrode is pointed. The droplets are equal to or smaller than the diameter of the electrode. While in the process of transferring through the welding arc, the metal particles do not interrupt the flow of current and there is a nearly constant spray of metal.

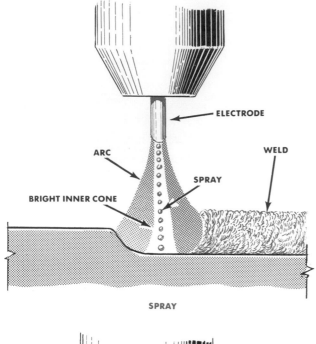

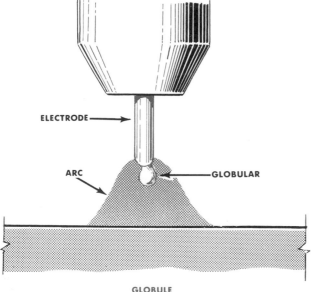

Fig. 19-6. Types of metal transfer. (Airco)

Spray transfer requires a high current density. With the higher current, the arc becomes a steady quiet column having a well defined narrow incandescent cone-shape core within which metal transfer takes place. See Fig. 19-6 top. The use of argon or a mixture of argon and oxygen is also necessary for spray transfer. Argon produces a pinching effect on the molten tip of the electrode, permitting only small droplets to form and transfer during the welding process.

With high heat input, heavy wire electrodes will melt readily and deep weld penetration becomes possible. Since the individual drops are small the arc is stable and can be directed where required. The fact that the metal transfer is produced by directional force which is stronger than gravity, spray transfer is effective for out-of-position welding. *It is particularly adapted for welding heavy gage metal.* It is not too practical for welding light gage metal because of the resulting burn-through.

Globular transfer. This type of transfer occurs when the welding current is low or below what is known as the transition current. The transition range extends from the minimum value where the heat melts the electrode to the point where the high current value induces spray transfer. Only a few drops are transferred per second at low current values, whereas many small drops are transferred when high current values are used.

In globular transfer the molten ball at the tip of the electrode tends to grow in size until its diameter is two or three times the diameter of the wire before it separates from the electrode and transfers across the arc to the workpiece. See Fig. 19-6 bottom. As the globule moves across the arc it assumes an irregular shape and rotary motion because of the physical forces of the arc. This frequently causes the globule to reconnect with the electrode and workpiece, causing the arc to go out and then reignite. The result is poor arc stability, poor penetration, and excessive spatter.

Because of these characteristics a globular type transfer is usually not very effective for most Mig welding operations. Its use is generally restricted to where a low heat input is desired and for welding thin sections.

Short circuiting transfer (short arc). The short circuiting transfer permits welding thinner sections with greater ease. It is extremely practical for welding in all positions, especially for vertical, horizontal and overhead welding where normally puddle control is a little more difficult.

With this process a shallow weld penetration is obtained. See illustrations in Fig. 19-7A-E. It is

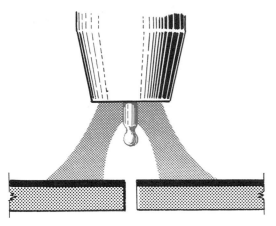

Fig. 19-7A. Start of the SHORT ARC cycle. High temperature electric arc melts advancing wire electrode into a drop of liquid metal. Wire is fed mechanically through the welding torch. Arc heat is regulated by the power supply. (Linde Co.)

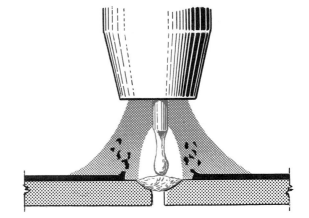

Fig. 19-7C. Electrode makes contact with work piece, creating short circuit. Arc is extinguished, allowing it to cool. Frequency of arc extinction in SHORT ARC varies from 20 to 200 times per second, according to job requirements. (Linde Co.)

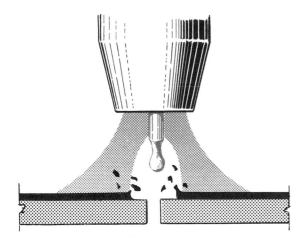

Fig. 19-7B. Molten electrode moves toward the work piece. Note cleaning action. Argon gas mixture, developed for SHORT ARC, shields molten wire and seam, insuring regular arc ignition, preventing spatter and weld contamination. (Linde Co.)

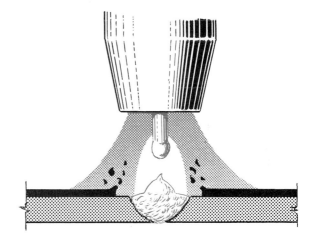

Fig. 19-7D. Drop of molten wire breaks contact with electrode, causing arc to re-ignite. Electrode is broken by pinch force, a squeezing power common to all current carriers. Amount and suddenness of pinch is controlled by power supply. (Linde Co.)

generally considered to be the most practical at current levels below 200 amperes with fine wire of 0.045″ or less in diameters. The use of fine wire produces weld pools that remain relatively small and are easily managed, making all-position welding possible.

As the molten wire is transferred to the weld, each drop touches the weld puddle before it has broken away from the advancing electrode wire.

The circuit is shorted, and the arc is then extinguished.

Electromagnetic pinch force squeezes the drop from the wire. The short circuit is broken and the arc re-ignites. Shorting occurs from 20 to 200 times a second according to preset controls. Shorting of the arc pinpoints the effective heat. The result is a small, relatively cool weld puddle which reduces burn-through. Intricate

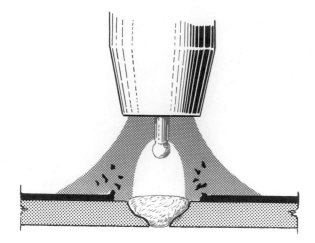

Fig. 19-7E. With arc renewed, SHORT ARC cycle begins again. Because of precision control of arc characteristics and cool, uniform operation, SHORT ARC produces perfect welds on metals as thin as 0.030″ on carbon or stainless steel. (Linde Co.)

welds are possible in most all of the positions, as indicated in Figs. 19-7C and D.

In *short-arc welding*, the shielding gas mixture consists of 25 percent carbon dioxide, which provides increased heat for higher speeds, and 75 percent argon which controls spatter. However, considerable usage is now being made of straight CO_2 where bead contour is not particularly important but good penetration is very essential.

MIG WELDING EQUIPMENT

The Gas-Metal Arc welding equipment consists of four major units: power supply, wire feeding mechanism, welding gun, and gas supply. See Fig. 19-8.

Power Supply

The recommended machine for Mig welding is a rectifier or motor generator supplying direct current with normal limits of 200 to 250 amperes for all position welding. Direct current reverse polarity (DCRP) is used for optimum efficiency. DCRP contributes to better melting, deeper penetration, and excellent cleaning action.

Constant current versus constant potential power supply. In Mig welding, heat is generated by the flow of current through the gap between the end of the wire electrode and the workpiece. A voltage forms across this gap which varies with the length of the arc. To produce a uniform weld, the welding voltage and arc length must be maintained at a constant value. This can be accomplished by (1) feeding the wire into the weld zone at the same rate at which it melts, or (2) melting the wire at the same rate it is fed into the weld zone.

With the conventional constant current welding machine used for many years in shielded metal-arc (stick) welding, the power source produces a constant current over a range of welding voltages. The current has a steep drooping volt-ampere characteristic. See Fig. 19-9. Remember, the volt-amp characteristic actually shows what occurs at the terminal of the welder (electrode) as the load on the power source varies. It indicates how voltage changes in its relationship to amperage between the open circuit stage (static electric potential but no current flowing) and the short circuit condition (electrode touching the work).

When an arc is struck with a power source having a drooping arc voltage the electrode is shorted to the workpiece. The highest voltage potential is present when the circuit is open and no current is flowing. This provides the maximum initial voltage to start the arc. As soon as the arc is struck, the amperage shoots up to maximum and the voltage drops to minimum. Then as the electrode is moved, the voltage rises to maintain the arc and the amperage drops to its normal working level. During welding, the

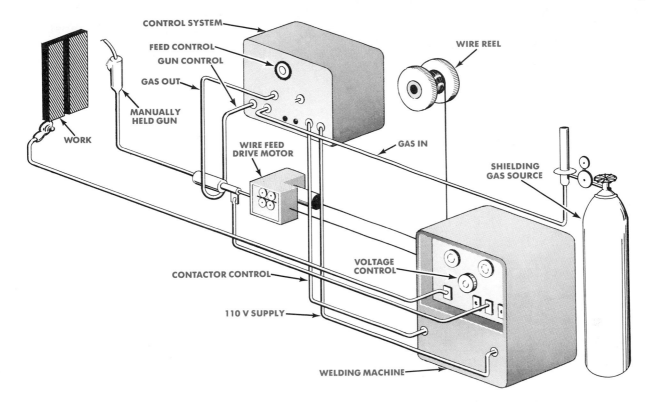

Fig. 19-8. Mig welding unit. (Hobart Brothers Co.)

Fig. 19-9. The conventional constant current welding machine has a drooping volt-ampere characteristic. (Miller Electric Manufacturing Co.)

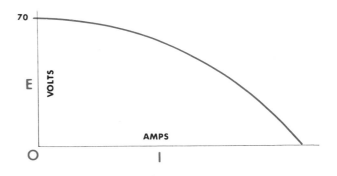

voltage varies directly and amperage inversely with the length of the welding arc. Consequently, the operator can keep reasonable control over the heat input to the work.

When a conventional power source is used for Mig welding, the wire feed speed must be adjusted to narrow limits to prevent the wire from burning back to the nozzle or plunging into the weld plate. Although the operator can, by means of electronic speed controls, adjust the wire speed for a predetermined arc length, neverthe-

less, whenever the nozzle-to-work distance changes, the arc length (voltage) changes. Thus, if the nozzle to work distance increases, the arc length increases. The result is a nonuniform weld.

With the need for better arc control, the *constant voltage (potential) power supply* was developed. Constant potential welding power supply has a nearly flat volt-ampere characteristic. See Fig. 19-10. This means that the preset voltage level can be held throughout its range. Although

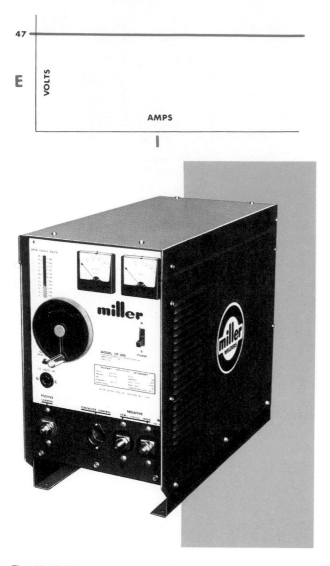

Fig. 19-10. The constant potential power supply has a nearly flat volt-ampere characteristic. (Miller Electric Manufacturing Co.)

be set on the power supply and any variations in nozzle-to-work distance will not produce changes in the arc length. For example, if the arc length becomes shorter than the pre-selected value, there is an automatic increase of current and the wire speed automatically adjusts itself to maintain a constant arc length. Similarly, if the arc becomes too long, the current decreases and the wire begins to feed faster.

Stated in another way, when the wire is fed into the arc at a specific rate, a proportionate amount of current is automatically drawn. The constant potential welder therefore provides the necessary current required by the load imposed on it. When the electrode wire is fed faster, the current increases; if it is fed slower the current decreases.

Because of this self-correcting feature, less operator skill is necessary to achieve good welds. There are only two basic controls: a rheostat on the welding machine to regulate the voltage and a rheostat on the wire feed mechanism to control the speed of the wire feed motor.

There is no current control on a constant voltage type machine; the welding current output is determined by the wire feeder.

Slope control. Some power supply units designed for Mig welding have provisions for controlling the slope. The incorporation of slope control gives the machine greater versatility. Thus by altering the flat shape of the slope it is possible to control the pinch force on the consumable wire which is particularly important in the short circuiting transfer method of welding. With better control of the short circuit, the weld puddle can be kept more fluid with better resulting welds. Slope control also helps to decrease the sudden current surge when the electrode makes its initial contact with the workpiece. By slowing down the rate of current rise, the amount of spatter can be reduced.

its static voltage potential at open circuit is lower than a machine with drooping characteristic, it maintains approximately the same voltage regardless of the amount of current drawn. Accordingly there is unlimited amperage to melt the consumable wire electrode.

The power supply becomes self-correcting with respect to arc length. The operator can change the wire feed speed over a considerable range without affecting the stubbing or burning back the wire. In other words, the arc length can

Wire Feeding Mechanism

The wire feeding mechanism automatically drives the electrode wire from the wire spool to the gun and arc. See Figs. 19-11 and 19-12. Control on the panel can be adjusted to vary the wire feeding speed. In addition, the control panel

Fig. 19-11. Typical wire feeding unit for Mig welding. (Miller Electric Manufacturing Co.)

Fig. 19-12. Types of control panels on wire feed mechanisms.

usually includes a welding power contactor and a solenoid to energize the gas flow. On units designed for welding with a water-cooled gun, a control is also available to turn on and shut off the water flow.

The wire feeder can be mounted on the power supply machine or it can be separate from the welding machine and mounted elsewhere to facilitate welding over a large area.

Welding Gun

The function of the welding gun is to deliver the wire, shielding gas, and welding current to the arc area. The manually operated gun is either water or air-cooled. An air-cooled gun is especially designed to weld light gage metals that require less than 200 amperes with argon as a shielding gas. However, such a torch can usually function at higher amperage (300A) with CO_2 because of the cooling effects of this gas. A water-cooled gun generally is best when welding with currents that are higher than 200 amperes.

Guns are either of the push or pull type. The pull gun has drive rolls that pull the welding wire from the wire feeder, and the push gun has the wire pushed to it by drive rolls in the wire feeder itself. The pull gun handles small diameter wires; the push gun moves heavier diameter wires. The pull type is also used to weld with soft wires such as aluminum and magnesium while the push gun is considered more suitable for welding with hard wires such as carbon and stainless steels and where currents are often in excess of 250 amperes.

Both guns have a trigger switch that controls the wire feed and arc as well as the shielding gas and water flow. When the trigger is released the wire feed, arc, shielding gas, and water, if a water-cooled torch is used, stop immediately. With some equipment a timer is included to

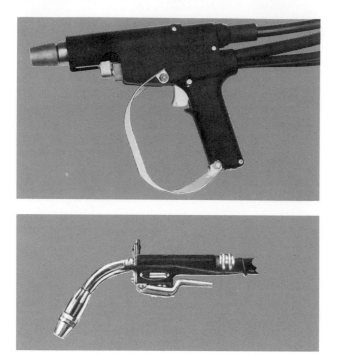
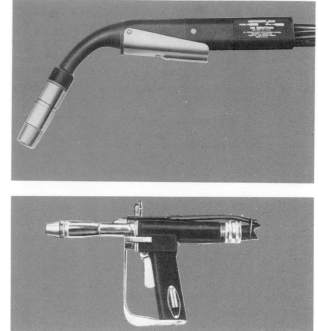

Fig. 19-13. Types of guns used for Mig welding.

permit the shielding gas to flow for a predeter-mined time to protect the weld until it solidifies.

Guns are available with a straight or curved nozzle. See Fig. 19-13. The curved nozzle pro-vides easy access to intricate joints and difficult to weld patterns.

Shielding Gas[1]

In any gas shielded arc welding process, the shielding gas can have a large effect upon the properties of a weld deposit. Therefore, welding is done in a controlled atmosphere. In shielded metal-arc welding, this is accomplished by plac-ing a coating on the electrode which produces a non-harmful atmosphere when it disintegrates in the welding arc. In the case of Mig welding, the same effect is accomplished by surrounding the arc area with gases supplied from an external source. See Table 19-1.

The air in the arc area is displaced by the shielding gas. The arc is then struck under the blanket of shielding gas and the welding is accomplished. Since the molten weld metal is exposed only to the shielding gas it is not con-taminated and strong dense weld deposits are obtained. The reason for shielding the arc area is to prohibit air from coming in contact with the molten metal.

By volume, air is made up of 21 percent oxygen, 78 percent nitrogen, 0.94 percent argon, and 0.04 percent other gases (primarily carbon dioxide). The atmosphere will also contain a certain amount of water depending upon its humidity. Of all of the elements that are in the air, the three which cause the most difficulty as far as welding is concerned are oxygen, nitrogen, and hydrogen.

Oxygen is a highly reactive element and com-bines readily with other elements in the metal or alloy to form undesirable oxides and gases. The oxide-forming aspect of the oxygen can be over-come with the use of deoxidizers in the steel weld metal.

The deoxidizers, such as manganese and sili-con, combine with the oxygen and form a light slag which floats to the top of the weld pool. If

1. Hobart Brothers Co.

TABLE 19-1. SHIELDING GASES FOR MIG WELDING.

MATERIAL	PREFERRED GAS	REMARKS
Aluminum alloys	Argon	With DC reverse polarity removes oxide surface on work piece
Magnesium aluminum alloys	75% He 25% A	Greater heat input reduces porosity tendencies. Also cleans oxide surface
Stainless steels	Argon + 1% O_2	Oxygen eliminates under-cutting when DC reverse polarity is used
	(Argon + 5% O_2)	When DC straight polarity is used 5% O_2 improves arc stability
Magnesium	Argon	With DC straight polarity removes oxide surface on work piece
Copper (deoxidized)	75% He, 25% A	Good wetting and increased heat input to counteract high thermal conductivity. Light gages
	(Argon)	
Low-carbon steel	Argon + 2% O_2	Oxygen eliminates under-cutting tendencies also removes oxidation
Low-carbon steel	Carbon dioxide (spray transfer)	High quality low current out of position welding low spatter
	Carbon dioxide (buried arc)	High speed low cost welding accompanied by spatter loss
Nickel	Argon	Good wetting decreases fluidity of weld metal
Monel	Argon	Good wetting decreases fluidity of weld metal
Inconel	Argon	Good wetting decreases fluidity of weld metal
Titanium	Argon	Reduces heat-affected zone, improves metal transfer
Silicon bronze	Argon	Reduces crack sensitivity of this hot short material
Aluminum bronze	Argon	Less penetration of base metal Commonly used as a surfacing material

Note: ( ) = Second choice

the deoxidizers are not provided, the oxygen will combine with the iron and form compounds which can lead to inclusions in the weld material, and lower its mechanical properties. On cooling, the free oxygen in the arc area combines with the carbon of the alloy material and forms carbon monoxide. If this gas is trapped in the weld metal as it cools, it collects in pockets which cause pores or hollow spaces in the weld deposit.

Of all of the elements in the air, nitrogen causes the most serious problems in welding steel materials. When iron is molten, it is able to take a relatively large amount of nitrogen into solution. At room temperature, however, the solubility of nitrogen in iron is very low. Therefore, in cooling, the nitrogen precipitates or comes out of the iron as nitrites. These nitrites cause a high yield and tensile strength, and increased hardness, but a pronounced decrease in the ductility and impact resistance of the steel materials. The loss of ductility often leads to cracking in and near the weld metal. Since air contains approximately 78 percent nitrogen by volume, therefore, if the weld metal is not protected from the air during welding, very pronounced decreases in weld quality will occur. In excessive amounts, nitrogen can also lead to gross porosity in the weld deposit.

Hydrogen is also harmful to welding. Very small amounts of hydrogen in the atmosphere produce an erratic arc. Of more importance is the effect that hydrogen has on the properties of the weld deposit. As in the case of nitrogen, iron can hold a relatively large amount of hydrogen when it is molten but upon cooling it has a low solubility for hydrogen. As the metal starts to solidify, it rejects the hydrogen. Hydrogen that becomes entrapped in the solidifying metal collects at certain points and causes large pressures or stresses to occur. These pressures lead to minute cracks in the weld metal which can later develop into large cracks. Hydrogen also causes defects known as fish eyes and underbead cracking.

The effects of oxygen, nitrogen, and hydrogen make it essential that they be excluded from the weld area during welding. This is done by using inert gases for shielding. The inert gases consist of atoms which are very stable and do not react readily with other atoms. In nature there are only six elements possessing this stability and each of these elements exists as a gas.

The six inert gases are helium, neon, argon, krypton, xenon and radon. Since the inert gases do not readily form compounds with other elements, they are very useful as shielding atmospheres for arc welding. Of the six inert gases only helium and argon are important to the welding industry. This is because they are the only two which can be obtained in quantities at an economical price.

Carbon dioxide gas can also be used for shielding the weld area. Although it is not an inert gas, compensations can be made for its oxidizing tendencies and it can readily be employed for shielding the weld. Characteristics of this gas will be explained in detail later.

Argon. Argon gas has been used for many years as a shielding medium for fusion welding. Argon is obtained by the liquification and distillation of air. Air contains approximately 0.94 percent argon by volume or 1.3 percent by weight. This seems like a small quantity, but calculations show that the amount of air covering one square mile of the earth's surface contains approximately 800,000 pounds (364 metric tonnes) of argon.

In manufacturing argon, air is put under great pressure and refrigerated to very low temperatures. Then, the various elements in air are boiled off by raising the temperature of the liquid. Argon boils off from the liquid at a temperature of −302.4°F (−185.9°C). For welding the purity of the argon is approximately 99.995 percent. When greater purity is required, the gas can be chemically cleaned to the purity of 99.999 percent.

Argon has a relatively low ionization potential. This means that the welding arc tends to be more stable when argon is used in the shielding gas. For this reason argon is often used in conjunction with other gases for arc shielding. The argon gives a quiet arc and thereby reduces spatter. Since argon has a low ionization potential, the arc voltage is reduced when argon is added to the shielding gas. This results in lower power in the arc and therefore lower penetration. The combination of lower penetration and reduced spatter makes the use of argon desirable when welding sheet metal.

Straight argon is seldom used for arc shielding except in welding such metals as aluminum, copper, nickel, and titanium. When welding steel the use of straight argon gas leads to undercutting and poor bead contour. Also, the penetration with straight argon is shallow at the bead edges and deep at the center of the weld. This can lead to lack of fusion at the root of the weld.

Argon plus oxygen. In order to reduce the poor bead contour and penetration pattern obtained with argon gas when welding on mild steel, it has been found that the addition of oxygen to the shielding gas is desirable. Small amounts of oxygen added to the argon produce significant changes. Normally, the oxygen is added in amounts of 1, 2, or 5 percent. Using gas metal-arc welding wires, the amount of oxygen which can be employed is limited to 5 percent. Additional oxygen might lead to the formation of porosity in the weld deposit.

Oxygen improves the penetration pattern by broadening the deep penetration finger at the center of the weld bead. It also improves bead contour and eliminates the undercut at the edge of the weld that is obtained with pure argon. Argon-oxygen mixtures are very common for welding low alloy steels, carbon steels, and stainless steel.

Carbon dioxide. Unlike argon or helium gases which are made up of single atoms, the carbon dioxide gas is made up of molecules. Each molecule contains one carbon atom and two oxygen atoms. The chemical formula for the carbon dioxide molecule is CO_2. Often, carbon dioxide is referred to simply as C-O-TWO.

At normal temperatures, carbon dioxide is essentially an inert gas. However, when subjected to high temperatures, carbon dioxide will disassociate into carbon monoxide and oxygen. In the high temperature of the welding arc this disassociation takes place to the extent that 20 to 30 percent of the gases in the arc area are oxygen (O_2). Because of this oxidizing characteristic of the CO_2 gas, the wires used with this gas must contain deoxidizing elements. The deoxidizing elements have a great affinity for the oxygen and readily combine with it. This prevents the oxygen atoms from combining with the carbon or iron in the weld metal and producing low quality welds. The most common deoxidizers used in wire electrodes are manganese, silicon, aluminum, titanium, and vanadium.

Carbon dioxide is manufactured in most plants from flue gases which are given off by the burning of natural gas, fuel oil, or coke. It is also obtained as a by-product of calcining operations of lime kilns, from the manufacturing of ammonia, and from the fermentation of alcohol. The carbon dioxide given off by the manufacturing of ammonia and the fermentation of alcohol is almost 100 percent pure.

The purity of carbon dioxide gas can vary considerably depending upon the process used to manufacture it. However, standards have been set up for the purity that must be obtained if it is to be used for arc welding. The purity specified for welding grade CO_2 is a minimum dew point of minus 40°F. This means that gas of this purity will contain approximately 0.0066 percent moisture by weight.

Carbon dioxide gas eliminates many of the undesirable characteristics that are obtained when using argon for arc shielding. With the carbon dioxide a broad, deep penetration pattern is obtained. This makes it easier for the operator to eliminate weld defects such as lack of penetration and lack of fusion. Bead contour is good and there is no tendency towards undercutting. Another advantage is its relatively low cost compared to other shielding gases.

The chief drawback of the CO_2 gas is the tendency for the arc to be somewhat violent. This can lead to spatter problems when welding on thin materials where appearance is of particular importance. However, for most applications this is not a major problem and the advantages of CO_2 shielding far outweigh its disadvantages.

CO_2 is used primarily for mild steel welding, although it has some application in the formulation of other shielding gas mixtures.

Helium. Helium is an inert gas and may be compared to argon in that respect. There the similarity ends. Helium has an ionization potential of 24.5 volts. It is lighter than air and has high thermal conductivity. The helium arc plasma will expand under heat (thermal ionization) reducing the arc density.

With helium there is a simultaneous change in arc voltage where the voltage gradient of the arc length is increased by the discharge of heat from the arc stream or core. This means that more arc energy is lost in the arc itself and is not transmitted to the work. The result is that, with helium, there will be a broader weld bead with relatively shallower penetration than with argon. (For Tig welding, the opposite is true.) This also accounts for the higher arc voltage, for the same arc length, that is obtained with helium as opposed to argon.

Helium is derived from natural gas. The process by which it is obtained is similar to that of argon. First the natural gas is compressed and cooled. The hydrocarbons are drawn off, then nitrogen, and finally the helium. This is a process of liquifying the various gases, each at its separate distillation temperature, until at $-452°F$ ($-269°C$), the helium is produced.

Helium has sometimes been in short supply due to governmental restrictions and, therefore, has not been used as much as it might have been for welding purposes. It is difficult to initiate an arc in a helium atmosphere with the tungsten arc process. The problem is less acute with the gas metal arc process.

Helium is used primarily for the non-ferrous metals such as aluminum, magnesium and copper. It is also used in combination with other shielding gases.

Argon CO_2. For some applications of mild steel welding, welding grade CO_2 does not provide the arc characteristics needed for the job. This will usually manifest itself, where surface appearance is a factor, in the form of intolerable spatter in the weld area. In such cases a mixture of argon-CO_2 has usually eliminated the problem. Some welding authorities believe that the mixture should not exceed 25 percent CO_2. Others feel that mixtures with up to 80 percent CO_2 are practical.

The reason for wanting to use as much CO_2 as possible in the mixtures is cost. By using a cylinder of each type of gas, argon and CO_2, the mixture percentages may be varied by the use of flowmeters. This method precludes the possibility of gas separation such as may occur in pre-mixed cylinders. When it is considered that pre-mixed argon-CO_2 gas is sold at the price of pure

argon then it makes good sense to mix your own. The price of CO_2 is approximately 15 percent that of argon in most areas of the country.

Argon-CO_2 shielding gas mixtures are employed for welding mild steel, low alloy steel and, in some cases, for stainless steels.

Argon-Helium-CO_2. This mixture of shielding gases is used primarily for welding austenitic stainless steels. The combination of gases provides a unique characteristic to the weld. It is possible to make a weld with very little build-up of the top bead profile. The result is excellent for those applications where a high crowned weld is detrimental rather than a help. This gas mixture has found considerable use in the welding of stainless steel pipe.

Gas Flow and Regulation

For most welding conditions, the gas flow rate will approximate 35 cu ft per hour. This flow rate may be increased or decreased depending upon the particular welding application. (See Tables 19-5 thru 19-10, p. 221, 223 through 225.)

The data presented in these tables are not intended as absolute settings but only as a point in making the starting settings. Final adjustments must often be made on a trial and error basis. Actually the correct settings will be governed by the type and thickness of metal to be welded, position of the weld, kind of shielding gas used, diameter of electrode and type of joint.

The proper amount of gas shielding usually results in a rapidly crackling or sizzling arc sound. Inadequate gas shielding will produce a popping arc sound with resultant weld discoloration, porosity, and spatter.

Gas drift may occur with high weld travel speeds or from unusually drafty or windy conditions in the weld area. Since one or more of these factors may cause the gas to drift away from the arc, the result is inadequate gas coverage. See Fig. 19-14. The gas nozzle should be adjusted for proper coverage and outside influences should be eliminated by proper windbreakers or shields. See Fig. 19-15.

Correct positioning of the nozzle with respect to the work will be determined by the nature of the weld. The gas nozzle may usually be placed

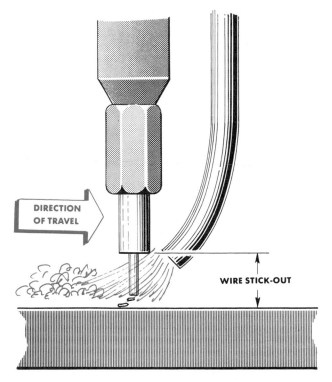

DIRECTION OF TRAVEL

WIRE STICK-OUT

Fig. 19-14. Gas drift with inadequate gas coverage. (Hobart Brothers Co.)

DIRECTION OF TRAVEL

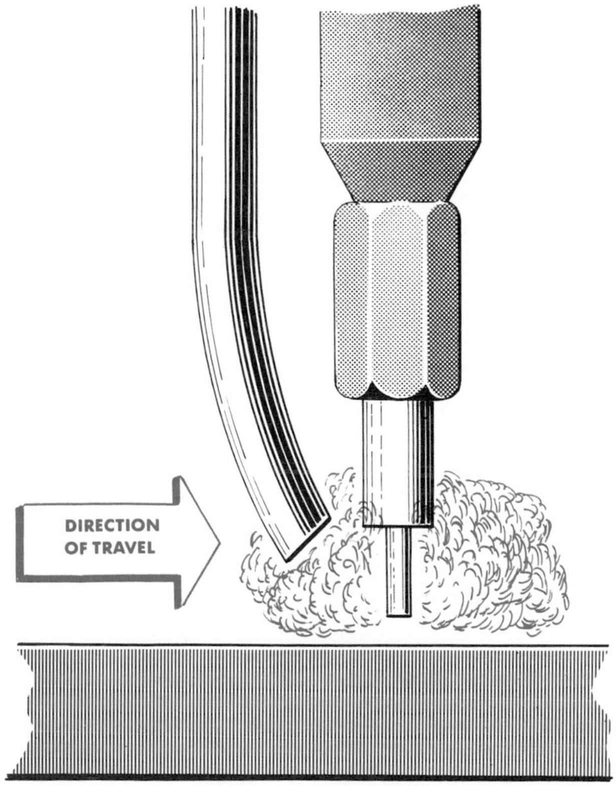

Fig. 19-15. Adequate gas coverage. (Hobart Brothers Co.)

up to 2″ from the work. Too much space between nozzle and work reduces the effectiveness of a gas shield while too little space may result in excessive weld spatter which collects on the nozzle and shortens its life.

Wire for Mig Welding

Filler wire for Mig welding should be similar in composition to the base metal. Several common wires and their suitability for a particular type of welding are listed in Table 19-2. These designations are based on the AWS classification system. Thus for mild steel wires, the *E* identifies it as an electrode, the next two digits show the tensile strength in psi per thousand, the *S* indicates a solid bare wire, and the final symbols specify a particular classification based on chemical composition of the wire.

Wires are usually available in spools of several different sizes as well as in 36″ rod lengths for Tig welding.

Best results are obtained by using the proper diameter wire for the thickness of the metal to be welded and the position in which the welding is to be done. See Tables 19-5 to 19-10.

Basic wire diameters are 0.020″, 0.030″, 0.035″, 0.045″, ¹/₁₆″ and ¹/₈″. Generally wires of 0.020″, 0.030″ or 0.035″ are best for welding thin metals. These wires are sometimes referred to as micro wires. The use of micro wires permits increased welding speeds and improves the appearance and quality of the welds. See Table 19-3. Micro-wire welding is especially adaptable for joining thin materials (20 gage), although it also can be used to weld low and medium-carbon steels, and low-alloy/high-strength steels of medium thickness. Medium thickness metals normally require 0.045″ or ¹/₁₆″ diameter electrodes. For thick metals, ¹/₈″ electrodes are usually recommended. However, the position of welding is a factor which must be considered in electrode selection. Thus for vertical or overhead welding, smaller diameter electrodes will be more satisfactory than larger diameter wires.

Wire feed. The amperage of the welding current used limits the speed of the wire feed to a definite range. However, it is possible to make adjustments of the wire feed within the range.

TABLE 19-2.   FILLER WIRES FOR GAS SHIELDED—ARC WELDING

	MILD STEEL WIRES
E-60S-1	Silicon deoxidized wire for low and medium-carbon steels. Can be used either with CO_2, argon, or argon-CO_2 mixtures. Performs best on killed steels
E-60S-2	Premium quality wire containing Al, Zr, and Ti in addition to silicon and manganese deoxidizers. Can be used with CO_2 or argon-CO_2 or argon-O_2. Recommended for pipe welding and heavy vessel construction
E-60S-3	Used for higher quality welding either with CO_2, argon-O_2, or argon-CO_2 mixtures. Produces medium quality welds in rimmed steels and high quality welds in semi-killed steels
E-70S-1B	Low-alloy wire for carbon steels, low-alloy steels, and high strength low-alloy steels
E-70S-3	General purpose welding of low to medium-carbon steels. Has a silicon content high enough to permit its use in either CO_2, argon-oxygen mixtures or mixture of the two
E-70S-6	Contains higher manganese and silicon levels and has more powerful deoxidizing characteristics for welding over rust and scale or where stringent cleaning practices cannot be followed
E-70S-5	Contains aluminum and is designed for single or multipass welding of rimmed, semikilled, or killed mild steels. Suitable to weld steels having rusty or dirty surfaces and normally used with CO_2 gas

	ALUMINUM WIRES
ER-1100	To weld aluminum of similar composition
ER-4043	
ER-5183	
ER-5554, 5556	
ER-5654	

	STAINLESS STEEL WIRES
ER-308L	For welding types 304, 308, 321, 347
ER-308L-Si	For welding types 301, 304
ER-309	For welding types 309 and straight chromium grades when heat treatment is not possible. Also for 304-clad
ER-310	For welding types 310, 304-clad and hardenable steels
ER-316	For welding 316
ER347	For welding types 321, and 347 where maximum corrosion resistance is required

	COPPER AND COPPER-BASE ALLOY WIRES
E-CuSi (Silicon Bronze)	Special wires for welding copper and copper-based alloys
E-CuAl-A1 (Aluminum Bronze)	
E-Cu (Deoxidized Copper)	
E-CuAl-A2 (Aluminum Bronze)	
E-CuAl-B (Aluminum Bronze)	

TABLE 19-3. MICRO-WIRE WELDING.
(Manual travel, single pass, flat fillet welds)

MATERIAL THICKNESS (inches)	ELECTRODE SIZE	WELDING CONDITIONS		GAS FLOW (cfh)	TRAVEL SPEED (ipm)
		DCRP (arc volts)	(amperes)		
0.025	0.030	15–17	30–50	15–20	15–20
.031	.030	15–17	40–60	15–20	18–22
.037	.035	15–17	65–85	15–20	35–40
.050	.035	17–19	80–100	15–20	35–40
.062	.035	17–19	90–110	20–25	30–35
.078	.035	18–20	110–130	20–25	25–30
.125	.035	19–21	140–160	20–25	20–25
.125	.045	20–23	180–200	20–25	27–32
.187	.035	19–21	140–160	20–25	14–19
.187	.045	20–23	180–200	20–25	18–22
.250	.035	19–21	140–160	20–25	10–15
.250	.045	20–23	180–200	20–25	12–18

Shielding gas: CO_2, welding grade
Wire stick-out—1/4″ to 3/8″
Hobart Brothers Co.

For a specific amperage setting, a high speed of wire feed will result in a short arc. A low speed contributes to a long arc. Also a higher speed must be used for overhead welding than speeds for flat position welding. (See Tables 19-5 to 19-10, pages 221, 223 through 225.)

Wire stick-out. Wire stick-out refers to the distance the wire projects from the nozzle of the gun. See Fig. 19-16, *top.* Stick-out influences the welding current since it changes the preheating in the wire. When the stick-out increases the preheating increases, which means that the power source does not have to furnish as much welding current to melt the wire at a given feed rate. Since the power source is self-regulating, the current output is automatically decreased. Conversely, if the stick-out decreases, the power source is forced to furnish more current to burn off the wire at the required rate.

For most Mig welding applications, the wire stick-out should measure from 3/8″ to 3/4″. On micro wires a shorter wire stick-out ranging from 1/4″ to 3/8″ is recommended. An excessive amount of wire stick-out results in increased wire preheating; this tends to increase the deposit rate.

Too much wire stick-out may also produce a ropy appearance in the weld bead. Too little stick-out will cause the wire to fuse to the nozzle tip, which decreases the life of the tip. As the amount of wire stick-out increases, it may become increasingly difficult to follow the weld seam, particularly with a small diameter wire. The tip should be either flush with the gas nozzle or recessed in the nozzle. See Fig. 19-16, *bottom.* An extended tip is seldom used and then for very low amperages.

The wire, in a near-plastic state between the tip and arc, tends to move (whip) around, describing a somewhat circular pattern. Decreasing the amount of wire stick-out and straightening the welding wire tend to decrease the amount of wire whip.

Welding Current

A wide range of current values can be used with each wire diameter. This permits welding various thicknesses of metal without having to change wire diameter. The correct current to use for a particular joint must often be determined by trial. The current selected should be high

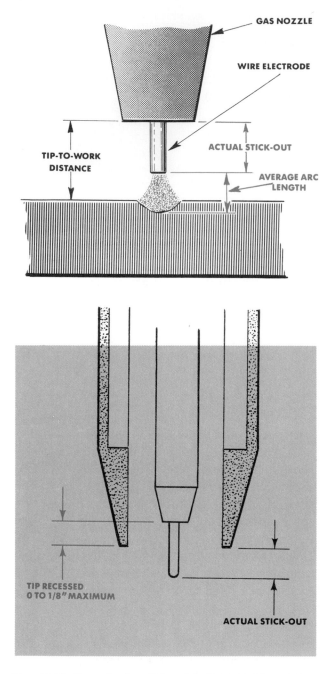

Fig. 19-16. Correct wire stick-out is important to achieve sound welds. (Hobart Brothers Co.)

enough to secure the desired penetration without cold lapping (cold shuts) but low enough to avoid undercutting and burn through. (See Tables 19-5 to 19-10, pages 221, 223 through 225.)

The term current is often related to current density. Current density is the amperage per square inch of cross sectional area of the electrode. Thus at a given amperage the current density of 0.030″ diameter electrode is higher than with 0.045″ diameter electrode. Current density is calculated by dividing the welding current by the electrode area.

Each type and size of electrode wire has a minimum and maximum current density. For example, if the welding current falls below the minimum, a satisfactory weld cannot be made.

The success of Mig welding is due to the concentration of a high current density at the electrode tip. Whereas the arc stream of Mig is sharp and deeply penetrating, metallic arc (stick electrode) is soft and widespread. Consequently, the width-to-depth ratio of gas metal arc will be less than with stick electrode. See Fig. 19-17.

Joint Edge Preparation and Weld Backing

Preparation of the edge of each member to be joined is recommended to aid in the penetration and control of weld reinforcement. For Mig welding, beveling the edges is usually desirable for butt joints thicker than ¼″ if complete root penetration is desired. For thinner sections, a square butt joint is best.

To a considerable extent the same conventional joint design recommended for other arc welding processes can be used for gas metal-arc welding. However, some joint modifications are often incorporated to compensate for the operating characteristics of gas metal-arc welding. Thus the arc in gas metal-arc welding is more penetrating and narrower than the arc in shielded metal-arc welding. Consequently, groove joints can have smaller root faces and root openings. Also since the nozzle does not have to be placed within the groove a narrower included angle can be provided. By reducing the joint area less weld metal is required; this lowers material and labor costs.

In the Mig process, weld backing is helpful in obtaining a sound weld at the roots. Backing prevents molten metal from running through the joint being welded, especially when complete weld penetration is desired.

There are several types of material used for backing: steel and copper blocks, strips, and

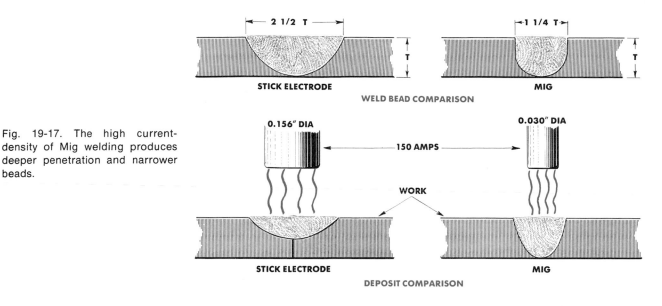

Fig. 19-17. The high current-density of Mig welding produces deeper penetration and narrower beads.

bars; carbon blocks, plastics, asbestos, and fire clay. Some of these serve to conduct heat away from the joint and also to form a mold or dam for the metal. The most commonly used backing for Mig welding is copper or steel.

Positioning Work and Welding Wire

The proper position of the welding torch and weldment is important. In Mig welding, the flat position is preferred for most joints because this position improves the molten metal flow, bead contour, and gives better gas protection. However, on gaged material, it is sometimes necessary or advantageous to weld with the work inclined 10° to 20°. The welding is done in the downhill position. This has a tendency to flatten the bead and increase the travel speed.

The alignment of the welding wire in relation to the joint is very important. The welding wire should be on the center line of the joint for most butt joints, if the pieces to be joined are of equal thickness. If the pieces are unequal in thickness, the wire may be moved toward the thicker piece. The recommended position of the welding gun for fillet and butt welds is shown in Fig. 19-18.

Either a pulling or pushing technique may be used with little or no weaving motion. See Fig. 19-19. Some weaving is desirable for poorly fitted edge joints. The pulling or drag technique is usually best for light gage metals and the pushing technique for heavy materials. In the pulling technique (backhand) the gun points away from the direction of travel, whereas in the pushing motion (forehand) the gun points forward in the direction of travel.

Generally the penetration of beads deposited with a pulling technique is greater than with a pushing technique. Furthermore, since the welder can see the weld crater easier in a pulling action he can produce high quality welds more consistently. On the other hand, forehand welding permits the use of higher welding speeds and produces less penetrating and wider welds.

For welding circular seams, the wire should be shifted off-center, approximately $1/3$ of the work radius, as shown in Fig. 19-20. This will allow the metal to solidify by the time it reaches the top of the circle. A shift of more than $1/3$ of the work radius will cause the weld metal deposit to run ahead of the weld bead.

Preliminary Welding Checks for Mig Welding

Before starting to weld, it is always a good practice to check the following:

1. All electric power controls are in the OFF position.

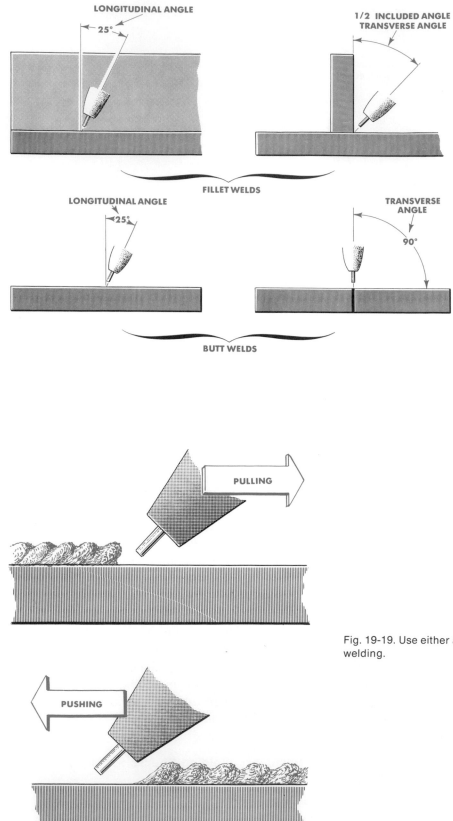

Fig. 19-18. Correct nozzle angle for Mig welding.

Fig. 19-19. Use either a pushing or a pulling technique in Mig welding.

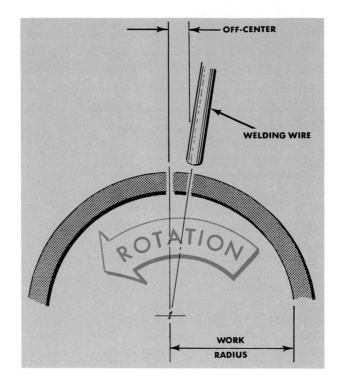

Fig. 19-20. Position of electrode wire in making a circular weld. (Hobart Brothers Co.)

2. All hose and cable connections from the gun to the feeder are in good condition, properly insulated, and connections have been correctly made and secured.

3. Correct nozzle for the diameter wire.

4. Wire is properly threaded through gun.

5. Apertures of contact tube and nozzle are clean. Blow out gun occasionally because sometimes the gun becomes loaded with dust which restricts proper wire feed and flow of protective gases.

6. Wire speed and feed have been predetermined and adjusted on the feeder control.

7. Shielding gas and water coolant sources are on and adjusted for desired output.

8. Wire stick-out is correct.

9. Contact tip is in proper shape. Tips eventually wear out, especially under high usage, and must be replaced for good welds.

WELDING PROCEDURE

In general, Mig welding procedure follows a definite sequence regardless of the kind of welding that is being done. Basically the following steps are involved:

1. Set the voltage, wire feed, and gas flow to the prescribed conditions for the required type of welding. (See Tables 19-7 to 19-10, pages 223 through 225.) During welding the wire speed rate may have to be varied to correct for too much or too little heat input.

2. Adjust the wire for the proper amount of stick-out.

3. Start the arc and move the gun along the seam at a uniform speed, keeping the gun at the correct angle. If the arc is not started properly the filler wire may stick to the work or actually freeze to the tip. Should this happen, shut off the machine and free the wire.

4. Move the gun along the seam with a pushing or pulling motion. As the gun is moved keep the wire at the leading edge of the puddle. Also be sure the wire is centered in the gas pattern to insure adequate shielding.

5. Release the trigger when reaching the end of the weld. This stops the wire feed and interrupts the welding current. However, always keep the gun over the weld until the gas stops flowing

in order to protect the puddle until it solidifies.

6. Shut down the welding unit when welding is completed. Follow this sequence:

   a. Turn OFF wire speed control.

   b. Shut OFF gas flow at cylinders.

   c. Squeeze the welding gun trigger to bleed the lines.

   d. Hang up the welding gun.

   e. Shut OFF welding machine.

During any welding operation certain welding condition may have to be changed. Some of the more specific welding variables with their required changes are shown in Table 19-4.

TABLE 19-4.   CORRECTING WELDING VARIABLES.

CHANGE DESIRED	ACTION REQUIRED
1. Deeper penetration	Increase welding current, or decrease wire stick-out, or use smaller wire size
2. Shallower penetration	Decrease welding current, or increase wire stick-out, or use larger wire size
3. Larger bead	Increase welding current, or decrease travel speed, or increase wire stick-out
4. Smaller bead	Decrease welding current, or increase travel speed, or decrease wire stick-out
5. Flatter, wider bead	Increase arc voltage, or decrease wire stick-out
6. Faster deposition rate	Increase welding current, or increase wire stick-out, or use smaller wire size
7. Slower deposition rate	Decrease welding current, or decrease wire stick-out, or use larger wire size

Arc Starting

Starting an electrical arc for a welding process involves three major factors: electrical contact, arc voltage, and time. To assure good arc starts,

it is necessary for the electrode wire to make good electrical contact with the work. The electrode must exert sufficient force on the workpiece to penetrate impurities.

Arc initiation becomes increasingly more difficult as wire stick-out increases. A reasonable balance of volts and amperes must be maintained in order to assure the proper arc and to deposit the metal at the best electrode melting rate.

The response time of a power source and circuit is generally fixed by equipment design. Most power sources have an optimum response time for wires.

The arc may be generated by the *run-in start method* or *scratch start method*, depending on the type of equipment used. Some welding units provide circuiting for both by incorporating separate toggle switches on the feeder panel.

In the run-in method the gun is aimed at the workpiece without touching the workpiece. The gun trigger is depressed and this immediately energizes the wire and starts the arc. See Fig. 19-21 for run-in method.

With the scratch method, the end of the welding wire must be scratched against the workpiece to start the arc.

A practice often used to insure a good weld start, is to strike the arc about one inch ahead of where the actual weld is to begin. Then the arc is

Fig. 19-21. Start the arc by pressing gun trigger. (Hobart Brothers Co.)

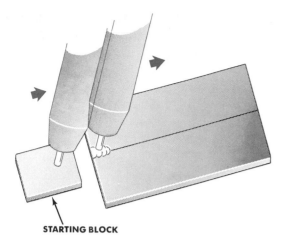

STARTING BLOCK

Fig. 19-22. Strike the arc ahead of the starting point of the weld.

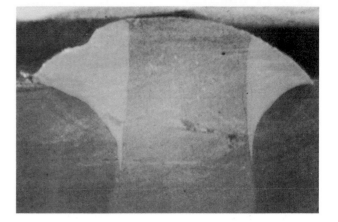

Fig. 19-23. Example of cold lap.

brought back quickly to the weld starting point. Another way is to strike the arc on a starting block outside the weld seam. See Fig. 19-22.

When finishing a weld a similar technique is used by reversing the direction of travel and at the same time increasing the speed. This helps to taper the width of the molten pool before breaking the arc. By following such a procedure there is less chance to leave a crater in the last deposited bead.

Once the arc is started, the gun is held at the correct angle and moved at a uniform speed.

Possible Weld Defects

Mig welding like any other form of welding must be controlled properly to produce consistently high quality welds. The beginner, in practicing Mig welding, should analyze each completed weld to avoid repeated weld defects. The following are a few of the more common defects which may be encountered during the early stages of the learning process.

Cold lap. Cold laps usually occur when the arc does not melt the base metal sufficiently, causing the slightly molten puddle to flow into the unwelded base metal. See Fig. 19-23. Very often if the puddle is allowed to become too large, this too will result in cold laps. For proper fusion, the arc should be kept at the leading edge of the puddle. When directed in such a

manner, the molten puddle is prevented from flowing ahead of the welding arc. Also remember that the size of the puddle can be reduced by increasing the travel speed or reducing the wire speed feed.

Surface porosity. Generally, surface porosity is the direct result of atmospheric contamination. See Fig. 19-24. It is caused by having the shielding gas set too low or too high. If it is too low the air in the arc area is not fully displaced; if the gas flow is too much, an air turbulence is generated which prevents complete shielding. On occasion, porosity will occur if the welding is being done in a windy area. Without some protective wind shield the gas envelope may be blown away, exposing the molten puddle to the contaminating air.

Crater porosity or cracks. The chief cause of crater defects is removing the gun and shielding gas before the puddle has solidified. See Fig. 19-25. Other possible causes of crater porosity or cracks are moisture in the gas, dirt, oil, rust or paint on the base metal, or excessive tip-to-work distance.

Insufficient penetration. Lack of penetration is due to a low heat input in the weld area or failing to keep the arc properly located on the leading edge of the puddle. See Fig. 19-26. If the heat input is too slow, increase the wire feed speed to get a higher amperage.

Excessive penetration. Too much penetration or *burn-through* is caused by having exces-

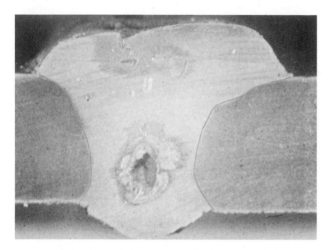

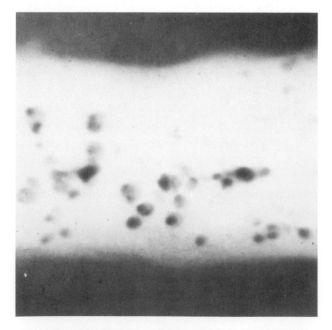

Fig. 19-24. Examples of surface and subsurface porosity. (Hobart Brothers Co.)

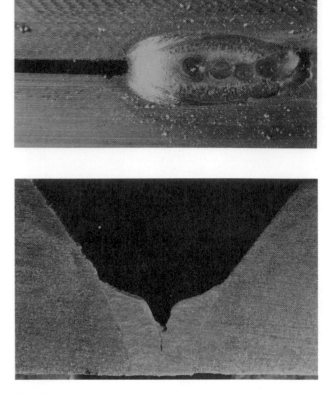

Fig. 19-25. Examples of crater porosity and cracks. (Hobart Brothers Co.)

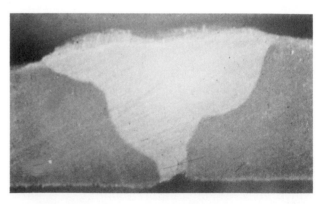

Fig. 19-26. Example of insufficient penetration. (Hobart Brothers Co.)

sive heat in the weld zone. See Fig. 19-27. By reducing the wire speed feed, the amperage is lowered and there will be less heat. Excessive penetration can also be avoided by increasing

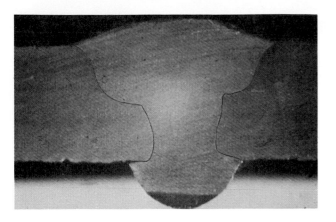 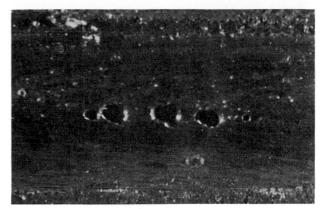

Fig. 19-27. Examples of burn-through and excessive penetration. (Hobart Brothers Co.)

Fig. 19-28. Examples of whiskers through a weld joint. (Hobart Brothers Co.)

the travel speed. If the root opening in the joint is too wide too much burn-through may result. Usually improper joint design can be remedied by increasing the stick-out and by weaving the welding gun.

Whiskers. Whiskers are short lengths of electrode wire sticking through the weld joint. See Fig. 19-28. They are caused by pushing the wire past the leading edge of the puddle. The small section of wire then protrudes inside the joint and becomes welded to the deposited metal. The best way to remedy this is to reduce travel speed, increase slightly the tip-to-work distance, or reduce the wire feed speed.

MIG WELDING COMMON METALS

Mig welding has become one of the most universally accepted processes for joining all types of metals. The ease with which sound welds can be produced by Mig welding has in many instances revolutionized welding practices in numerous industries. One of its particularly outstanding features is the ease with which production welding can be mechanized, thereby substantially reducing manufacturing costs.

Generally speaking, the same type of equipment and welding technique apply to joining all metals. A few specific characteristics for welding several common metals are included in the paragraphs that follow.

Carbon Steels

Both the spray arc and short arc produce excellent welds in carbon steels. For spray-arc welding a mixture containing 5 percent oxygen with argon is generally recommended. The addition of oxygen provides a more stable arc, minimizes undercutting and permits greater speeds.

Considerable amount of steel welding is done with a mixture of argon and CO_2. A straight CO_2 gas is sometimes used, especially for high-speed production welding. However, with CO_2 the arc is not a true spray arc.

For short arc welding of carbon and low-alloy steels a 25 percent carbon dioxide and 75 percent argon mixture is preferred. The dioxide mixture improves arc stability and minimizes spatter.

Thin steel plates 0.035″ to 1/8″ in thickness may be butt-welded with square edges. Usually an opening of 1/16″ or less is recommended. For wider openings the short arc is better since relatively large gaps are more easily bridged without excessive penetration. Plates 3/16″ and 1/4″ may be square butt-welded with a 1/16″ to 3/32″ root opening but usually two passes are necessary. For quality welds some beveling is desired. Plates 1/4″ and greater require single or double-V grooves with 50 to 60 degree included angles. U-grooves having a root spacing of 1/32″ to 3/32″ are necessary on plates thicker than one inch. See Fig. 19-29.

Table 19-5 lists specific requirements for Mig welding carbon steels. On multipass welds the sequence of bead deposits is similar to metallic arc welding. See Chapter 9.

Practice welding—lap and T-joints:

1. Arrange two pieces of 3/16″ or 1/4″ steel plates to form a lap joint.

2. Set voltage regulator, wire feed speed, gas flow, and wire stick-out. See Tables 19-3 and 19-5 for these conditions.

3. Position the gun with a wire stick-out of 3/8″ to 3/4″, squeeze the gun trigger and tack both ends of the lap joint.

4. Hold the gun so the nozzle bisects (cuts in two) the joint angle and leans 5 to 10 degrees

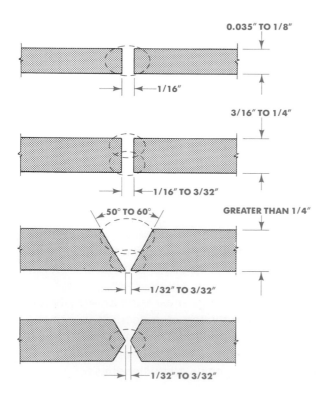

Fig. 19-29. Joint spacing for Mig welding carbon steels.

TABLE 19-5.   MIG WELDING—CARBON STEEL.

PLATE THICKNESS (inches)	JOINT AND EDGE PREPARATION	WIRE dia (inches)	GAS FLOW (cfh)	DCRP CURRENT (amps)		WIRE FEED (ipm)
0.035				55	16*	117
.047				65	17*	140
.063	Non-positioned fillet or lap	.030	10–15	85	17*	170
.078				105	18*	225
.100				110	18*	225
1/8				130	19*	300
1/8	Butt (square edge)	1/16		280	—	165
3/16	Butt (square edge)	1/16		375	—	260
3/16	Fillet or lap	1/16		350	—	230
1/4	Double V butt (60° included angle, no nose)			375 (1st pass) / 430 (2nd pass)	27	83 (1st) / 95 (2nd)
5/16	Double V butt (60° included angle, no nose)		40–50	400 (1st pass) / 420 (2nd pass)	28	87 (1st) / 92 (2nd)
5/16	Non-positioned fillet			400		87
1/2	Double V butt (60° included angle, no nose)	3/32		400 (1st pass) / 450 (2nd pass)		87 (1st) / 100 (2nd)
1/2	Non-positioned fillet			450	28	100
3/4	Double V butt (90° included angle, no nose)			450 (all 4 passes)	29	100
3/4	Positioned fillet			475	30	110
1	Fillet			450 (all 4 passes)	28	100

GAS FLOW column: Mixture (75%A + 25% CO_2) — upper rows; Mixture (95% argon + 5% O_2) — lower rows.

*short arc

Linde Co.

in the direction of travel. Use a pulling motion in which the gun is pointing away from the direction of travel. See Fig. 19-30.

5. Run a fillet weld on one side and do the same on the other side.

6. Repeat the same procedure on a T-joint.

Practice welding—butt joint:

1. Tack two pieces to form a square butt joint with a 1/16″ root opening.

2. Incline the gun about 5 degrees in the direction of travel and use a pushing motion. See Fig. 19-31.

Practice welding—horizontal, vertical, and overhead positions: Practice welding lap, T, and butt joints in the horizontal, vertical, and overhead positions. In the vertical position practice both uphill and downhill welding.

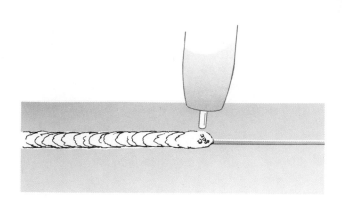

Fig. 19-32. Keep gun with wire in leading edge of puddle.

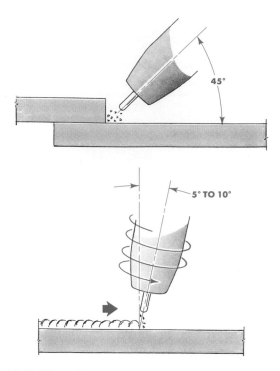

Fig. 19-30. Mig welding a lap joint using a pulling technique.

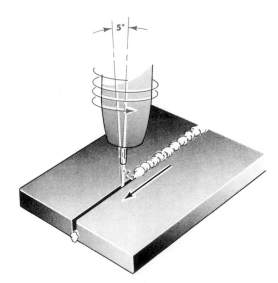

Fig. 19-31. Mig welding a butt joint using a pushing motion.

Remember that travel speed controls the weld bead size. If you want a smaller bead, increase the rate of travel. For a larger bead reduce the rate of travel. Keep the arc stream near the leading edge of the puddle. See Fig. 19-32. If undercutting occurs, use a slight side to side weaving motion.

Aluminum

The design of seams for aluminum is similar to that for steel. However, narrower joint spacing and lower welding currents are recommended due to the higher fluidity of the metal.

Argon gas is preferred for Mig welding plates up to 1″ in thickness since it provides better metal transfer and arc stability with less spatter. Sometimes in position welding of 1100 and 3003 aluminum the addition of a small amount of oxygen to the argon improves coalescence (flow of metals together).

When welding plates between 1″ and 2″ a mixture of 50 percent argon and 50 percent helium will often prove advantageous. This provides a higher heat input associated with helium and good cleaning action obtained with argon.

The *short-arc welding* of aluminum produces a colder arc than the spray type arc and thereby permits the weld puddle to solidify rapidly.

This action is especially advantageous in vertical, overhead, and horizontal welding, and in welding lighter materials. In vertical welding, a downhill technique is preferred.

Spray-arc welding aluminum is especially suitable for thick sections. With spray arc more heat is produced to melt the wire and base metal. As a rule vertical, horizontal and overhead welds are more difficult to make than with the short arc.

TABLE 19-6.   MIG WELDING—ALUMINUM (SPRAY-ARC).

PLATE THICKNESS (inches)	TYPE OF JOINT	WIRE dia (inches)	ARGON FLOW (cfh)	DCRP (amperes)	VOLTAGE (volts)	APPROXIMATE WIRE FEED (ipm)
0.040	Fillet or tight butt	0.030	30	40	15	240
.050	Fillet or tight butt	.030	15	50	15	290
.063	Fillet or tight butt	.030	15	60	15	340
.093	Fillet or tight butt	.030	15	90	15	410

Linde Co.

TABLE 19-7.   MIG WELDING—ALUMINUM (SPRAY-ARC).

PLATE THICKNESS	PREPARATION	WIRE DIAMETER (inches)	ARGON FLOW (cfh)	DCRP (amperes)	VOLTAGE
0.250	Single V butt (60° included angle) sharp nose back-up strip used	3/64	35	180	24
	Square butt with back-up strip	3/64	40	250	26
	Square butt with no back-up strip	3/64	35	220	24
.375	Single V butt (60° included angle) shop nose, back-up strip used	1/16	40	280	27
	Double V butt (75° included angle, 1/16" nose). No back-up. Back chip after root pass	1/16	40	260	26
	Square butt with no back-up strip	1/16	50	270	26
.500	Single V butt (60° included angle) sharp nose. Back-up strip used	1/16	50	310	27
	Double V butt (75° included angle 1/16" nose). No back-up. Back chip after root pass	1/16	50	300	27

Linde Co.

Practice welding—lap, T, and butt joints:

1. Tack two pieces of aluminum to form proper joint.

2. Set voltage, wire feed, and gas flow according to conditions in Tables 19-6 and 19-7.

3. Angle gun nozzle about 15° with a wire stick-out of 1/2" to 3/4".

4. Use a pushing motion.

Stainless Steel

Copper back-up strips should be used to weld stainless steel up to 1/16" in thickness. Precaution must be taken to prevent air from reaching the underside of the weld while the puddle is solidifying since the oxygen and nitrogen will weaken the weld. To prevent air from contacting the underside of the weld an argon back-up gas is often used.

Spray arc welding with 1/16" diameter wire and high current produces good welds. DCRP with a 1 or 2 percent argon-oxygen mixture is recommended for most stainless steel welding.

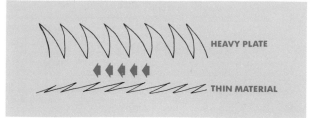

Fig. 19-33. Gun motion for Mig welding stainless steel.

The forehand or pushing technique is generally used for welding stainless steel. On plates 1/4" or more in thickness the gun should be moved back and forth with a slight side to side movement. Thin materials are best welded with just a slight back-and-forth motion along the joint. See Fig. 19-33. As a rule the short arc produces better welds on thin materials when overhead or vertical welding is required.

Tables 19-8 and 19-9 list specific requirements for Mig welding stainless steels.

TABLE 19-8.   GENERAL WELDING CONDITIONS FOR SPRAY-ARC WELDING OF STAINLESS STEEL.

PLATE THICKNESS (inches)	JOINT AND EDGE PREPARATION	WIRE Dia	GAS FLOW	CURRENT DCRP (Amps)	WIRE FEED (ipm)	SPEED (imp)	WELDING PASSES
0.125	Square butt with backing	1/16	35	200–250	110–150	20	1
.250	Single V butt 60° inc. angle no nose	1/16	35	250–300	150–200	15	2
.375	Single V butt 60° inc. angle 1/16" nose	1/16	(O₂–1)	275–325	225–250	20	2
0.500	Single V butt 60° inc. angle 1/16" nose	3/32	(O₂–1)	300–350	75–85	5	3–4
.750	Single V butt 90° inc. angle 1/16" nose	3/32	(O₂–1)	350–375	85–95	4	5–6
1.000	Single V butt 90° welded angle 1/16" nose	3/32	(O₂–1)	350–375	85–95	2	7–8

Linde Co.

TABLE 19-9. GENERAL WELDING CONDITIONS FOR SHORT ARC WELDING OF STAINLESS STEEL.

PLATE THICKNESS	JOINT AND EDGE PREPARATION	WIRE dia (inches)	GAS FLOW (cfh)	CURRENT DCRP (amps)	VOLTAGE	WIRE FEED (ipm)	WELDING SPEED (ipm)	PASSES
0.063	Non-positioned fillet or lap	0.030	15–20	85	15	184	18	1
.063	Butt (square edge)	.030	0₂–2	85	15	184	20	1
.078	Non-positioned fillet or lap	.030	0₂–2	90	15	192	14	1
.078	Butt (square edge)	.030	0₂–2	90	15	192	12	1
.093	Non-positioned fillet or lap	.030	0₂–2	105	17	232	15	1
.125	Non-positioned fillet or lap	.030	0₂–2	125	17	280	16	1

*Voltage values are for C-25 gas or 02-2 gas. For 90% HE—10% C-25, voltage will be 6 to 7 volts higher
Linde Co.

TABLE 19-10. CONDITIONS FOR WELDING COPPER.

THICKNESS (inches)	CURRENT DCRP (amps)	VOLTS	TRAVEL (ipm)	WIRE dia (inches)	WIRE FEED (ipm)	JOINT DESIGN
1/8	310	27	30	1/16	200	Square butt, steel back-up strip required
1/4 (1)	460	26	20	3/32	135	Square butt
1/4 (2)	500				150	
3/8 (1)	500	27	14	3/32	150	Double bevel, 90° included angle,
3/8 (2)	550				170	3/16″ nose
1/2 (1)	540	27	12	3/32	165	Double bevel, 90° included angle,
1/2 (2)	600		10		180	1/4″ nose

Linde Co.

Practice welding—lap, T, and butt joints:

1. Arrange two pieces of stainless steel to form the required joint.

2. Set voltage, wire feed, and gas flow according to conditions shown in Table 19-8.

3. Use a pushing gun motion.

Copper

Mig welding of copper is usually confined to the deoxidized types. Welding electrolytic copper is not advisable because such welds exhibit low strength.

Argon is preferred as the shielding gas for thin materials. For materials 1″ or more in thickness a mixture of 65 percent helium and 35 percent argon is recommended.

Steel back-up blocks are required for welding sheets 1/8″ or less in thickness. Although no preheat is necessary for materials of this thickness, some preheating (400°F or 204°C) is advisable on sections 3/8″ or more in thickness.

See Table 19-10 for special Mig welding conditions of copper.

Points to Remember

1. Mig (GMAW) welding is often referred to by the manufacturer's trade name as Micro-wire, Aircomatic, Sigma, and Millermatic Welding.

2. Mig welding is faster than stick electrode welding and is much easier to learn.

3. Spray transfer type of welding is particularly adapted for welding heavy gage metals.

4. Short circuiting transfer welding is best for welding light gage metals.

5. For optimum efficiency, DCRP current is required for Mig welding.

6. For Mig welding, a constant potential power supply with a nearly flat volt-ampere characteristic produces the best results.

7. As a general rule, an air-cooled gun is satisfactory when welding with amperage around 200 and a water-cooled gun for welding heavy metals requiring higher amperages.

8. The use of CO_2 as a shielding gas is most effective and less expensive when welding steel.

9. Argon or a mixture of argon and oxygen will produce the most effective results in welding aluminum and stainless steel.

10. The rate of gas flow for welding most metals is approximately 35 cu ft/hr. However, this rate may have to be varied somewhat, depending on the type, electrode size, and thickness of metal.

11. The effectiveness of the shielding gas is often governed by the distance of the gun from the workpiece. Generally the gas nozzle should not be spaced more than 2″ from the workpiece.

12. The use of correct diameter wire electrode is necessary for good welds. Check recommendations for correct electrode diameters.

13. The correct current for welding must often be determined by trial. Check recommendations for starting current.

14. Be sure the wire feed is set for the amperage which is to be used for welding.

15. For most Mig welding applications, the wire stick-out should be about $3/8″$ to $3/4″$.

16. Keep the gun properly positioned to insure uniform weld with proper penetration.

17. Cold laps will occur if the arc does not melt the base metal sufficiently.

18. Check the weld for surface porosity. Surface porosity is usually caused by improper gas shielding.

19. Do not remove the gun from the weld area until the puddle has solidified, otherwise cracks may develop.

20. Remember, insufficient or excessive penetration is the result of failure to control heat input.

QUESTIONS FOR STUDY AND DISCUSSION

1. How does Mig welding differ from Tig welding?

2. What are some of the specific advantages of Mig welding?

3. What is the difference between spray and globular metal transfer?

4. Why is globular transfer ineffective for welding heavy gage metals?

5. What is meant by short-circuiting transfer? For what type of welding is this the most effective means?

6. Why is DCRP current best for doing Mig welding?

7. What results can be expected if DCSP current is used?

8. How does a constant potential power supply unit differ from the conventional constant current welding machine?

9. What is the advantage of using a constant potential power supply unit for Mig welding?

10. What is meant by slope control?

11. How is the electrode wire fed to the welding gun?

12. When should a water-cooled or air-cooled gun be used?

13. What are the elements that make up air?

14. Why is oxygen generally a harmful element in welding?

15. Why does nitrogen cause the most serious problems in welding?

16. When is argon or a mixture of argon and oxygen considered the ideal gas for shielding purposes?

17. When is CO_2 better for shielding purposes instead of an inert gas?

18. How is it possible to determine if the gas flow is proper for shielding?

19. What is likely to happen if the gas flow is allowed to drift from the weld area?

20. What factors must be taken into consideration in selecting the correct diameter (size) electrode?

21. What is meant by current density?

22. Why is it impossible to make a proper weld if the current density falls below the required minimum?

23. What determines the rate at which the wire feed should be set?

24. Why is correct wire stick-out important?

25. What position should the gun be held for horizontal fillet welding?

26. What position should the gun be held for flat fillet welding?

27. What determines whether a pulling or pushing technique of the gun should be used?

28. How does the run-in start differ from the scratch start method in starting a weld?

29. What is the probable cause for the formation of cold laps in a weld?

30. What should be done to avoid surface porosity in a weld?

31. How can crater porosity or cracks be avoided?

32. If weld penetration is insufficient, what should be done?

33. What causes whiskers in a weld?

CHAPTER 20 related Mig welding processes

In addition to the Mig welding process described in Chapter 19, several other gas shielded arc techniques have been developed to meet the ever increasing industrial demands for joining metals more effectively and at lower cost. Some of the more common ones are described in this chapter.

Buried-Arc CO_2 Welding Process

Buried-arc CO_2 welding is a high-energy, fast-weld method in which the end of the wire electrode is held either level or below the surface of the work with practically a zero arc length. See Fig. 20-1. This process is designed for high-speed welding of mild steel. Although it can be employed for manual welding its greatest application is in mechanized welding. The process is widely used in many industries where the fabrication of parts requires deep penetration and fast deposition of weld metal without critical control of bead contour. In most instances welding wire diameters range from 0.045″ to $\frac{1}{8}$″.

Regular Mig welding equipment is utilized for the buried-arc process. The shielding gas is pure carbon dioxide which provides additional economy over the more expensive argon. The metal transfer is globular, but since the wire is buried and a high density current is used, a deep cavity is formed. This deep cavity traps the molten globules that normally would be ejected sideways through the arc. Thus, splatter which otherwise would be severe is minimized and does not affect the welding process.

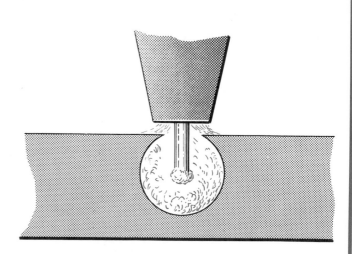

Fig. 20-1. Buried-arc CO_2.

Pulsed-Spray Arc Welding

The pulsed-spray process is an extension of spray-transfer welding to a current level much below that required for continuous spray transfer. The pulsing current used may be considered as having its peak current in the spray-transfer current range and its minimum value in the globular transfer current range.

The need for current values less than the transition level becomes apparent when attempting to weld under heat transfer conditions which are inadequate for spray transfer. For example, when welding out-of-position, high current will result in a molten pool which cannot be retained in position unless the material being

228

welded has adequate thermal-conductivity (coupled with the joint type and plate thickness).

The same factors explain the burn-through obtained when a weld on thin material is attempted with too high a welding current. While smaller diameter electrodes have lower transition currents, the basic limitation cannot be avoided. The net result is that the spray transfer process is very applicable to flat position use but is rather more limited in its use for out-of-position and thin material welding.

Pulsed-spray transfer is achieved by pulsing the current back and forth between the spray transfer and globular transfer current ranges. Fig. 20-2 illustrates, on the left, the current-time relationships for two power sources, A and B, with A putting out a current in the globular transfer range and B putting out a current in the spray-transfer range. Fig. 20-2 also shows, on the right, the two outputs combined to produce a simple pulsed output by electrically switching back and forth between them.

Transfer is restricted to the spray mode. Globular transfer is suppressed by not allowing sufficient time for transfer by this mode to occur. Conversely, at the high-current level, spray transfer is insured by allowing more than sufficient time for transfer to occur.

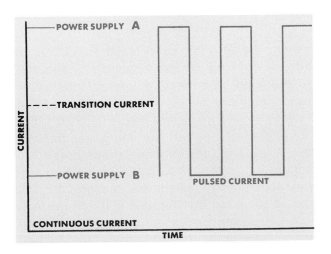

Fig. 20-2. Illustration of how a switching system can convert two steady-state DC output currents in a simple pulsing-current output waveform. (Airco)

For the given electrode deposited by the pulsed-spray method, all the advantages of the spray-transfer process are available at average current levels from the minimum possible with continuous spray transfer down to values low in the globular transfer range.

Features of pulsed-spray welding. Pulsed-spray welding method provides many features not previously available.

1. The heat input range bridges the gap between, and laps over into, the heat input ranges available from the spray and short-circuiting arc processes. Into its lower heat input range the pulsed-spray process brings the advantages of the continuous spray-transfer process. Also, due to lower heat input, the use of spray transfer is extended greatly into poor heat transfer areas, mainly related to welding out-of-position and on thinner materials.

2. The area of overlap with the spray-transfer process occurs because, having a higher transition current, a larger diameter electrode leaves the continuous spray and enters the pulsed-spray range at a higher current than a smaller electrode. Further, the use of a larger diameter electrode can be continued down to a current value considerably below the transition current associated with using a smaller diameter electrode.

The pulsed-spray process will not displace the short-circuiting arc process in those areas where the short-circuiting arc process is properly applicable and more economical.

3. The pulsed-spray method produces a higher ratio of heat input to metal deposition, permits the use of a completely inert gas shield where necessary, and is essentially free from spatter.

4. The pulsed-spray process is characterized by a uniformity of root penetration which approaches that possible with the gas tungsten-arc process; because of this feature, the process may permit deletion of weld backing in some cases.

Power source for pulsed spray. As Fig. 20-3 illustrates, the power source combines a standard, three phase, full-wave unit with a single phase, half-wave unit, both of the constant-potential type. The three phase unit is termed the background unit and the single phase unit is

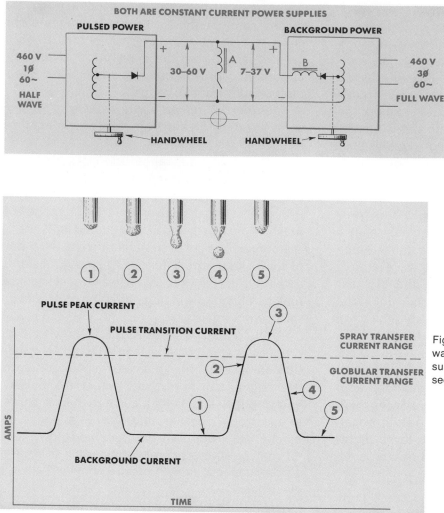

Fig. 20-3. Block diagram of the essential features of the pulsed-current power supply. (Airco)

Fig. 20-4. Illustration of the output current waveform of the pulsed-current power supply; also showing the metal transfer sequence. (Airco)

termed the pulsing unit. These units are connected in parallel but commutate (to form a unidirectional current) in operation. The waveform of the pulsing current output is schematically illustrated in Fig. 20-4.

The units are made to switch back and forth in operation by means of the varying output voltage of the pulsing unit. The diode rectifiers in each unit alternately permit or block the passage of current depending upon whether there is a positive or negative voltage difference across their terminals. When the pulse is OFF or its voltage is less than the background voltage, the diode rectifiers of the background unit pass the full value of the instantaneous current. Conversely, when the pulse voltage exceeds the background voltage, blocking the background diode rectifiers, the pulse diode rectifiers pass the full value of the instantaneous current.

Two chokes are used. (See Fig. 20-3.) The choke labeled A performs a commutation function. When the pulse voltage drops below the background voltage, it sustains the welding current momentarily, giving the background unit time to respond to the demand for current. Choke B, in series with the output of the background unit, filters the background current and prevents undesirable arc outages at low background current levels. The pulsed-current power source is shown in Fig. 20-5.

Fig. 20-5. Pulsed-current power supply designed for pulsed-spray arc welding. (Airco)

Operation of the pulsed-current power source is similar to that of conventional constant potential sources. With the arc OFF, the value of the pulse peak voltage, which depends upon the electrode type and diameter (Table 20-1) is selected and remains constant. This setting is made by rotating the pulse peak voltage handwheel while pressing a button which converts the average voltage meter to a peak voltage meter. The electrode feeder is set at the value which will produce the required current and is determined from Table 20-2 for the type and diameter of electrode to be used. The arc is then initiated and the background voltage handwheel rotated to produce the proper arc length. So long as the type and diameter of electrode remain the same, all further power source adjustments are made by means of the background voltage handwheel. The meters on the power supply read the average voltage and the average current which are the values familiar to every welding operator.

Tubular Wire Welding (Flux-core)

Tubular Wire welding is a gas metal-arc welding process in which a continuous fluxed core wire instead of a solid wire serves as the elec-

TABLE 20-1.   TYPICAL PULSED CURRENT POWER SUPPLY SETTINGS.

ELECTRODE TYPE	ELECTRODE DIA (inches)	PULSE PEAK RANGE (volts)	AVERAGE CURRENT RANGE (amps)	AVERAGE VOLTAGE (volts)
Mild and low-alloy steel	0.035	34–36	55–130	18–20
	0.045	37–39	90–180	19–23
	1/16	42–44	110–250	20–25
Stainless steel	0.035	33–35	55–130	18–20
	0.045	36–38	90–180	19–23
	1/16	41–43	110–250	20–25
Aluminum	1/16	34–36	80–250	20–30

TABLE 20-2.   AVERAGE CURRENT VS. ELECTRODE FEED SPEED FOR TYPICAL DIAMETERS OF STEEL AND ALUMINUM ELECTRODES WHEN USING PULSED CURRENT UNIT.

ELECTRODE FEED (ipm)	CORRESPONDING AVERAGE CURRENT (amps) MILD STEEL AND STAINLESS STEEL*			ALUMINUM†
	0.035″ dia	0.045″ dia	1/16″ dia	1/16″ dia
70	—	70	115	70
90	—	90	175	90
105	50	105	215	105
125	60	125	—	125
135	70	135	—	135
155	80	155	—	155
185	90	185	—	185
220	110	220	—	220
235	120			
255	130			
275	140			
300	150			
325	160			
345	170			
365	180			
380	190			
425	210			
500	—			

*Argon + 290 O_2
†Argon

trode. The wire can be used on any automatic or semi-automatic Mig welding equipment and is used principally in combination with CO_2 as a shielding gas. See Fig. 20-6. The wire is frequently referred to by the manufacturer's trade name such as *Fluxcor* (Airco) and *FabCo* (Hobart).

The common AWS flux-cored wires are E-70T-1, E-70T-2, E-70T-3, and E-70T-4. The identity of the symbols is similar to those for solid wires except the letter T which designates a tubular wire.

The flux ingredients in the wire include ionizers to stabilize the arc, deoxidizers to purge the deposit of gas and slag, and other metals to produce high strength, ductility and toughness in weld deposits. The flux generates a gas shield, which is augmented by the regular CO_2 shield, and a slag blanket that retards the cooling rate and protects the weld deposit as it solidifies.

Fig. 20-6. Mig welding equipment is used to weld with flux core wire. (Airco)

TABLE 20-3.   FLUX-CORED ARC WELDING CONDITIONS—DCRP.

	MATERIAL THICKNESS (inches)[1]	CURRENT DCRP (amps)	ARC VOLTAGE	WIRE FEED ipm	SHIELDING GAS FLOW CFH[2]	TRAVEL SPEED ipm	NO. OF PASSES	WIRE STICK-OUT (inches)
flux-cored	1/8	300–350	24–26	100–120	35–40	25–30	1	3/4 to 1 1/2
arc welding	3/16	350–400	24–28	120–150	35–40	25–35	1	3/4 to 1 1/2
of steel	1/4	350–400	24–28	120–150	35–40	20–30	1	3/4 to 1 1/2
using 3/32″	3/8	475–500	28–30	180–210	35–40	15–20	1	3/4 to 1 1/2
diameter electrode	1/2	400–450	25–28	150–170	35–40	18–20	2–3	3/4 to 1 1/2
wire size	5/8	400–450	25–28	150–170	35–40	14–18	2–3	3/4 to 1 1/2
flat and horizontal	3/4	400–450	25–28	150–170	35–40	14–18	5–6	3/4 to 1 1/2
positions								

[1]For groove and fillet welds. Material thickness also indicates fillet size. Use V groove for 1/4 inch and thicker. Double V for 1/2 inch and thicker.
[2]Welding grade CO_2.

TABLE 20-3. (Cont.)

ELECTRODE SIZE		FLAT POSITION[3]		HORIZONTAL POSITION[3]		VERTICAL POSITION[3]	
(inches)	(mm)	(ampere)[4]	(voltage)[5]	(ampere)[4]	(voltage)[5]	(ampere)[4]	(voltage)[5]
0.045	1.2	150–225	22–27	150–225	22–26	125–200	22–25
1/16	1.6	175–300	24–29	175–275	25–28	150–200	24–27
5/64	2.0	200–400	25–30	200–375	26–30	175–225	25–29
3/32	2.4	300–500	25–32	300–450	25–30	—	—
7/64	2.8	400–525	26–33	—	—	—	—
1/8	3.2	450–650	28–34	—	—	—	—

[3]Applies to groove, bead or fillet welds in position shown.
[4]Ampere range can be expanded. Higher currents can be used, especially with automatic travel.
[5]Voltage range can be expanded. It will increase when larger stick-out tip to work distance is employed.
Hobart Bros.

Tubular wire is designed for high current-densities and deposition rates which when combined with high duty-cycles result in sharply increased production speeds. It is especially intended for application in large fillet single and multi-pass welds in either a horizontal or flat position using DCRP current. Because of its deep penetrating qualities into the weld root, tubular wire fillet welds of smaller leg size will have the same strength as stick fillet welds of larger size. For instance, double welded butt joints up to 1/2″ thick can be welded without edge preparation.

The actual operation of flux core wire welding is similar to other Mig welding processes. See Table 20-3.

Vapor-Shielded Arc Welding[1]

Vapor-shielded arc welding, known as *Innershield,* is a welding process introduced by Lincoln Electric Company. This process uses a vapor instead of a gas to shield the arc and molten metal. (A *vapor* in its natural state is a liquid or a solid, as distinct from a *gas,* which is in its natural state.) The vapor is generated by a continuous tubular electrode, containing vapor-producing materials. The electrode also serves as a filler rod.

Equipment. Equipment for the vapor-shielded arc consists of a DC power source, a

1. Courtesy, The Lincoln Electric Co.

Fig. 20-7. Semi-automatic Innershield welder.

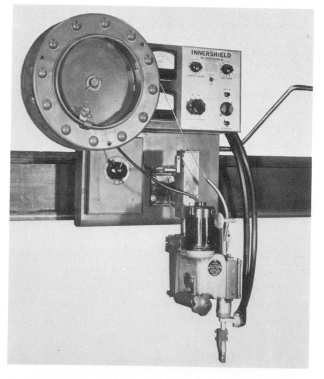

Fig. 20-8. A fully automatic Innershield welder. (The Lincoln Electric Co.)

control station for adjusting amperage and voltage, and a continuous wire feed mechanism. The equipment can be fully automatic or semi-automatic. See Figs. 20-7 and 20-8.

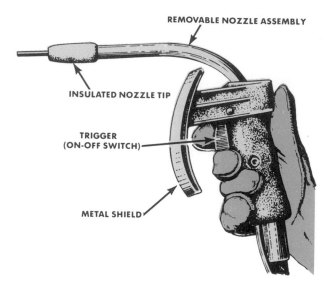

Fig. 20-9. The gun used in welding with a semi-automatic Innershield welder. (The Lincoln Electric Co.)

With semi-automatic equipment, the operator moves a gun-type nozzle along the weld seam. See Fig. 20-9. To weld, he simply pulls the trigger which starts the welding current and wire feed. When he wants to stop welding, he merely releases the trigger.

Filler wire. The tubular mild steel wire which serves as a filler rod, contains all of the necessary ingredients for shielding, deoxidizing, and fluxing. Some of the ingredients are metallic salts and oxides which melt before the electrode. These metallic salts vaporize and expand when they meet the intense heat of the arc. The vaporized salts expand until they reach a distance surrounding the arc where the temperature corresponds to their condensing temperature. The vaporized salts form a thick vapor shield around the arc and molten metal. This shield prevents contamination by the surrounding atmosphere. Other materials in the electrode have de-

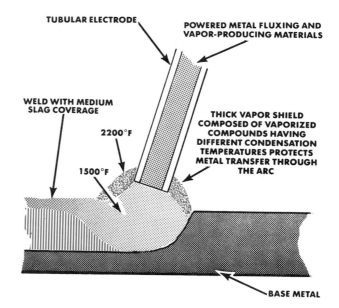

Fig. 20-10. Action of the tubular electrode in Innershield welding. (The Lincoln Electric Co.)

oxidizing and alloying functions in creating high grade weld metal. See Fig. 20-10.

Submerged-Arc Welding

Submerged-arc welding is a process wherein an electric arc is submerged or hidden beneath a granular material. The electric arc provides the necessary heat to melt and fuse the metal. The granular material, called flux, completely surrounds the electric arc thus shielding the arc and the metal from the atmosphere. A metallic wire is fed into the welding zone underneath the flux.

The welding process can be either semi-automatic or fully automatic. In the semi-automatic, a special hand welding gun is used. See Fig. 20-11. Any regular Mig welding DC power source can be adapted for submerged arc welding.

The difference between submerged-arc welding and other forms of gas metal-arc welding is that no inert shielding gas is required. The gun hopper is simply filled with flux, pointed over the weld area and the gun trigger depressed. As soon as the trigger is pulled, the wire is energized and the arc is started. At the same time the flux begins to flow. The actual welding operation is now carried out in the same manner as in the procedure for Mig welding.

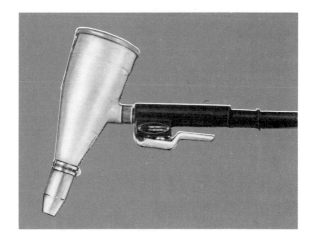

Fig. 20-11. Hand gun used for submerged-arc welding. (Hobart Brothers Co.)

As the metallic wire is fed into the weld zone the feeding hopper deposits the granulated flux over the weld puddle and completely shields the welding action. The arc is not visible since it is buried in the flux, thus there is no flash or spatter. That portion of the granular flux immediately around the arc fuses and covers the molten metal but after it has solidified it can be tapped off easily with a ball peen hammer.

With the fully automatic process, the welding unit is arranged to move over the weld area at a controlled speed. On some arrangements, the

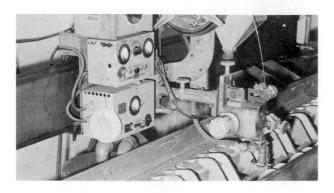

Fig. 20-12. An automatic submerged-arc welder is being used here to weld a cylindrical housing.

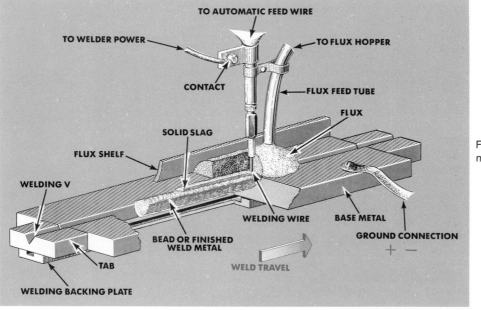

Fig. 20-13. Cutaway view of submerged arc welding. (Linde Co.)

welding head moves and the work remains stationary. In others the head is stationary and the work moves. See Figs. 20-12 and 20-13.

Submerged-arc welding is usually best adapted where relatively thick sections are to be joined and deep penetration is required. For example, it is possible to weld 3″ plate in a single pass. Little, if any, edge preparation is necessary—as a rule none on material under one-half inch in thickness. Generally back-up support is essential for heavy welding plate.

Welding is done on a horizontal or nearly horizontal plane.

Points to Remember

1. Buried-arc CO_2 is a high energy, fast welding process used for fabricating parts that require fast deposition and deep penetration.

2. Keep the wire completely buried in the weld cavity when using the buried Arc-CO_2 welding process.

3. The pulsed-spray process permits welding thin materials and making out-of-position welds with less danger of burn-through.

4. Operation of a pulsed-current power source is the same as the conventional constant potential welding machine.

5. Flux-core wire is used principally in combination with CO_2 and is intended for high current densities.

6. Use a DCRP current with flux-core wire.

7. Vapor-shielded arc welding uses a vapor instead of a gas for shielding purposes. The vapor is generated by a continuous tubular electrode containing vapor-producing materials, such as salts and oxides.

8. In submerged-arc welding, the electric arc is completely hidden beneath a flux.

9. Submerged-arc welding has its greatest application in welding thick metals where deep penetration is required.

10. No inert shielding gas is required for submerged-arc welding since the flux completely surrounds the electric arc.

QUESTIONS FOR STUDY AND DISCUSSION

1. How does the buried-arc CO_2 process differ from regular Mig Welding?

2. Where does the buried-arc CO_2 process have its greatest application?

3. What is pulsed-spray arc welding?

4. Pulsed-spray arc welding is particularly adapted for what kinds of operations?

5. What is tubular wire welding?

6. In combination with CO_2, Flux-cored wire is designed for what types of welding?

7. How does vapor-shielded arc welding such as Innershield differ from other gas-shielded arc welding processes?

8. In vapor-shielded arc welding, how is the vapor generated for the cover shield?

9. What is submerged-arc welding and by what material is it shielded?

10. What are some of the special advantages of submerged-arc welding?

oxy - acetylene welding

CHAPTER 21 equipment

The oxy-acetylene welding process has to a large extent been superseded by other types of welding for industrial production purposes. However, oxy-acetylene welding still has its place in the welding family. Undoubtedly its greatest use is in general maintenance, auto body shops, and in repairing small parts where other welding processes would be too expensive to use both in terms of material and equipment set-up costs.

The process of oxy-acetylene welding is possible because of the principle that, when acetylene is mixed with oxygen in correct proportions and ignited, the resulting flame is one of the hottest known. This flame, which reaches a temperature of 6300°F (3482°C), melts all commercial metals so completely that metals to be joined actually flow together to form a complete bond without application of any mechanical pressure or hammering. In most instances, some extra metal in the form of a wire rod is added to the molten metal in order to build up the seam slightly for greater strength. On very thin material the edges are usually flanged and just melted together. In either case, if the weld is performed correctly, the section where the bond is made will be as strong as the base metal itself.

With the oxy-acetylene flame, such metals as iron, steel, cast iron, copper, brass, aluminum, bronze, and other alloys may be welded. In many instances dissimilar metals can be joined, such

Fig. 21-1. Flame hardening with oxy-acetylene flame permits hardening just those surfaces of the gear teeth that wear more in service. (Linde Co.)

as steel and cast iron, brass and steel, copper and iron, brass and cast iron.

The oxy-acetylene flame is also employed for a variety of other purposes, notably for cutting metal, case hardening, and annealing. It is used for several types of metallic spray guns which spray fine particles of molten metal on worn

238

surfaces that need refacing or building up. As a matter of fact it can be used in practically any situation which involves joining metal parts. See Fig. 21-1 for flame hardening application.

Oxygen for Welding

The atmosphere, which we commonly refer to as air, is composed of approximately 20 percent oxygen, the rest being nitrogen and a small percentage of rare gases such as helium, neon, and argon. To obtain the oxygen in a state that makes it usable for welding, it is necessary to separate it from the other gases.

The two general methods that may be used to produce oxygen are the electrolytic and the liquid-air methods. In the electrolytic method, the fact is utilized that water is a chemical compound consisting of oxygen and hydrogen. By sending a current through a solution of water containing caustic soda, oxygen is given off at one terminal plate, and hydrogen at the other, thereby making possible the separation of the two gases. Because of the greater cost in manufacturing oxygen in this manner, the liquid-air method is more commonly used to produce commercial oxygen.

In a plant where oxygen is made by the liquid-air method, the air is drawn from the outside into huge containers known as washing towers, where the air is washed and purified of carbon dioxide. A solution of caustic soda is used to wash the air, this solution being run from a nearby tank and circulated through the towers by means of centrifugal pumps.

As the air leaves the washing towers, it is compressed and passed through *oil-purging cylinders* in which oil particles and water vapor are removed. From this point the air goes into drying cylinders. These cylinders contain dry, caustic potash which dries the air and removes any remaining carbon dioxide and water vapor. On the top of each cylinder, special cotton filters are provided to prevent any particles of foreign matter from being carried into the high-pressure lines.

The dry, clean, compressed air then goes into rectifying or liquification columns where the air is cooled and then expanded to approximately atmospheric pressure. The process of changing the extremely cold air under high pressure to the lower atmospheric pressure causes the air to liquefy.

The separation of the nitrogen from the oxygen becomes possible at this stage because of the difference in the boiling point between the nitrogen (−320°F or −195.5°C) and the oxygen (−296°F or −182°C). The nitrogen, having the lower boiling point, evaporates first, leaving the liquid oxygen in the bottom of the condenser. The liquid oxygen next passes through a heated coil which changes the liquid into a gaseous form. From here the gas goes into a storage tank, flowing through a gas meter which registers the amount of gas entering the storage tank. The gas is then drawn from this tank and compressed into receiving cylinders.

The oxygen cylinder. Oxygen cylinders are made from seamless, drawn steel and tested with a water pressure of 3360 psi. The cylinders are equipped with a high-pressure valve which can be opened by turning the hand wheel on top of the cylinder. See Fig. 21-2. *This valve should always be opened by hand and not with a wrench. The hand wheel must be turned slowly to permit a gradual pressure load on the regulator, and then opened as far as the valve will turn, to full gas pressure.* Unless the valve is wide open the high oxygen pressure may cause the oxygen to leak around the valve, resulting in considerable waste.

There are three common sizes of oxygen cylinders. The large size, which is the one popularly used by industrial plants and shops consuming large quantities of gas, is filled with 244 cubic feet (cu ft) of oxygen. The medium sized cylinder contains 122 cu ft, and the small cylinder 80 cu ft at standard conditions.

Cylinders are charged with oxygen at a pressure of 2200 psi at a temperature of 70°F (21°C). Gases expand when heated and contract when cooled, so the oxygen pressure will increase or decrease as the temperature changes. For example, if a full cylinder of oxygen is allowed to stand outdoors in near freezing temperature, the pressure of the oxygen will register less than 2200 psi. However, this does not mean that any of the oxygen has been lost; cooling has only reduced the pressure of the oxygen.

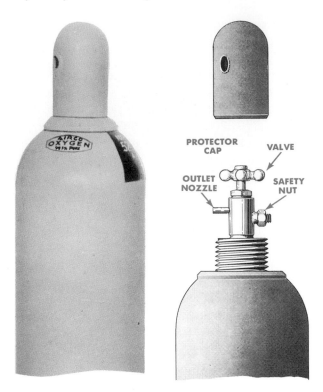

Fig. 21-2. Oxygen cylinders are available in three sizes, one of which is shown at left. The view at the right shows the top of the cylinder with the cap removed.

Since the pressure of gas will vary with the temperature, all oxygen cylinders are equipped with a safety nut that permits the oxygen to drain slowly in the event the temperature increases the pressure beyond the safety load of the cylinder. Thus if a cylinder were exposed to a hot flame, the safety nut would relieve the pressure before the cylinder would reach the point where it would explode.

A *protector cap* which screws onto the neck ring of the cylinder is furnished to protect the valve from any damage. This cap must always be in place when the cylinder is not in use.

Acetylene for Welding

Acetylene is a gas formed by the mixture of calcium carbide and water. The commercial generator in which the gas is made consists of a huge tank containing water. A specified quantity of carbide is dumped into a hopper and raised to the top of the generator. The carbide is then

allowed to fall into the water, and upon coming in contact with the water, bubbles of gas are given off. This gas is collected, purified, cooled, and slowly compressed into cylinders.

Acetylene is a colorless gas, with a very distinctive nauseating odor. It is highly combustible when mixed with oxygen or air. Although it is stable under low pressures, it becomes very unstable if compressed to more than 15 psi.

CAUTION: acetylene becomes dangerous if used beyond a 15 pound pressure!

The acetylene cylinder.   To insure safety in storing acetylene, the cylinder is packed with a porous material. This material is saturated with acetone, which is a chemical liquid that dissolves or absorbs large quantities of acetylene under pressures greater than 15 psi without changing the nature of the gas. See Fig. 21-3.

The acetylene cylinder is equipped with a fusible plug to relieve any excess pressure if the cylinder should be subjected to undue heat, or any other mechanical pressure.

Fig. 21-3. Acetylene cylinders come in three common sizes—60, 100 and 300 cu ft. A typical cylinder is shown above as it normally appears (left) and in a cutaway view (right). (Linde Co.)

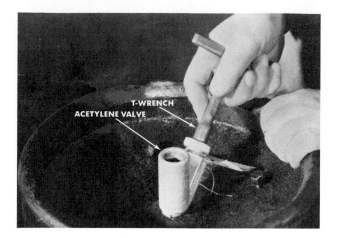

Fig. 21-4. The acetylene cylinder valve is operated with a T-wrench.

Fig. 21-5. The acetylene (top illustration) manifold and the oxygen (bottom) manifold makes a convenient and efficient system when a great deal of welding is done. (Airco)

The cylinder valve is operated by means of a T-wrench. *This valve should never be opened more than one and one-half turns.* A slight opening is advisable since it permits closing the valve in a hurry in case of an emergency. See Fig. 21-4.

Acetylene Generator. When considerable welding is done, such as in industry or in a school welding shop, the oxygen cylinders are frequently connected to a manifold with pipe lines carrying the gas to the welding stations as shown in Fig. 21-5. Although it is also possible to manifold acetylene cylinders, the acetylene generator, as illustrated in Fig. 21-6, is often more economical to operate than resorting to a manifold system.

Two kinds of acetylene generators are made—the low-pressure and the high-pressure type. The low-pressure generator furnishes acetylene under pressures of less than 1 psi. With this kind of generator only the injector type of blowpipe can be used. (See the following section entitled *The Welding Torch.*) Today the low-pressure generator is rarely installed.

The high-pressure generator supplies acetylene up to pressures of 15 psi. This generator is the one that is more commonly used.

The operation of a generator is relatively simple. The unit feeds controlled amounts of calcium carbide into the water. When these ingredients are mixed, acetylene gas is produced.

Various devices are incorporated in the generator to insure safe operation. However, there are definite official codes which must be followed as to where a generator is to be located and how it is to be used.

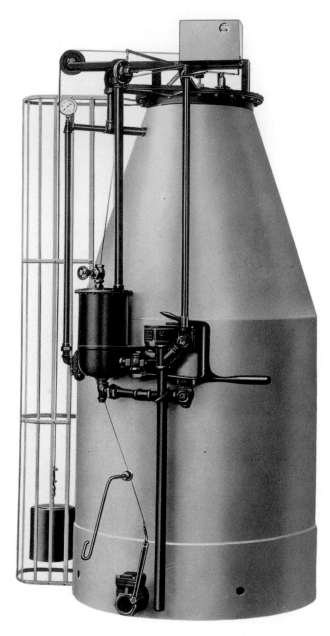

Fig. 21-6. Many welding shops manufacture their own acetylene in a generator such as this. (Airco)

Fig. 21-7. Correct way to move a cylinder. The rolling motion not only is less dangerous, but is also an easy way to handle a heavy cylinder.

Safety in Handling Cylinders

To move a cylinder, rotate it on its bottom edge. Place the palm of one hand over the protector cap and tilt the cylinder toward you. Start the tank rolling by pushing it with the other hand as shown in Fig. 21-7.

Here are a few precautions that must be followed when handling oxygen and acetylene:

1. Never use the valve-protector caps for lifting cylinders.

2. Do not allow cylinders to lie in a horizontal position.

3. Never permit grease or oil to come in

contact with cylinder valves. Although oxygen is in itself nonflammable, if it is allowed to come in contact with any flammable material it will quickly aid combustion.

4. Avoid exposing cylinders to furnace heat, radiators, open fire, or sparks from the torch.

5. Never transport a cylinder by dragging, sliding, or rolling it on its side. Avoid striking it against any object that might create a spark. There may be just enough gas escaping to cause an explosion.

6. If cylinders have to be moved, be sure that the cylinder valves are shut off.

7. Never tamper with or attempt to repair the cylinder valves. If valves do not function properly or if they leak, notify the supplier immediately.

8. Keep valves closed on empty cylinders.

9. If cylinder valves cannot be opened by hand, do not use a hammer or wrench—notify the supplier.

10. When not in use, keep cylinders covered with cylinder caps.

WELDING APPARATUS

A set of welding apparatus consists of a torch with an assortment of different-sized tips; two lengths of hose, one red for the acetylene and the other green or black for the oxygen; two pressure regulators; two cylinders, one containing acetylene and the other oxygen; a pair of dark-colored goggles; and a sparklighter. See Fig. 21-8.

As a rule, the cylinders are chained to a two-wheel truck to permit moving the equipment to any desired place. If the cylinders are positioned near the work bench, they should be chained to some fixed object.

CAUTION: Securing cylinders is important; otherwise they may tip over and cause an explosion or ruin the regulators. See Fig. 21-9.

The Welding Torch

The torch, or blowpipe, as it is sometimes called, is a tool which mixes acetylene and oxygen in the correct proportions and permits the mixture to flow to the end of a tip where it is burned. Although blowpipes vary to some extent

Fig. 21-8. A welding unit on a hand truck permits transporting the unit to where ever the welding job is to be done. (Linde Co.)

in design, basically they are all made to provide complete control of the flame during the welding operation.

The two main types of blowpipes are the *injector* and the *medium (equal) pressure*. The injector torch is designed to use acetylene at very low pressures, from 1 psi to zero. The equal pressure type requires acetylene pressures of 1

Fig. 21-9. Always chain cylinders to prevent them from tipping over.

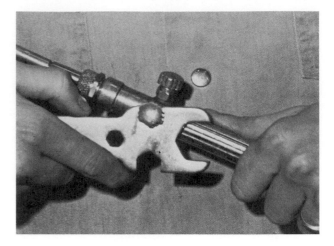

Fig. 21-10. Tighten the needle valve lock nuts with a correct fitting wrench.

to 15 psi. Both will operate when acetylene is supplied from cylinders or medium-pressure generators. Whereas the injector blowpipe will function with a low-pressure generator, the equal-pressure torch cannot be used on this type of generator.

With the injector blowpipe, the oxygen, as it passes through a small opening in the injector nozzle, draws acetylene into the oxygen stream. One advantage of this blowpipe is that small fluctuations in the oxygen supplied to it will produce a corresponding change in the amount of acetylene drawn, thereby making the proportions of the two gases constant while the torch is in operation. In the equal-pressure type the acetylene and oxygen are fed independently to a mixing chamber, after which they flow out through the tip. See Fig. 21-11. The equal pressure type of blowpipe is the one most generally used.

Both kinds of torches are equipped with two needle valves, one regulates the flow of oxygen and the other the acetylene. On the rear end of the torch there are two fittings for connecting the two hoses. In order to eliminate any danger of interchaning the hoses, the oxygen fitting is made with a right-hand thread and the acetylene with a left-hand thread.

Care of the torch.   When the welding is completed, always suspend the torch securely so it will not fall. The needle valves are especially delicate, and care must be taken never to drop the torch so they will strike some hard object. Occasionally the needle valves will turn too freely, making it difficult to secure and keep the adjustment for the proper mixture. When this occurs, give the safety-lock nuts on the stem of the needle valves a slight turn with a correct fitting wrench as shown in Fig. 21-10.

Welding Tips

To make possible the welding of different thicknesses of metal, blowpipes are equipped with an assortment of different size heads, or tips, as shown in Fig. 21-12. The size of the tip is governed by the diameter of its opening which is marked on the tip. The system of digits used depends largely on the manufacturer of the welding equipment. The most common system consists of numbers which range from 0 to 15. With this system, the larger the number the greater the tip diameter.

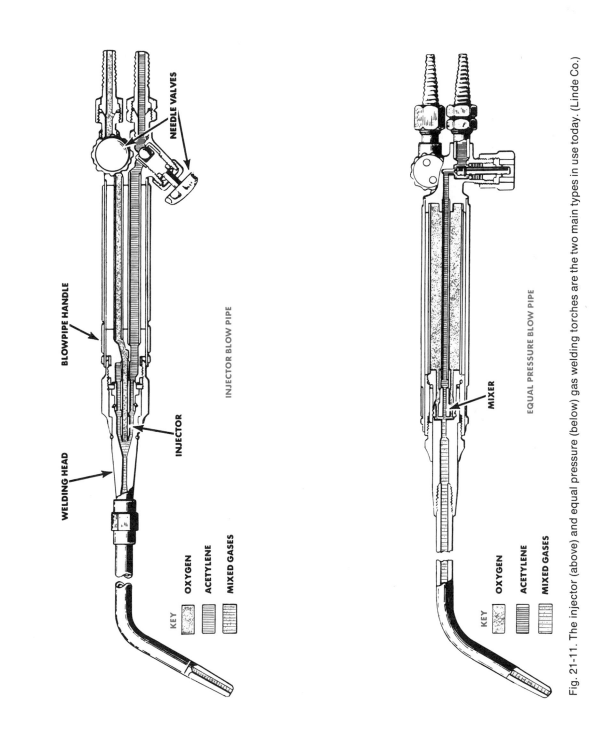

NEEDLE VALVES

BLOWPIPE HANDLE

INJECTOR BLOW PIPE

INJECTOR

WELDING HEAD

KEY

OXYGEN

ACETYLENE

MIXED GASES

MIXER

EQUAL PRESSURE BLOW PIPE

KEY

OXYGEN

ACETYLENE

MIXED GASES

Fig. 21-11. The injector (above) and equal pressure (below) gas welding torches are the two main types in use today. (Linde Co.)

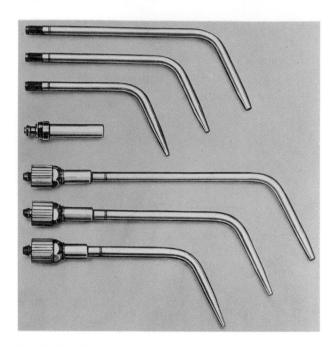

Fig. 21-12. The size of the welding tip is governed by the diameter of the opening. (Air Reduction Sales Co.)

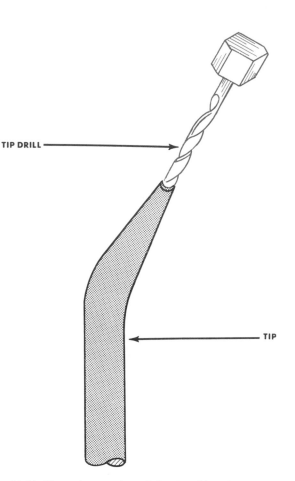

Fig. 21-14. Clean the opening of the tip with a tip reamer.

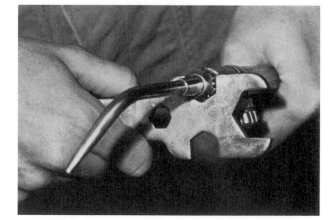

Fig. 21-13. Change the tip with the proper tip wrench.

Care of welding tips. When you have to remove a tip, change it with a wrench especially made for this purpose as in Fig. 21-13.
CAUTION: Under no conditions should pliers be used to remove a welding tip. Pliers quickly ruin the nut on the tip, rendering it nearly useless.
CAUTION: Never insert a tip while the torch is hot. Heat expands the threads, causing the tip to *freeze onto the torch after it has cooled. Attempting to force a tip loose from the torch later may result in breaking off the threaded section, thereby ruining the tip. Moreover you also have the problem of removing the broken section from the opening of the torch. Keep the tip clean at all times to produce satisfactory welds.*

Frequent use of the torch will cause a formation of carbon in the passage of the tip. Remove this carbon by inserting a tip reamer. See Fig. 21-14. Brushing the end of the tip with a fine grade of sandpaper is also advisable at times.

Regulators

The oxygen and acetylene pressure regulators perform two functions. They reduce the cylinder pressure to the required working pressure and produce a steady flow of gas under varying cylinder pressures. To illustrate, assume that the oxygen in the cylinder is under a pressure of 1800 psi and a pressure of 6 pounds is needed at the torch. The regulator must maintain a constant pressure of 6 pounds even if the cylinder pressure drops to 500 pounds.

There are two types of regulators, the *two-stage* and the *single-stage*. With the two-stage regulator, the reduction of the cylinder pressure to that required at the torch is accomplished in two stages. In the first stage, the gas flows from the cylinder into a high-pressure chamber. A spring and diaphragm keeps a predetermined gas pressure in this chamber. For oxygen such a pressure is usually set at 200 psi and for acetylene 50 psi. From the high-pressure chamber the gas then passes into a second reducing chamber. Control of the pressure in the reducing chamber is governed by an *adjusting screw.* The principal advantage of the two-stage regulator is that it provides a more uniform flow of gas.

The single-stage regulator is not as expensive as the two-stage type. With this regulator, there is no intermediate chamber through which the gas passes before it enters the low pressure chamber. The gas from the cylinder flows into the regulator and is controlled entirely by the adjusting screw. The disadvantage of this regulator is that as the cylinder pressure drops, the regulator pressure also falls, thereby necessitating occasional readjustment of the pressure. With the two-stage regulator the pressure remains constant until the gas in the cylinder is exhausted.

Both types of regulators have two gages. One gage indicates the actual pressure in the cylinder and the other shows the working or line pressure used at the torch. The oxygen high pressure gage is usually graduated to 3000 pounds. As a rule this gage also has a second scale which is calibrated to register the content of the cylinder in cubic feet. The acetylene high pressure gage is graduated to 350 to 400 pounds as shown in Fig. 21-15.

The oxygen working pressure gage is graduated in divisions between 50 to 60 pounds and the working pressure gage on the acetylene regulator in divisions of 15 to 30 pounds.

WORKING PRESSURE GAGE   **CYLINDER PRESSURE GAGE**

ADJUSTING SCREW

OXYGEN

ACETYLENE

Fig. 21-15. The oxygen and acetylene regulators control the flow of gas to be used for welding.

The most important thing to remember when using a regulator is to *make absolutely certain that the adjusting screw is released (turned out) before the cylinder valve is opened.* If the adjusting screw is not released and the cylinder valve opened, the tremendous pressure of the gas in the cylinder, forced on to the working pressure gage, may result in damage to the regulator.

Care of regulators. Regulators are sensitive instruments and must at all times be regarded as such. It takes only a slight jar to put a regulator out of commission. Be extremely careful in handling the regulator while removing it from the cylinder. *Never allow a regulator to remain on a bench top or floor for any length of time,* as someone may come along and carelessly move it, which may result in breaking the glass on the gages or damaging the regulator itself. Here are a few more rules that should be followed:

1. Always check the adjusting screw before the cylinder valve is turned on, and release this screw when the welding has been completed.

2. *Never use oil on a regulator.* Use only soap or glycerine to lubricate the adjusting screw.

3. *Do not try to interchange the oxygen and acetylene regulators.*

4. If a regulator does not function properly, shut off the supply of gas and have a qualified repairman check it.

5. If a regulator creeps (does not remain at set pressure), have it repaired immediately. Creeping will be noticed on the working pressure gage after the needle-valves on the torch are closed. A creeping regulator usually requires a change of valve seat or stem.

6. If the gage pointer fails to go back to the pin when the pressure is released, this condition should be repaired. The trouble is probably due to a sprung mechanism brought about by allowing the pressure to enter a gage suddenly.

7. Always keep a tight connection between the regulator and the cylinder. If the connection leaks after a reasonable force has been used to tighten the nut, close the cylinder valve and remove the regulator. Clean both the inside of the cylinder valve seat and the regulator inlet-nipple seat. If the leak persists, the seat and threads are probably marred, in which case the regulator will have to be returned to the manufacturer for repair.

Oxygen and Acetylene Hose

A special non-porous hose is used for welding. To prevent mistakes in connecting them, *the oxygen hose is either green or black* and *the acetylene hose red.* If oxygen were to pass through an old acetylene hose, a dangerous combustible mixture might result.

A standard connection is used to attach the hose to the regulator and torch. This connection consists of a nipple which is forced into the hose and a nut that connects the nipple to the regulator and torch. The acetylene nut may be distinguished from the oxygen by the groove that runs around the center, indicating a left-hand thread as illustrated in Fig. 21-16. A clamp of some type is used to squeeze the hose around the nipple to prevent it from working loose.

Care of welding hose. The acetylene and oxygen hose is an important part of the welding equipment. First of all, as previously mentioned, all the connections on the hose must be perfectly tight. These connections should be tightened with close-fitting wrenches to prevent damaging the nuts.

Avoid dragging the hose around on a greasy floor, since grease or oil eventually will soak into it. Be careful in pulling the hose around sharp objects and especially over hot metal. Prevent anyone from stepping or dropping anything on the hose. When the welding has been complet-

ACETYLENE   OXYGEN

Fig. 21-16. Hose connections (above) and welding hose (below).

ed, roll up the hose and suspend it in such a manner that it will not drop to the floor.

Here are a few additional precautions:

1. All new hose is dusted with talcum powder inside. This powder should be blown out before using the hose.

2. Do not use ordinary wire to bind hose to a connection. Use regular hose binders or clamps.

3. When splicing hose, use standard brass splicing nipples—never use copper tubing.

4. Long lengths of hose tend to kink.

5. Do not repair leaking hose with tape. Splice in a new piece or discard.

The Sparklighter

The sparklighter, as illustrated in Fig. 21-17, is the implement used for igniting the torch. While welding, form the habit of always employing a sparklighter to light a torch. *Never use matches.* The use of matches for this purpose is very dangerous because the puff of the flame produced by the ignition of the acetylene flowing from the tip is likely to burn your hand.

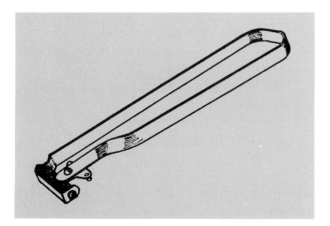

Fig. 21-17. Sparklighter.

Goggles

An oxy-acetylene flame produces light rays of great intensity and also heat rays which, if meeting the naked eye, may eventually prove destruc-

Fig. 21-18. Always wear proper goggles when welding.

tive to the eye tissues. Consequently, always wear goggles having suitable, approved, colored glass. The density of the colored lenses should be such that no more of the light rays are shut off than are necessary to the welder. One test which may be performed to check the suitability of the colored lenses is to use one pair of goggles for a few minutes while welding and then remove them, noting whether or not white spots dance before your eyes. If these white spots are noticeable the lenses do not provide sufficient protection, and a darker shade should be used.

For most gas welding, goggles with shade numbers of 4, 5 and 6 are recommended. Goggles, Fig. 21-18, also protect the eyes from flying sparks and pieces of molten metal that may splatter around.

Protective Clothing

It is a good idea to wear an apron, shop coat, or coveralls while welding with oxy-acetylene equipment. Sparks will invariably shoot away from the molten metal, and, without some suitable covering, numerous holes will be burned in your clothes. *Under no circumstances should a sweater be worn.* A small spark falling on this garment may burst into a rapid-spreading flame that may have dire consequences. Some type of skull cap is also desirable to prevent any hot metal particles from falling on the hair.

The beginner should wear a pair of lightweight gloves to avoid possible burns. Occasionally the hot end of a rod or a piece of metal, which has been momentarily laid aside to cool, is picked up by mistake, and without gloves serious burns are apt to result. However, oxy-acetylene welders frequently do not wear gloves.

Special Gas Welding Processes

Although oxy-acetylene is very common for certain types of welding, other gas welding processes also are used such as oxy-Mapp, oxy-hydrogen and air-acetylene. The principal difference between these and oxy-acetylene primarily is in the type of gas employed in the burning mixture, otherwise the welding technique is the same.

Oxy-Mapp welding. Mapp gas is a Dow Chemical Company product having many of the physical properties of acetylene but it lacks the shock sensivity of acetylene. The gas is the result of rearranging the molecular structure of acetylene and propane. Although propane itself is very stable, its limiting factor for welding is its low flame temperature. Acetylene on the other hand produces a very high flame temperature but is very unstable. By combining the two gases and changing their molecular structure we have a very stable fuel with a flame temperature nearly comparable to acetylene, without the dangers of acetylene.

Generally, with Mapp gas a slightly larger tip is required because of its greater gas density and slower flame propagation rate. The only significant difference is in the flame appearance. A neutral flame for welding will have a longer inner cone than with oxy-acetylene gas.

Since Mapp gas is not sensitive to shock it can be stored and shipped in lighter cylinders. Acetylene to be kept safe must be stored in cylinders filled with a porous filler and acetone. Where empty acetylene cylinders weigh around 220 pounds Mapp cylinders weigh only 50 pounds. Normally a filled cylinder of acetylene weighs 240 pounds while a filled cylinder of Mapp gas weighs 120 pounds. See Fig. 21-19.

Oxy-hydrogen welding. This combination generates a low temperature flame used primarily for welding thin sections of metal where low temperatures are required. One of the unusual characteristics of oxy-hydrogen is that the flame is practically non-luminous. Consequently, difficulty is often experienced in adjusting for a neutral flame. To avoid welding with what might be an oxidizing flame, the practice is to adjust

Fig. 21-19. Mapp gas cylinder and valve. (Mapp Industrial Gas Co.)

the regulator so it produces a greater flow of hydrogen at first.

Air-acetylene welding. The air-acetylene flame is generated by burning a mixture of acetylene with air. As the acetylene flows to the torch under pressure it draws in the right amount of air for proper combustion. The temperature of the air-acetylene flame is even lower than that of the oxy-hydrogen and thereby is used mostly for soldering and brazing very light metals. It has wide applications in the plumbing industry for joining copper tubing.

Points to Remember (Cautions)

1. Handle oxygen and acetylene cylinders with care. Never expose them to excessive heat and prevent oil and grease from coming in contact with them.

2. Be sure the adjusting screw on a regulator is fully released before opening a cylinder valve.

3. Always hang up a torch when not in use to prevent it from dropping to the floor and being bent or damaged.

4. Never remove tips with pliers. If a tip has to be cleaned, insert a piece of soft wire or tip drill in the opening.

5. Do not lubricate the adjusting screw on a regulator with oil. Use soap or glycerine.

6. Never interchange the hoses. Avoid dragging them over greasy floors.

7. Always wear proper goggles when welding, as well as suitable protective clothing.

8. Never light a torch with a match.

9. Never use oxygen or acetylene from the tank to blow dirt and dust from clothing.

QUESTIONS FOR STUDY AND DISCUSSION

1. What safety devices are used to prevent cylinders from exploding when subjected to intense pressure?

2. What is the purpose of the protector cap on a cylinder?

3. How much should the cylinder valve be opened on an acetylene cylinder? On the oxygen cylinder?

4. Why is it dangerous to allow grease or oil to come in contact with the oxygen cylinder valve?

5. What is the function of the needle valves on a welding torch?

6. Why are the oxygen and acetylene hose fittings made with different screw threads?

7. How are sizes of welding tips indicated?

8. Why should a close-fitting wrench be used in removing the various welding tips?

9. What is the objection to using pliers for removing welding tips?

10. What is a tip drill, and when and why should it be used?

11. What is meant by a two-stage pressure regulator?

12. Why should the adjusting screw on a regulator be fully released before opening a cylinder valve?

13. What precautions should be observed in handling a pressure regulator?

14. Why is it a poor practice to light a torch with a match?

15. Why is Mapp gas sometimes used instead of acetylene?

16. Oxy-hydrogen and air-acetylene are often used for what operations?

CHAPTER 22 setting-up and operating

One of the first things you should learn in starting your oxy-acetylene welding operations is to assemble a welding outfit. A certain sequence must be followed if the equipment is to be properly and safely connected. Once you have learned to do this then you need to know how to light the torch and adjust the flame. This chapter explains how these operations may be performed.

ASSEMBLING THE WELDING OUTFIT

To assemble a welding outfit follow these directions:

Chain cylinders. Fasten the cylinders securely to a truck or to some other fixed object where they are to be located. Remove the protector cap from each cylinder and examine the outlet nozzles closely. Make sure the connection seat or screw threads are not damaged. A damaged screw thread may ruin the regulator nut, while a poor connection seat will cause the gas to leak.

Crack cylinder valves. Particles of dirt frequently collect in the outlet nozzle of the cylinder valve. If not cleaned out, this dirt will work into the regulator when the pressure is turned on. To avoid any possibility of dirt clogging up any passage, open the valves of each cylinder for just an instant. Then with a clean cloth, wipe out the connection seat. See Fig. 22-1.

Fig. 22-1. Crack cylinder valves to blow out dirt which may be lodged in the outlet nozzles.

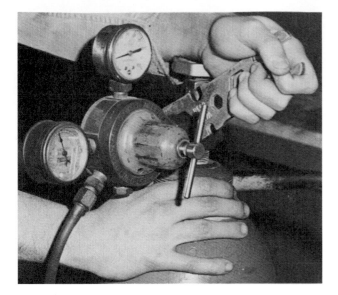

Fig. 22-2. Fasten the regulators to the cylinders with a proper fitting wrench.

Fig. 22-3. Connect the hoses to the regulator.

Fig. 22-4. Blow out each hose.

Attach regulators. Connect the oxygen pressure regulator to the oxygen cylinder and the acetylene regulator to the acetylene cylinder as shown in Fig. 22-2. Use a close-fitting wrench to tighten the nuts, and avoid stripping the threads. Always use a proper wrench to tighten the nuts, as a loose-fitting wrench will eventually ruin the corners of the regulator nuts.

Connect hoses to cylinders. Connect the green or black hose to the oxygen regulator and the red hose to the acetylene regulator as illustrated in Fig. 22-3. Check the adjusting screw on each regulator to make certain that it is released, and open the cylinder valves. Then blow out any dirt that may be lodged in the hoses by opening the regulator adjusting screws and promptly closing them again when the hoses have been cleaned out. See Fig. 22-4.

Connect hoses to the torch. Connect the hoses to the respective fitting on the torch, the red hose to the needle valve fitting, marked *AC,* and the green or black hose to the oxygen needle valve, marked *OX,* as shown in Fig. 22-5. *Remember, acetylene hose connections always have left-hand threads and oxygen hose connections have right-hand threads.*

Testing for leaks. All new welding apparatus needs to be tested for leaks before being operated. Thereafter it is advisable to periodically apply this same test in order to insure that no leakage has developed. A leaky apparatus is very danger-

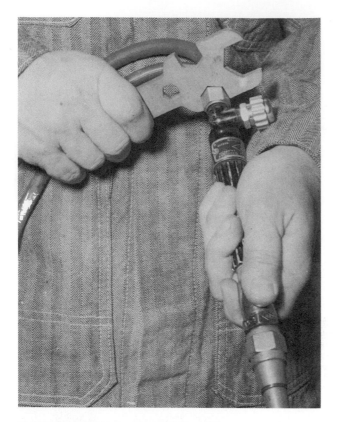

Fig. 22-5. Connect the hoses to the torch.

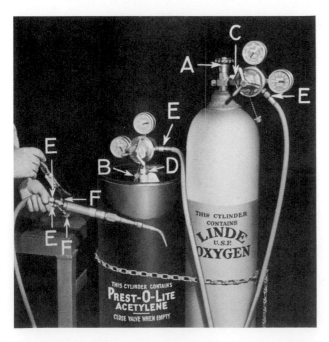

Fig. 22-6. Test these points with soapy water for leakage. (Linde Co.)

ous, since a fire may develop. Furthermore, leaks mean wasted gas.

To test for leaks, open the oxygen and acetylene cylinder valves and with the needle valves on the torch closed, adjust the regulators to give about normal working pressure. Test the following points as shown in Fig. 22-6:

A—Oxygen cylinder valve
B—Acetylene cylinder valve
C—Oxygen regulator inlet connection
D—Acetylene regulator inlet connection
E—Hose connections at the regulators and torch
F—Oxygen and acetylene needle valves

If there is a leak in the acetylene cylinder valve that cannot be stopped by closing the valve or tightening the packing gland, or if the leak is at the fuse plug, remove the cylinder out of doors away from possible sources of ignition and notify the supplier immediately.

Apply soapy water with a brush to test for leaks. Under no circumstances should any other means be used for this operation. If bubbles form, a leak is present.

To remedy any leakage, tighten up the connections a little with a wrench. If this does not stop the leak, shut off the gas pressure, open the connections, and examine the screw threads. Also, old hose should be tested very closely, since it has a tendency to become porous.

Selecting the Proper Welding Tip

The size of the tip will depend upon the thickness of the metal to be welded. If very light sheet iron is to be welded, a tip with a small opening is used, while a large-sized tip is needed for thick metal. Table 22-1 lists the sizes of tips for various thicknesses of metals and the approximate working pressures required.

It is very important to use the correct tip, with the proper working pressure. If too small a tip is employed, the heat will not be sufficient to fuse the metal to the proper depth. When the tip is too large, the heat is too great, thereby burning holes in the metal. A good weld must have the right penetration and smooth, even, overlapping ripples. Unless the conditions are just right, it is

TABLE 22-1.   TIP SIZES VS METAL THICKNESS.

TIP NUMBER	THICKNESS (inches)	OXYGEN PRESSURE (pounds)	ACETYLENE PRESSURE (pounds)
00	1/64	1	1
0	1/32	1	1
1	1/16	1	1
2	3/32	2	2
3	1/8	3	3
4	3/16	4	4
5	1/4	5	5
6	5/16	6	6
7	3/8	7	7
8	1/2	7	7
9	5/8	7 1/2	7 1/2
10	3/4 & up	9	9

Airco.

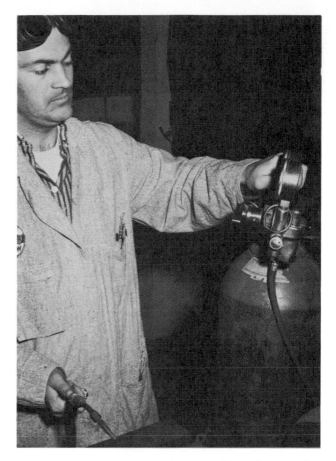

Fig. 22-7. Do not stand in this facing position when opening a valve—it is dangerous!

impossible for the torch to function the way it should and, consequently, the weld will be poor.

Lighting the Torch

1. Select a tip for welding ¹⁄₈″ or ¹⁄₁₆″ metal and screw it into the torch.

2. Open the oxygen and acetylene cylinder valves and set the working pressure to correspond to the size of tip that is to be used. *Stand aside, do not face the regulator when opening the cylinder valve.* A defect in the regulator may cause the gas to blow through, shattering the glass and blowing it into your face. See Fig. 22-7. *Make it a practice to stand to one side of the regulator, as in Fig. 22-8, and turn the cylinder valve slowly.* Remember, oxygen and acetylene are charged in the tanks under a high pressure and if the gas is permitted to come against the regulator suddenly, it may cause some damage to the equipment. Open the acetylene cylinder valve approximately one complete turn and the oxygen all the way. Next turn the oxygen and acetylene regulator adjusting valves to the required working pressures.

Then turn the acetylene needle valve on the torch approximately three quarters of a turn. With the sparklighter held about one inch away from the end of the tip, ignite the acetylene as it

Fig. 22-8. Stand to one side of the regulator when opening a cylinder valve.

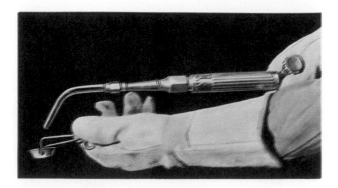

Fig. 22-9. Some people prefer to hold the torch in the right hand and the sparklighter in the left hand. Others find it easier to reverse this procedure. Do what comes naturally.

leaves the tip as shown in Fig. 22-9. Do this as rapidly as possible to avoid unnecessary wasting of gas. If not enough acetylene is turned on, the flame will produce considerable smoke; therefore, quickly turn on more acetylene until the flame has a slight tendency to jump away from the tip.

CAUTION: Never use a match to light a torch. This procedure brings your fingers too close to the tip and the sudden ignition of the acetylene is very apt to burn them.

When igniting a torch, keep the tip facing downward and preferably against the bench top. Lighting the torch while it is facing outward or upward may result in burning someone nearby as the ignited flame spurts out.

CAUTION: If other people are welding around the same area, never reach over to the other follow for a light. He or she may be welding, and bringing your torch near will not only disturb the work, but might burn his or her face or hands.

CAUTION: Make no attempt to relight a torch from the hot metal when welding in an enclosed box, tank, drum, or other small cavity. There may be just enough unburned gas in this confined space to cause an explosion as the acetylene from the tip comes in contact with the hot metal. Instead, *move the torch into the open,* relight it in the usual manner, and make the necessary adjustments before resuming the weld.

Adjusting the Flame

With the acetylene burning, gradually open the oxygen needle valve until a well-defined white cone appears near the tip, surrounded by a

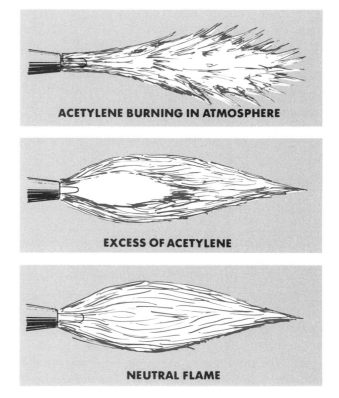

ACETYLENE BURNING IN ATMOSPHERE

EXCESS OF ACETYLENE

NEUTRAL FLAME

Fig. 22-10. Steps in adjusting for a neutral flame.

second, bluish cone that is faintly luminous. This is known as a *neutral flame* because there is an approximate one-to-one mixture of acetylene and oxygen resulting in a flame which is chemically neutral. The brilliant white cone should be from $1/16''$ to $3/4''$ long, depending on the tip size. See Fig. 22-10. The neutral flame is used for most welding operations.

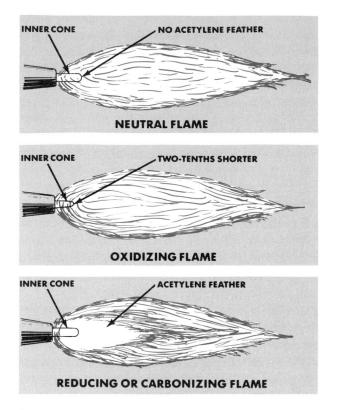

INNER CONE — NO ACETYLENE FEATHER

NEUTRAL FLAME

INNER CONE — TWO-TENTHS SHORTER

OXIDIZING FLAME

INNER CONE — ACETYLENE FEATHER

REDUCING OR CARBONIZING FLAME

Fig. 22-11. Three types of oxy-acetylene welding flames. (Linde Co.)

Any variation from the one-to-one mixture will alter the characteristics of the flame. When an excess amount of oxygen is forced into the mixture, the resulting flame is said to be *oxidizing.* This flame resembles the neutral flame slightly, but has a shorter and more pointed inner cone with an almost purple color rather than brilliant white. It is sometimes used for brazing. See Fig. 22-11.

If the mixture consists of a slight excess of acetylene, the flame is *carbonizing,* or *reducing.* This flame can easily be identified by the existence of three flame zones instead of the usual two found in the neutral flame. The end of the billiant white cone is no longer as well defined, and it is also surrounded by an intermediate white cone, which has a feathery edge in addition to the usual bluish outer envelope. See Fig. 22-11 for this feathery edge.

Testing the Flames

You will probably have a little difficulty at first in making the correct flame adjustment, but if the characteristics of the carbonizing and oxidizing flames are understood, the adjustment for a neutral flame will be relatively easy. As an aid in becoming familiar with the effects of these various flames, try the following experiments:

Carbonizing flame. Obtain a piece of scrap metal. Light the acetylene and turn on the oxygen until a white cone appears on the end of the tip enveloped by another fan-shaped cone which has a feathered edge.

Put on a pair of goggles and apply this flame to the metal, holding the point of the white cone close to the metal. You will notice that as the metal melts it has a tendency to boil. This is an indication that carbon is entering the molten metal. You will also discover, after the metal has cooled, that the surface is pitted and very brittle.

Oxidizing flame. Now, open the oxygen needle valve. The white cone becomes short and the color changes to a purplish hue. The flame burns with a decided roar.

Apply this flame to a piece of metal, allowing the cone to come in contact with the surface. You will find that as the metal melts there are numerous sparks given off and a white foam or scum forms on the surface. After this piece cools, the metal will be shiny.

Neutral flame. Now, adjust the needle valve until the flame is balanced. Apply this neutral flame to a piece of metal. The molten metal flows smoothly like syrup, with very few sparks, clean and clear.

Flame Characteristics

A flame may be harsh or quiet. The harsh type is induced by forcing too much pressure of both gases to the tip. This flame is undesirable, since it has a tendency to depress the molten surface and cause the metal to splatter around the edges of the puddle. Such a flame is noisy, and its use makes it extremely difficult to bring about perfect fusion, with smooth, uniform ripples.

The quiet flame is just the opposite of the harsh flame and is achieved by the correct pressure of gases flowing to the tip. The flame is not a forcing, noisy flame but one that permits a continuous flow of the molten puddle without any undue amount of splatter.

To secure a soft neutral flame, see that (1) the tip is absolutely clean, and (2) the mixture is correct. Even if the proportion of acetylene and oxygen is right, a good weld is difficult to achieve unless the opening in the tip permits a free flow of gases. Any foreign matter in the tip will simply restrict the source of heat necessary to melt the metal.

Flame Control

Once the flame has been properly set, it does not mean that further adjustments are unnecessary. From time to time, as the welding progresses, it is necessary to observe the flame cone to be certain that the mixture has not altered. Changes in the flame will occur as a result of some slight fluctuation in the flow of the gases from the regulators. A slight turn of one needle valve or the other will quickly readjust the flame.

In the course of welding, a torch may occasionally start popping. This noise is an indication that there is an insufficient amount of gases flowing to the tip. Such popping can be stopped by opening both the oxygen and acetylene needle valves on the torch to a greater extent. Another reason for popping is the overheating of the molten pool by lingering or keeping the flame too long in one position and not melting enough rod into the pool.

Backfire and Flashback

When the flame goes out with a loud pop, it is called a *backfire.* A backfire may be caused (1) by operating the torch at lower pressures than required for the tip used, (2) by touching the tip against the work, (3) by overheating the tip, or (4) by an obstruction in the tip. If a backfire should occur, shut the needle valves and after remedying the cause, relight the torch.

A *flashback* is a condition that results when the flame flashes back into the torch and burns inside with a shrill hissing or squealing noise. If this should happen, close the needle valves

immediately. A flashback generally is an indication that something is wrong. Perhaps it is a clogged tip, or the improper functioning of the needle valves, or even an incorrect acetylene or oxygen pressure. In any case, investigate the cause before relighting the torch.

Shutting Off the Torch

Following is the correct sequence of steps for shutting OFF a torch:

1. Close the acetylene needle valve first.

The acetylene needle valve is closed first, since shutting off the flow of this gas will immediately extinguish the flame, whereas if the oxygen is shut off first, the acetylene will continue to burn, throwing off smoke and soot.

2. Close the oxygen needle valve.

3. If the entire welding unit is to be shut down, shut off both the acetylene and the oxygen cylinder valves.

4. Remove the pressure on the working gages by opening the needle valves until the lines are drained. Then promptly close the needle valves.

5. Release the adjusting screws on the pressure regulators.

Points to Remember

1. Be sure to fasten the cylinders securely so they will not fall over.

2. Crack the cylinder valves before attaching regulators.

3. Periodically test the welding outfit for leaks. Use soapy water only.

4. Use the correct size tip for the thickness of metal to be welded.

5. Always stand to one side of the regulator when opening cylinder valves and be sure the regulator adjusting valves are fully released.

6. Never use a match to light a torch; use a regulation sparklighter.

7. Adjust the torch to a soft, neutral flame for welding unless the nature of the metal calls for a different type of flame.

8. Keep the passage in the welding tip clean.

9. Avoid conditions that may cause a backfire or flashback.

QUESTIONS FOR STUDY AND DISCUSSION

1. Why should the cylinders be securely fastened before being used?

2. Why should the outlet nozzles be examined closely?

3. What is meant by the term *cracking the valve?* Why is this done?

4. List the proper order for setting up the welding apparatus.

5. Why should a close-fitting wrench be used in fastening all connections?

6. Why should a flame never be used in testing for leaks?

7. What is the proper method of testing for gas leaks?

8. What governs the size of the tip which should be used?

9. Why must the working pressure be correct for the size of tip that is to be used?

10. How can you determine the amount of acetylene to turn on at the needle valve control when lighting the torch?

11. What is meant by an *oxidizing flame?*

12. What is a *carbonizing flame?*

13. How can you distinguish these flames from the neutral flame?

14. How can you determine when you have a *neutral flame?*

15. Why should the acetylene needle valve be closed first in shutting off the torch?

16. What is the difference between a harsh and soft flame?

17. What are some of the conditions that may cause a *backfire?*

18. What is meant by a *flashback* when one is using an oxy-acetylene torch?

CHAPTER 23 the flat position

To master the skill of welding with an oxy-acetylene torch, you will have to practice a series of operations in a definite order. These operations involve carrying a puddle without a rod, laying beads with a filler rod, and welding various types of joints. All of this welding should be done with the metal lying in a flat position.

Carrying a Puddle without a Filler Rod

Obtain a piece of metal 1/16″ or 1/8″ in thickness and approximately 5″ in length. Be sure the surface is free of oil, dirt, and scale. Light the torch and adjust it for a neutral flame. Then proceed as follows:

Holding the torch. A torch may be held in either one of two ways, depending on which is the more comfortable for you. When welding light-gage metal, most operators prefer to grasp the handle of the blowpipe, with the hose over the outside of the wrist, in the same manner a pencil ordinarily is held. See Fig. 23-1. In the other grip, the torch is held like a hammer, with the fingers lightly curled underneath as shown in Fig. 23-2. In either case the torch should balance easily in the hand to avoid fatigue.

Position and motion of the torch. Hold the torch so the flame points in the direction you are going to weld and at an angle of about 45° with the completed part of the weld. See Fig. 23-3. If you are right handed, start the weld at the right edge of the metal and bring the inner cone of the

Fig. 23-1. This is one way to hold the torch.

neutral flame to within 1/8″ of the surface of the plate. The left-handed person reverses this direction. Hold the torch still until a pool of molten metal forms. Then move the puddle across the plate. As the puddle travels forward, rotate the torch to form a series of overlapping ovals as shown in Fig. 23-3.

Do not move the torch ahead of the puddle, but slowly work forward, giving the heat a

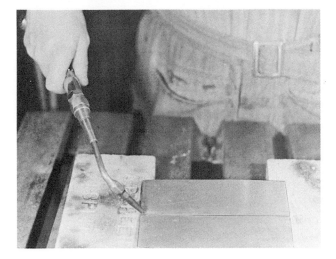

Fig. 23-2. Here is another way to hold the torch.

Fig. 23-4. Strip A and D shows correct procedure for carrying the puddle without a filler rod. Strip B shows excess heat. Strip C shows insufficient heat. (Linde Co.)

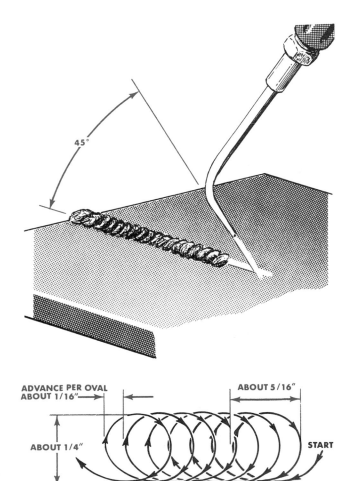

Fig. 23-3. Hold and rotate the torch in this manner while moving the puddle forward.

chance to melt the metal. If the flame is moved forward too rapidly, the heat fails to penetrate far enough and the metal does not melt properly. If the torch is kept in one position too long, the flame will burn a hole through the metal. See Fig. 23-4 for the various conditions.

Laying Beads with a Filler Rod

On some types of joints it is possible to weld the two pieces of metal without adding a filler rod. In most instances the use of a filler rod is advisable because it builds up the weld, thereby adding strength to the joint. The strength of a weld depends largely on the skill with which the rod is blended, or interfused, with the edges of the base metal.

The use of a filler rod requires coordination of the two hands. One hand must manipulate the torch to carry a puddle across the plate, while the other hand must add the correct amount of filler rod. To gain experience in coordinating these movements, practice laying a bead across a flat plate. To do this, obtain a flat plate $1/16''$ or $1/8''$ in thickness and then proceed as follows:

Selecting the filler rod. A welded joint should always possess as much strength as the base metal itself. If this is to be accomplished it is necessary to employ a welding rod that has the same properties as the base metal. It is a mistake to attempt to use just any kind of wire, because an inferior rod contains so many impurities that it is difficult to use and makes a weld that is weak and brittle. A good welding rod will flow smoothly and readily unite with the base metal without any excessive amount of sparking. A rod of poor weldability will spark profusely, flow irregularly, and leave a rough surface filled with punctures like pinholes.

Filler rods come in a variety of sizes ranging from 1/16″ to 3/8″ in diameter. The size of rod to use will depend largely upon the thickness of the metal. *The general rule is to use a rod with a diameter equal to the thickness of the base metal.* In other words, if a 1/16″ metal is to be welded, a 1/16″ diameter rod should be used.

A great many different kinds of rods are available for welding a variety of metals. For example, a mild-steel rod is used for welding mild steel, a cast-iron rod for cast iron, a nickel rod for nickel steel, a bronze rod for bronzing malleable cast iron and other dissimilar metals, an aluminum rod for aluminum welding, a copper rod for copper products, etc.

Manipulating the filler rod. Hold the rod at approximately the same angle as the torch but slant it away from the torch. It is advisable to bend the end of the rod at right angles, since this permits holding the rod so that it is not in a direct line with the heat of the flame. See Fig. 23-5.

Melt a small pool of the base metal and then insert the tip of the rod in this pool. Remember, to secure perfect fusion, the correct diameter rod is important. If the rod is too large the heat of the pool will be insufficient to melt the rod, and if the rod is too small the heat cannot be absorbed by the rod, with the result that a hole is burned in the plate.

As the rod melts in the pool, advance the torch forward. Concentrate the flame on the base metal and not on the rod. Do not hold the rod above the pool. If you do this the molten metal will fall through the air to the puddle. When this happens it combines with the oxygen of the air and part of it burns up, causing a weak, porous weld. Always dip the rod in the center of the molten pool.

Laying beads. Rotate the torch to form overlapping ovals and keep raising and lowering the rod as the molten puddle is moved forward. *An alternate torch movement* is the semi-circular motion as shown in Fig. 23-6. When the rod is not in the puddle, keep the tip just inside the outer envelope of the flame.

At the start of this exercise, you will find that the end of the rod may stick occasionally. This is because a beginner often experiences difficulty in holding the welding rod steady. Instead of inserting the rod in the middle of the puddle where the heat is sufficient to melt it readily, the beginner may insert it near the edge of the pool where the temperature is lower. The heat at the edge is not enough to melt the rod. When this happens, do not try to jerk it loose, since such an action will simply interrupt the welding. To loosen the rod, play the flame directly on the tip of the rod and the rod will be freed immediately. In

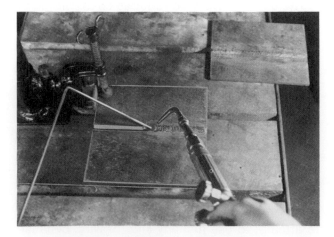

Fig. 23-5. Position for holding a rod. (Linde Co.)

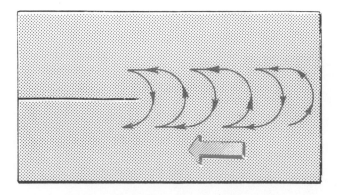

Fig. 23-6. Instead of the circular torch motion shown in Fig. 23-3, you can use this semicircular movement.

all probability, while the rod is being freed, the puddle will have solidified; therefore be sure to reform the puddle before moving forward.

Travel speed. To secure beads of uniform width and height you must keep the forward movement of the torch just right. If the puddle is carried forward too slowly, it becomes too large and you may even burn through the metal. If it is moved too rapidly, the rod is not actually fused with the base metal but merely stuck on the surface. Furthermore, it will be impossible to form ripples evenly.

When it appears that the puddle is getting too large, withdraw the flame slightly so only the outer envelope of the flame is touching the molten puddle. *Do not flip the flame off to one side, since such a movement allows the air to strike the hot metal and oxidize it.*

Continue to practice melting filler rod on scraps of metal until you are able to lay straight ripple beads 4″ to 5″ in length as shown by the example in Fig. 23-7.

Welding Butt Joints

After you have mastered the knack of carrying a puddle across the surface of a plate while adding filler rod, your next task is to fuse two pieces together. Follow these steps when practicing this exercise:

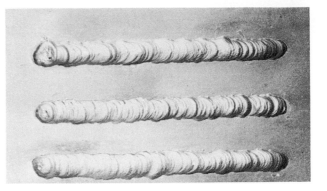

Fig. 23-7. Examples of straight beads laid with a filler rod.

Space plates. Obtain two pieces of $\frac{1}{16}$″ or $\frac{1}{8}$″ metal 4″ to 5″ in length and set them on two firebricks. Allow a gap of about $\frac{1}{16}$″ at the starting end of the joint and approximately $\frac{1}{8}$″ at the other. This is known as progressive spacing. The purpose of the space is to allow for expansion of the metal; otherwise the edges will overlap before the weld is completed as illustrated in Fig. 23-8. Furthermore, this space permits the flame to melt the edges all the way to the bottom of the plates.

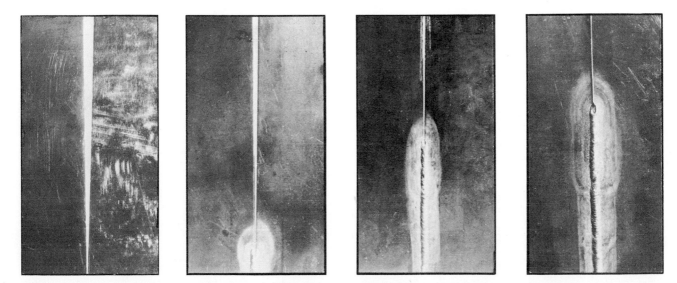

Fig. 23-8. This illustrates the manner in which the spacing closes up during welding. (Linde Co.)

Tack plates. If progressive spacing between the edges of a seam is not used, then the plates must be tacked at various intervals to restrict expansion forces. See Fig. 23-9.

To make a tack weld, simply apply the flame to the metal unit until it melts and then add a little filler rod.

Weld plates. Begin welding at the right end (or the left end if you are left-handed), using the same torch and rod movement as previously described. Work the torch slowly to give the heat a chance to penetrate the joint, and add sufficient filler rod to build up the weld about $1/16''$ above the surface. Be sure the puddle is large enough and the metal is flowing freely before you dip in the rod. Watch closely the course of the flame to make certain that its travel along both edges of the plates is the same, maintaining a molten puddle approximately $1/4''$ to $3/8''$ in width. Advance this puddle about $1/16''$ with each complete motion of the torch. Unless the molten puddle is kept active and flowing forward, correct fusion will not be achieved.

Keep the motion of the torch as uniform as possible. This will produce smooth, even ripples as shown in Fig. 23-10. At first such a task may appear very difficult but after a little practice a uniform motion of the torch will be mastered.

Test the weld. The welded piece should next be tested. Pick the metal up with a pair of pliers

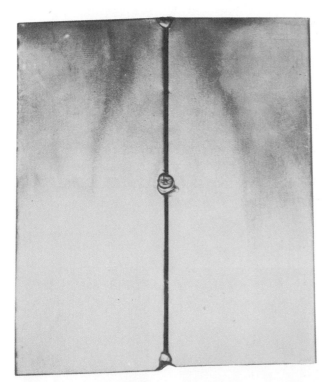

Fig. 23-9. Tack the plates before welding.

Fig. 23-10. A good ripple weld on a butt joint will look like this. (Linde Co.)

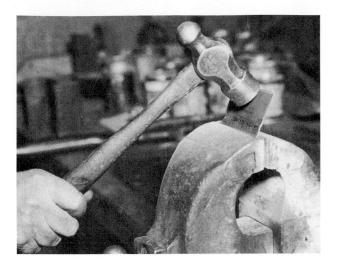

Fig. 23-11. Test the weld by striking it with a hammer, bending the metal against the weld.

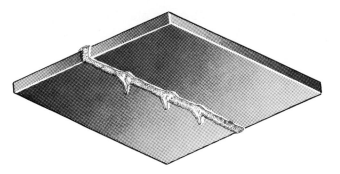

Fig. 23-12. The underside of this plate shows an uneven penetration.

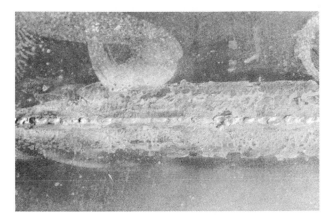

Fig. 23-13. If the penetration is correct, the underside of the plate will look like this.

and clamp it in a vise with the weld parallel to the top of the jaws and the welded joint just above the top of the vise. Strike the metal with a hammer, bending it against the weld as illustrated in Fig. 23-11. The test should be made so that the underside or bottom of the weld is put in tension and the top in compression.

Hammer the metal until the joint has been bent to an angle of 90° to 180°. If the fusion has been correct no cracks will show in the weld joint.

Check for defects. It is only natural to assume that your first few welds will break easily. Continue trying until you can make a straight, smooth weld that will not open when bent.

The following are some of the common defects that you may expect to find in your first few attempts to weld:

1. Uneven weld bead, caused by moving the torch too slowly or too rapidly.

2. Holes in the joint, caused by holding the flame too long in one spot.

3. A brittle weld, due to improper flame adjustment during welding.

4. Excessive metal hanging underneath the weld, showing too much penetration. See Fig. 23-12 for example.

5. Insufficient penetration, caused by moving the torch forward too rapidly. When the penetration is correct, the underside of the seam should show that fusion has taken place clear to the bottom edges as shown in Fig. 23-13.

6. Hole in the end of the joint, caused by not lifting the torch when the end of the weld has been reached.

7. Very often a joint appears to have correct penetration and still cracks open when tested. This may be caused by a number of reasons, such as:

a. Improper space allowances between the edges of the plates.

b. Filling the space between the metal plates with molten rod without sufficiently melting the edges of the plates to insure a good bond between the parent metal and rod.

c. Holding the torch too flat, causing the molten puddle to lap over an area that has not been properly melted to merge with it.

Welding a Flange Joint

A flange joint is used a great deal in sheetmetal work, particularly on material that is 20 gage or less. The flange should extend above the surface of the sheet a distance equal to about the thickness of the sheet. See Fig. 23-14.

To weld a flange joint, no filler rod is needed. The position of the torch should be the same as for welding a butt joint. Place the two pieces with the flange edges touching and tack weld them. Hold the torch flame on the right end (left end for a left-handed person) until a puddle forms and then move the puddle forward.

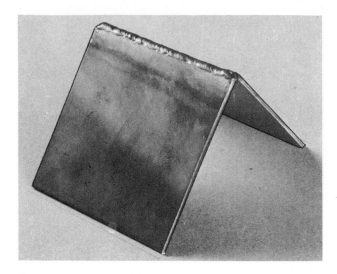

Fig. 23-15. A corner weld.

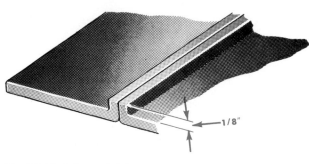

Fig. 23-14. A flange joint.

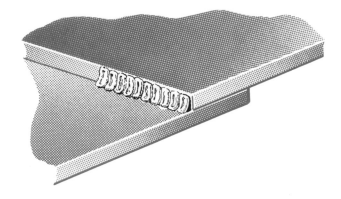

Fig. 23-16. A lap joint.

Welding a Corner Joint

The corner joint is used extensively in fabricating numerous products as well as in repair work. To weld such a joint, set up the plates as illustrated in Fig. 23-15 and tack them in position. The edges are then fused as in welding a flange joint; that is, without a filler rod. To build an edge, add a little filler rod. Test the completed weld by bending the piece out flat and then fold it a little farther over to put stress on the bottom of the weld to see if it holds.

Welding a Lap Joint

Obtain two pieces of $1/16''$ or $1/8''$ metal 4″ to 5″ in length. Lay the plates on the workbench with one piece of metal overlapping as in Fig. 23-16.

Weld the two plates, using a semi-circular movement of the torch and working from right to left, or the reverse for a left-handed person. Since it requires less heat to melt the edge of the upper plate than the flat surface of the lower one, direct more heat onto the lower plate. This may be accomplished by prolonging the travel of the flame on the lower surface and cutting short the flame on the edge of the upper metal. Weld one side of the plate, test it as shown in Fig. 23-17, and then practice on the reverse side. Repeat the exercise until a satisfactory weld is made every time a weld is made.

Fig. 23-17. Use a chisel and hammer to open up the back side of a lap joint. (Linde Co.)

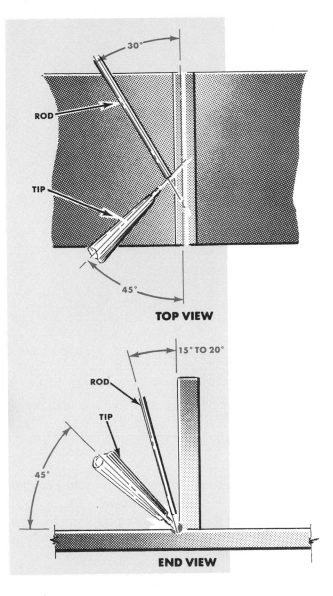

Welding a T-Joint

This joint is made by standing one piece of metal at right angles to a second plate. Proceed as follows to weld a T-joint:

Tack plates in position.   Obtain two pieces of $1/16''$ or $1/8''$ metal 4" to 5" in length and tack them in position to form a T-joint.

Weld the joint.   Start the welding at the right end if you are right-handed (the reverse applies if you are left-handed) and proceed to the left, using a semicircular torch movement.

Hold the torch so the tip forms an angle of about 45° to the flat plate and the same angle to the line of the weld. Point the rod toward the welding tip at an angle of approximately 30° to the line of weld and 15° to 20° to the horizontal plate. See Fig. 23-18.

Direct the flame evenly over both plates, keeping the inner cone of the flame about $1/8''$ away from the surface. In a weld of this type the tendency is to heat the vertical plate too much, causing an undercutting above the weld. To prevent undercutting, keep the molten puddle moving, adding the rod nearer the vertical plate.

Test the weld.   Pick up the welded piece with a pliers and fasten it in a vise as shown in Fig.

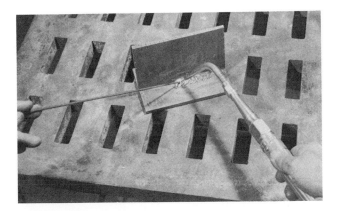

Fig. 23-18. Position of the torch and rod for welding a T-joint. (Linde Co.)

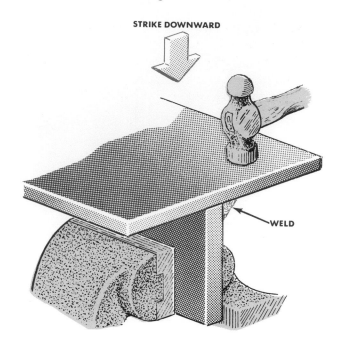

STRIKE DOWNWARD

WELD

Fig. 23-19. Testing a T-weld.

23-19. Hammer the flat plate downward, bending it against the weld.

Points to Remember

1. Move the torch just fast enough to keep the puddle active and flowing forward.

2. When reaching the end of a joint, raise the flame away momentarily to give the puddle a chance to solidify partly.

3. Use a filler rod that has the same composition as the constitution of the base metal.

4. Use a rod with a diameter equal to the thickness of the base metal.

5. Do not hold the rod too high above the pool so the molten metal falls drop by drop onto the puddle.

6. When welding with a filler rod, move the torch in a semicircular or circular motion.

7. Allow a space between plates to compensate for expansion forces.

QUESTIONS FOR STUDY AND DISCUSSION

1. What governs the rate at which the flame should be moved forward?

2. What happens if the torch is moved forward too slowly?

3. Why should the torch be raised when it nears the end of the weld?

4. What are some of the common defects of a beginner's weld?

5. How should the top and bottom surfaces appear when proper fusion has taken place?

6. If the metal does not melt readily, what is the probable cause for it?

7. Why is a filler rod used in welding?

8. How can you determine whether or not a rod is good or poor?

9. What determines the size of the rod that should be used?

10. What happens if the rod is too large for the size of metal that is being welded? If it is too small?

11. How should the torch be manipulated when using a filler rod on a butt weld?

12. Why is the length of the flame prolonged on the bottom plate of a lap joint?

13. What precautions should be observed in welding a T-joint?

14. If the metal does not melt freely, what should be done?

15. What should be the position of the torch for welding a T-joint?

16. What is the usual method for testing the weld?

CHAPTER 24 other welding positions

Welding with an oxy-acetylene torch cannot always be done with the work in a flat position. On some occasions the location of the piece will be such as to require horizontal, vertical, or overhead welding. Undoubtedly welding in a flat position is easier and somewhat faster; nevertheless after a little practice, welds in other positions can be performed without too much difficulty.

Vertical Welding

1. Obtain two pieces of ¹/₁₆″ or ¹/₈″ metal and tack them to form a butt joint. Then mount the plates in a jig (see Fig. 12-2.)

2. Hold the torch and rod at about the same angle as in flat welding as shown in Fig. 24-1. As the welding progresses you may have to vary the angle slightly to control the puddle.

3. Start the weld at the bottom and work upward, using a semicircular torch motion. Do not allow the puddle to become too large. If the puddle gets too big or too fluid, it will get out of control and run down the face of the weld. When you find that the weld is getting too hot, pull the flame away slightly so that it does not play directly on the puddle.

4. To prevent the puddle from getting too fluid, direct more of the flame on the filler rod. Continue to practice this weld until you can obtain beads that appear good and have correct penetration. Test each weld by bending it in a vise as previously described.

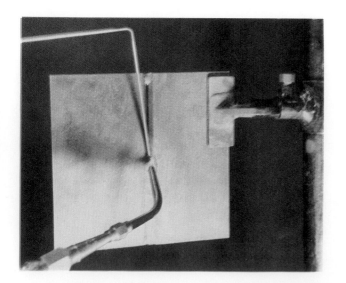

Fig. 24-1. Hold the torch and rod in this position for vertical welding. (Linde Co.)

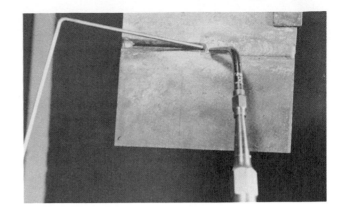

Fig. 24-2. Hold the torch and rod in this position for horizontal welding. (Linde Co.)

Horizontal Welding

1. Tack two pieces of $1/16''$ or $1/8''$ plates to form a butt joint. Allow a space between the edges for expansion.

2. Clamp the joint in a jig so the surfaces are in a vertical position with the line of weld running horizontally. See Fig. 24-2.

3. Start welding from the right edge (or left, if you are left-handed) using a semicircular torch motion. As the welding progresses, you will find that the metal has a tendency to build up much more on the edge of the lower plate. To overcome this, direct the flame longer on the lower edge of the plate without allowing the molten puddle to drop. Keep the tip of the filler rod nearer the upper plate. After each weld, be sure to test the joint by bending it in the vise.

Overhead Welding

The overhead weld is a little more difficult to perform because of the unusual position in which you are required to work and the skill needed to keep the molten puddle from dropping off the plates.

Overhead welding is possible because of the fact that molten metal has cohesive (sticky) qualities as long as the puddle is not permitted to get too large. In other words, molten metal will not fall if the puddle is not allowed to form in complete drops. The amount of heat directed on the seam must be very carefully regulated, since excessive heat will increase the flow of the molten metal. With the correct flame, proper torch manipulation, and practice, the overhead weld can be mastered quickly.

Practicing overhead welds. Tack two $1/16''$ or $1/8''$ plates together and set the work up on a suitable jig that will permit sufficient freedom for manipulating the torch. See Fig. 24-3. Use the same semicircular motion of the torch as previously described. To help keep the puddle shallow, move the filler rod slowly in a circular or swinging motion. The movement of the rod will distribute the molten puddle and prevent it from forming into large drops and falling off. Watch the flame very closely and, if the puddle has a tendency to run, pull the torch away slightly. After the weld is completed, subject it to the

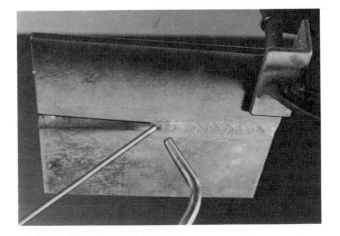

Fig. 24-3. Notice the positions of the torch and rod for overhead welding. (Linde Co.)

bending test. Continue practicing overhead welds until you can make joints that will not break when you apply the break test.

Points to Remember

1. Use a semicircular torch movement for vertical, horizontal, and overhead welding.
2. Do not allow the puddle to become too large as you will lose control of it and it will have a tendency to crack.
3. If the puddle has a tendency to become too fluid, raise the flame slightly.
4. In horizontal welding direct the flame more on the edge of the lower plate.
5. On overhead welds move the filler rod slowly in a circular or swinging motion.

QUESTIONS FOR STUDY AND DISCUSSION

1. What can be done to prevent the puddle from sagging in vertical welding?
2. At what angle should the torch be held for vertical welding?
3. How should the torch be manipulated for vertical, horizontal, and overhead welding?
4. In horizontal welding, why should the flame be directed more on the edge of the lower plate?
5. What should be done to prevent the puddle from becoming too fluid?
6. Why is overhead welding somewhat more difficult to perform?
7. How can the puddle be prevented from dropping off in overhead welding?
8. How should the filler rod be manipulated in overhead welding?

CHAPTER 25 heavy steel plate

Rolled steel stock is commonly labeled as sheet or plate, depending upon the thickness. Generally speaking, if the metal is $1/8''$ or less in thickness, it is referred to as *sheet*. If the thickness is over $1/8''$ it is designated as *plate*.

Although the technique for welding heavy plate is about the same as in welding light material, the problems associated with heavy plate welding are somewhat more complicated. For this kind of welding more attention has to be given to the manner in which the joints are prepared and to the amount of heat required for sufficient penetration.

Welding a Single-V Butt Joint

To have the greatest maximum strength, a weld must possess complete penetration. Full penetration in stock $1/8''$ or less in thickness is reasonably easy to achieve. When the thickness exceeds $1/8''$, penetration cannot be obtained if the edges are left square. Therefore, on such metal the edges must be beveled. The easiest method of beveling plates is by means of the cutting torch. See Chapter 33. Another way to prepare the edges is by grinding them on an emery wheel.

For plates up to $1/2''$ in thickness, the single-V, as shown in Fig. 25-1, is ample. In the single-V, the included angle should be between 60° and 90°. The bottom of the V can have a $1/16''$ or $1/8''$ square root face (unbeveled) or the edges feathered to a sharp point.

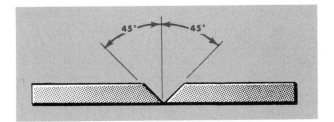

Fig. 25-1. For a single-V butt joint, bevel the edges like this.

To practice welding a single-V butt joint, proceed as follows:

1. Obtain two pieces of $1/4''$ plate and bevel the edges.

2. Separate the plates about $1/16''$ and tack weld them together. Use a filler rod that is approximately $3/16''$ in diameter.

3. Hold the torch so the flame will be at an angle of 60° from the vertical rather than the 45° angle used on light stock as previously described. See Fig. 25-2. Be sure you have the correct size tip for this weld. See Chapter 21.

4. Direct the flame on the V and, as the edges begin to melt, dip the tip of the rod in the puddle. Before adding filler rod make certain that the sides of the V are in a molten state all the way to the bottom of the V. Fill in the bottom of the V for a length of about $1/2''$ with the puddle extending upward to one-half the depth of the V. Then, while the puddle is still in a plastic state, swing

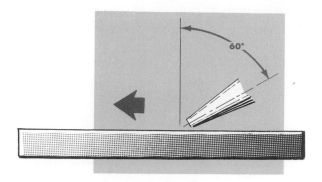

Fig. 25-2. For heavy plate welding hold the torch at a 60° angle.

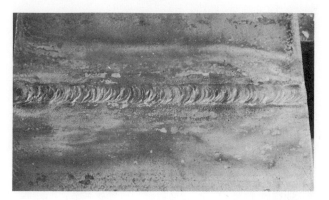

Fig. 25-4. A good ripple weld in heavy plate should look like this. (Linde Co.)

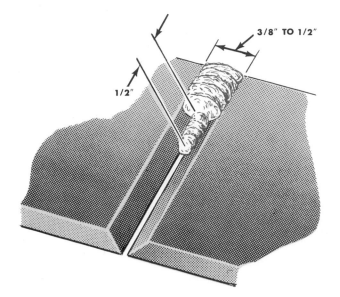

Fig. 25-3. Fill in the bottom of the V and then carry the puddle to the top.

Fig. 25-5. Cut off a specimen and grind off the weld surplus.

the torch in a semicircular motion and fill the V as shown in Fig. 25-3. The completed bead should be between $3/8''$ to $1/2''$ in width and project slightly above the surface of the plate. Return the flame to the bottom of the V, advance another $1/2''$, and again raise this section of the bead to the top of the V. Continue to do this until the weld is finished. See Fig. 25-4.

5. To test the weld, cut off several 1" strips as shown in Fig. 25-5. Grind off the surplus weld metal so that the top of the weld (face) is flush with the top of the plate. Place the specimen in a vise with the face of the weld towards you and bend the piece so the weld is under tension. A good weld should bend at least 90° with no indication of cracking or fracturing. See Fig. 25-6 for example.

Backhand Welding

All of the welding that has been described so far is known as forehand welding. As you know, in this method the welding progresses from right to left for the right-handed person with the flame between the welding rod and the completed portion of the weld.

Fig. 25-6. A good weld specimen should stand bending without cracking.

Fig. 25-7. In the backhand technique, the welding progresses from left to right for right handed persons. (Linde Co.)

In the *backhand* technique, the weld is carried from left to right (right to left for a left-handed person). The welding flame is directed backward at the completed portion of the weld and the rod is between the flame and the completed weld section. See Fig. 25-7. Since the flame is constantly directed on the edges of the V ahead of the puddle, no sidewise motion of the torch is necessary. As a result, a narrower V can be utilized than in forehand welding.

In backhand welding the puddle is less fluid. This results in a slightly different appearance of the weld surface. The ripples are heavier and spaced further apart.

To practice a backhand weld, proceed as in the following method.

1. Place two $1/4''$ thick plates on the bench with the beveled edges separated about $1/16''$ and tack them together.

2. Start the weld at the left (right, if left-handed) and bring the edges of the V to a molten puddle. While doing so, hold the end of the rod in the outer envelope of the flame so it will be ready to melt as soon as the puddle forms.

3. At the start, concentrate the flame a little more on the bottom of the V. As soon as the puddle is fluid enough, dip the rod into it. Once the puddle begins to move, direct the flame more on the filler rod and build up the puddle to the top of the V. As the molten metal begins to fill up the V, move the filler rod slightly from side to side to make sure that the weld metal fuses evenly with the edges of the base metal.

4. Test the joint in the same manner as described in making a single-V butt-weld with the forehand method.

Welding a Double-V Butt Joint

When a plate is $1/2''$ or more in thickness, it is much better to run a weld on both sides. For this purpose a double-V is required. See Fig. 25-8. Notice that in a double-V a $1/16''$ or $1/8''$ root face is provided in the center. Either forehand or backhand method can be used to make the weld.

In making a double-V weld it will be necessary to build up the weld in layers, because it is too difficult to control the molten puddle and secure good penetration in attempting to fill the V in

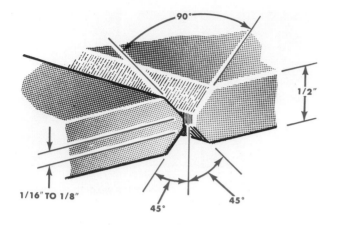

Fig. 25-8. For extremely heavy plates a double-V should be used.

one lap. The usual practice is to first run one layer near the bottom of the V on both sides of the plate. Then successive layers are added to fill the V. Extreme care must be taken when applying the successive layers to fuse each layer with the one already deposited and with the sides of the base metal V.

Welding High-Carbon Steel

Any metal having a carbon content of 0.80 percent carbon or more is considered to be a high-carbon steel. This includes all carbon machine steels, spring steel, and common tool steels.

Whether a steel is weldable or not depends upon the effects the heating and cooling cycle has on the physical properties of the weld area and the base metal itself. If the metal can be taken through the heating and cooling cycle without cracks developing in the weld zone or without seriously affecting the characteristics of the original metal, a steel is assumed to be weldable. See Chapter 3 for a more complete discussion of the effects of heat on steel.

To weld a high-carbon steel, it must be preheated first. This can be done by playing the oxy-acetylene flame uniformly over the metal until it reaches a faint, red color. The welding should be done with an excess acetylene flame with just enough heat to secure fusion between the weld metal and the base metal. For thin parts where there is likely to be considerable mixing of

base metal and weld metal, a low-carbon filler rod should be used. On heavier sections and especially when the parts are to be heat-treated again, a high-carbon rod is advisable.

Points to Remember

1. Plate thicker than $1/8''$ should be beveled before welding.

2. Plate $1/2''$ or more in thickness should have a double-V.

3. When beveling plate, the included angle should be between 60° and 90°.

4. Hold the torch at a 60° angle when welding heavy plate.

5. Be sure the bottom surfaces of the V have reached the proper temperature before adding filler rod.

6. The backhand welding technique is particularly adaptable for welding heavy plate.

7. In backhand welding do not swing the torch; move the filler rod instead.

8. On plates $1/2''$ or more in thickness do not try to fill the V in a single pass. Use several passes.

9. Remember that a high-carbon steel will lose its hardness if welded. To restore its hardness it must be put through a heat-treating process after welding.

10. Use a slightly excess acetylene flame for welding high-carbon steel.

11. Use a high-carbon filler rod on steels that are to be heat-treated after welding.

QUESTIONS FOR STUDY AND DISCUSSION

1. Why is some metal called sheet and other plate?

2. Why should the edges be beveled when the plate is $1/4''$ or more thick?

3. At what angle should the torch be held when welding heavy plate?

4. Why should the edges of heavy plate be spaced for welding?

5. How should test specimens of heavy plate be prepared for checking strength of weld?

6. How does the backhand welding technique differ from the forehand?

7. How should the torch and rod be handled in backhand welding? At what angle?

8. At what angle should heavy plates be beveled in preparation for welding?

9. Why is a double-V used on some plates?

10. In welding plates over $1/2''$ thick, why use more than one pass?

11. Why must extreme care be taken when welding high-carbon steel that has been heat-treated?

12. What kind of flame should be used in welding high-carbon steel?

13. What kind of a filler rod is needed to weld high-carbon steel?

14. How is high-carbon steel heated in preparation for welding?

CHAPTER 26 *stainless steel*

Stainless steels, particularly the 300 series (See Chapter 15), can readily be welded with the oxy-acetylene flame. However, better results are obtained with the gas shielded-arc processes.

Welding Considerations

Choosing the correct join design. For thin metal, the flange-type joint is probably the most satisfactory design. Slightly heavier sheets, up to $1/8''$ in thickness, may be butted together. For plates heavier than $1/8''$, the edges should be beveled to provide a V so fusion can be obtained entirely to the bottom of the weld.

Since stainless steels have a much higher coefficient of expansion with lower thermal conductivity than mild steels, there are greater possibilities for distortion and warping. Therefore whenever possible clamps and jigs should be used to keep the pieces in line until they have cooled.

Selecting the proper tip size. In welding stainless steel with oxy-acetylene, it is necessary to use a tip that is one or two sizes smaller than one ordinarily employed for welding mild steel. Since heat has a tendency to remain longer in the weld zone of stainless steel, a smaller flame reduces the possibility of destroying the properties of the metal.

Adjusting the flame. A neutral flame is very essential for welding stainless steel. Even a slightly oxidizing flame oxidizes the chromium in the steel, thereby materially reducing its corrosion-resistant qualities. An excessive carbonizing flame is also undesirable because it increases the carbon content in the weld area, making the weld weak and brittle. Since it is often difficult to maintain an exact balance of gases for a neutral flame, it is often better to use a slight excess of acetylene.

Selecting the filler rod. Special treated columbium 18-8 filler rods are essential for achieving satisfactory welding of stainless steel. As a matter of fact, it is often desirable to have a rod that contains 1 to $1^{1}/_{2}$ percent more columbium than the base metal. Such a rod allows for the slight oxidation losses that may occur during the welding operation. If special rods are not available, the next best thing is to cut strips from the base metal and use these as filler rods.

Using flux. Not only does the chromium in stainless steel oxidize easily, but the oxide formed during welding acts as an insulating barrier between the flame and the work. Hence a flux is needed to insure better control of the molten metal and to make possible a sound, clean, good-appearing weld.

Most fluxes are prepared by mixing the powder with water to a consistency of thin paste. The flux is then applied by brushing it on the seam, on the filler rod, or on both. It is also a good policy to coat the underside of the seam with flux, since this prevents oxidation and allows a more perfect union to be made along the bottom of the seam.

278

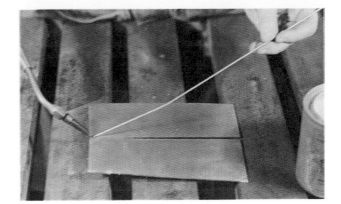

Fig. 26-1. For welding stainless steel, incline the pieces downward from the point where you start.

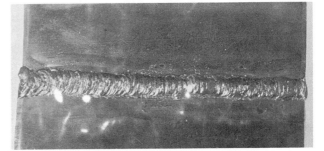

Fig. 26-2. A weld on stainless steel should look like this.

Welding Procedure

1. Use the forehand technique on light sheets and either the backhand or forehand method on heavy plate.

2. Keep the tip of the inner cone of the flame to within 1/16″ of the molten puddle. This will help prevent oxidation. Hold the torch at an angle of about 45° to the work. It is usually helpful to elevate the starting end of the joint so the line of the weld inclines slightly downward in the direction of the finishing end. In this way the flux, which fuses at lower temperatures than the steel, can flow forward and provide protection for the metal as it fuses. See Fig. 26-1.

3. When adding filler rod, hold it close to the cone of the flame. Upon withdrawing it from the puddle, remove it entirely from the flame until you are ready to dip it back into the puddle. A columbium-treated rod flows freely and care must be taken not to direct too much heat on it.

4. Weld from one side only and fill the seam in the first pass. See Fig. 26-2. Refrain as much as

possible from going back over a weld. Success in welding stainless steel depends on keeping the heat to a minimum. Retracting a hot weld produces excessive heat, which is likely to increase the loss of the corrosion-resistant elements in the metal.

Points to Remember

1. Whenever possible, use copper backing strips to weld stainless steel.

2. Clamp the pieces to reduce distortion and warpage.

3. Use a tip that is one or two times smaller than for welding mild steel.

4. Be sure the flame is fully neutral or slightly reducing.

5. Use columbium treated 18-8 filler rods.

6. Always use a flux especially designed for stainless steel.

7. On thin sheet use the forehand welding

technique. Use either the forehand or backhand method on heavy plate.

8. Incline the pieces slightly downward in the direction of the weld.

9. Withdraw the rod completely from the flame when it is not being dipped in the puddle.

10. Complete the weld in one pass.

QUESTIONS FOR STUDY AND DISCUSSION

1. Why is the gas-shielded arc more adaptable for welding stainless steel?

2. How should the edges be prepared on stainless steel pieces to be welded?

3. Why should greater precautions be taken with stainless steel to avoid distortion and warping in welding?

4. What kind of filler rod is needed to weld stainless steel?

5. Why is a flux required to weld stainless steel? What type of rod is used?

6. Where and how should flux be applied?

7. How should the torch be held when welding stainless steel?

8. How should the filler rod be applied?

9. Why complete the weld in one pass?

10. Why should the pieces be inclined slightly downward in the direction of the welding?

CHAPTER 27 gray cast iron

Gray cast iron may be fusion-welded, except that greater caution must be taken to offset expansion and contraction forces than if it were bronze welded. See chapter 30. Since gray cast iron is unusually brittle, it is very susceptible to temperature changes. Consequently, preheating is necessary and considerably more care is required to cool the parts after welding.

To maintain the gray iron structure throughout the weld area, the weld has to be made with the correct filler rod and the parts cooled slowly. If the casting is cooled rapidly, the weld area is likely to turn into white cast iron, thereby making the weld section not only extremely brittle but so hard that machining might be impossible.

Preparing the Edges

The edges of the casting should be beveled to have a 90° included angle, but the V should extend only to 1/8″ from the bottom of the break. Beveling in this manner makes it easier to build up a sound weld near the bottom and lessens the danger of melting through. Placing carbon back-up blocks underneath the joint also helps to prevent the molten cast iron from running out the seam.

Precautions must be taken to clean the surfaces of the joint before welding. The weld area should be cleaned at least one inch on both sides of the V as shown in Fig. 27-1. Improperly cleaned surfaces will result in porous spots and blowholes in the weld, even though sufficient flux is used.

Fig. 27-1. Clean the surface around the seam to be welded.

Preheating the Cast Iron

One important rule that must be followed for successful fusion-welding of cast iron is to preheat the entire casting to a dull red. Uniform preheating will equalize the expansion and con-

281

traction forces and thus minimize the possibility of cracks developing. On a small section, the heating can be carried out by playing the flame over the casting. A large casting may have to be placed in a preheating furnace. The temperature has to be watched very carefully on a heavy casting, especially if it has thin members. The thin members will heat more rapidly, so care must be taken not to get them too hot.

It is good practice to bring the entire casting up to a uniform temperature after the welding is completed. Then the casting can be covered with asbestos paper so it will cool slowly.

Using Filler Rod and Flux

A special cast iron filler rod having the same composition as the base metal is needed to weld cast iron. It is important that the rod contain the correct amount of silicon. During the welding process, the silicon in the weld area has a tendency to burn away. Hence if the rod has sufficient silicon, there will be a proper amount of this element in the weld area after the welding is completed.

A flux is essential in welding cast iron to keep the molten puddle fluid. Otherwise, infusible slag will mix with the iron oxide that forms on the puddle. When this happens, the weld will contain inclusions and blowholes.

Welding Procedure

1. Prepare the edges to be welded. Bevel the joint if necessary and remove all foreign matter from the surface.

2. Slowly heat the entire metal to a dull red.

3. Concentrate the flame near the starting point of the weld until the metal begins to melt. Keep the torch in the same position as in welding mild steel with the inner cone of the flame about 1/8″ to 1/4″ from the seam.

4. When the bottom of the V is thoroughly fused, move the flame from side to side, melting down the sides so the molten metal runs down and combines with the fluid metal in the bottom of the V. Rotate the torch in a circular motion to keep the sides and bottom of the V in a molten condition. If the metal gets too hot and tends to run away, raise the torch slightly.

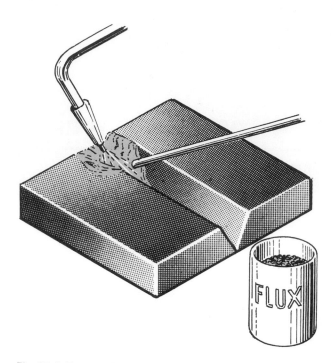

Fig. 27-2. Be sure the metal is fluid before dipping in the rod.

5. Once you have a molten puddle, bring the filler rod into the outer envelope of the flame and keep it there until the rod is fairly hot. Then dip it into the flux. Now insert the fluxed end of the rod into the molten puddle. The heat of the puddle will melt the rod. Never keep dipping the rod in and out of the puddle. As the rod melts, the molten metal will rise in the groove. When it has been built up slightly above the top surface move the puddle forward about one inch and repeat the operation. Be sure not to move the puddle before the sides of the V have been broken down, as this will force the molten puddle ahead on the cold metal. See Fig. 27-2.

6. When gas bubbles or white spots appear in the puddle or at the edges of the seam, add more flux and play the flame around the specks until the impurities float to the top. Skim these impurities off the puddle with the rod. By tapping the rod against the bench the impurities can be removed.

7. After the weld is completed, reheat the entire piece to a dull red. Then cover the metal with asbestos paper and allow it to cool slowly. A

Fig. 27-3. Appearance of properly welded cast iron.

proper cast iron weld should look like the specimen shown in Fig. 27-3.

8. To test your weld sample, place it in a vise with the weld flush with the top of the jaws. Strike the upper end with a heavy hammer until the piece breaks. If the metal has been welded properly, the break should not occur along the fused line but in the base metal.

Points to Remember

1. If possible, use carbon back-up blocks when welding cast iron.

2. Clean the surfaces at least one inch around the seam which is to be welded.

3. Heat the cast iron to a dull red before welding.

4. Use a good grade of cast iron filler rod.

5. Apply flux to the molten metal as the weld is being made.

6. Keep the torch moving in a circular motion to distribute the heat evenly.

7. Reheat the entire piece after the weld is completed and then cover it with asbestos paper so the weld will cool slowly.

QUESTIONS FOR STUDY AND DISCUSSION

1. Why should a rod having the same properties as the base metal be used for this type of welding?

2. Why should cast iron pieces be preheated before welding?

3. How can you tell when a casting has been preheated enough to weld satisfactorily?

4. Why should welded cast iron pieces be cooled slowly?

5. What are the various steps of preparation to be followed before welding a joint?

6. Why is a flux necessary in welding cast iron?

7. How is the flux manipulated in order to deposit it in the weld?

8. How should the rod be introduced into the puddle?

9. What precaution should be taken in moving the puddle forward?

10. If the metal has been properly welded, where should the break occur when the completed weld is tested?

CHAPTER 28 aluminum

Although the gas shielded-arc processes are the most practical for welding commercially pure aluminum, there are occasions when oxy-acetylene welding is used (See Fig. 28-1).

Welding Considerations

The following must be kept in mind when welding aluminum with a gas flame:

1. Aluminum has a relatively low melting point compared to other metals that are welded. Pure aluminum melts at 1220°F (660°C).

2. The thermal conductivity of aluminum is high, almost four times that of steel.

3. Due to its light color, there is practically no indication when the melting point is reached; it collapses suddenly into liquid.

4. Molten aluminum oxidizes very rapidly, forming a heavy coating on the surface of the seam, which necessitates the use of a good flux.

5. Aluminum when hot is very flimsy and weak, and care must be taken to support it adequately during the welding operation.

Choosing the Correct Joint Design

In general, the same principles of joint design for welding steel apply to aluminum. On thin material up to about 1/16″ thick, the edges should

Fig. 28-1. Fusion welding with a gas torch flame is sometimes used in welding aluminum. (Aluminum Company of America)

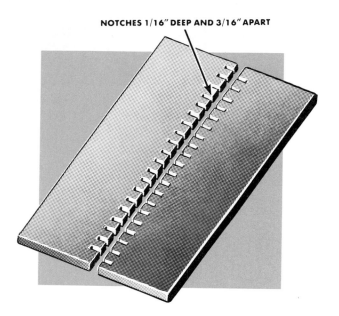

Fig. 28-2. Unbeveled butt joints on ¹/₁₆″ to ³/₁₆″ aluminum should be notched.

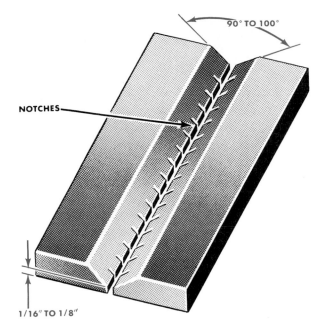

Fig. 28-3. For welding aluminum plate, the edges should be beveled and the base notched.

be formed to a 90° flange at a height equal to the thickness of the material. Flanges will prevent excessive warping and buckling and also serve as filler metal when the flange is melted in the welding operation. Usually no additional filler rod is necessary.

Aluminum from ¹/₁₆″ to ³/₁₆″ in thickness can be butt welded, providing the edges are notched with a saw or cold chisel. See Fig. 28-2. Notching minimizes the possibility of burning holes through the joint, permits full penetration, and prevents local distortion.

As a rule the lap joint is not recommended for welding aluminum because of the danger of flux and oxide being trapped between the surfaces of the joint. When this happens the aluminum is likely to corrode.

For welding heavy aluminum plate ³/₁₆″ or more in thickness, the edges should be beveled to form a 90° to 120° V. It is usually better to allow a ¹/₁₆″ to ¹/₈″ root at the base. This shoulder should be notched as shown in Fig. 28-3.

Aluminum that is greater than ³/₈″ in thickness should have a double-V with the center shoulder notched as illustrated in Fig. 28-4.

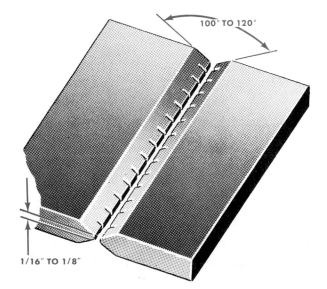

Fig. 28-4. Use a double-V on aluminum plate greater than ³/₈″ in thickness.

Using flux. The most important step in welding aluminum is to clean the edges to be joined thoroughly. All grease, oil, and dirt must be

Fig. 28-5. Be sure to clean the edges before welding aluminum object.

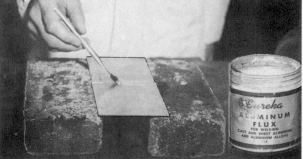

Fig. 28-6. Flux should be applied to the rod (top) and the metal (bottom) when welding heavy aluminum.

removed either with gasoline, naphtha, carbon-tetrachloride, or by rubbing the surface with steel wool or a wire brush. See Fig. 28-5.

Since all aluminum oxidizes very rapidly, a layer of flux must be used to insure a sound weld. The flux is sold as a powder, which is usually mixed with water to a consistency of a thin paste (approximately two parts of flux to one part of water).

If the welding does not require the addition of any filler rod, the flux is applied to the joint by means of a brush. When a filler rod is used, the rod also is coated with flux. On heavy sections it is advisable to coat the metal as well as the rod for greater ease in securing better fusion. See Fig. 28-6.

When the welding has been completed, it is very important that all traces of the flux be washed away. Otherwise, the remaining flux will subsequently cause corrosion. The flux is removed by washing the piece in hot water or by immersing the weld in a 10 percent cold solution of sulfuric acid, followed by rinsing in hot or cold water.

Selecting the filler rod. As in welding other metals, the proper selection of a filler rod is important for welding aluminum. The composi-

tion of the rod should compare to that of the metal to be welded. The three most common rods for welding non-heat-treatable aluminum are 1100, 4043, and 5356. The 4043 and 5356 rods are recommended when greater strength is required.

Welding rods are obtainable in sizes of $1/16''$, $1/8''$, $3/16''$, and $1/4''$ diameter. As a rule, a rod whose diameter equals the thickness of the metal should be used.

Preheating aluminum. All aluminum to be welded, including thin sheet, should be preheated as this will decrease the effects of expansion and minimize cracks. Aluminum plate $1/4''$ or

TABLE 28-1.   TORCH WELDING DATA FOR VARIOUS THICKNESSES OF ALUMINUM.

| ALUMINUM THICKNESS (inches) | OXYACETYLENE | | OXYHYDROGEN* | | HYDROGEN PRESSURE (psi) |
	OXYGEN PRESSURE (psi)	ACETYLENE PRESSURE (psi)	ACETYLENE PRESSURE (psi)	TIP ORIFICE dia (inches)	
1/16	1 +	0.021–0.031	1	0.031 –0.0465	1–3
1/8	1–2	.025– .038	1–2	.038 – .055	2–4
3/16	1–3	.031– .0465	1–3	.0465– .067	3–5
1/4	2–4	.038– .055	2–4	.055 – .076	4–6
3/8	5–7	.067– .086	5–7	.086 – .110	7–9
1/2	6–8	.076– .098	6–8	.098 – .1285	8–10

Reynolds Metals Co.

more in thickness should be preheated to a temperature of 300° to 500°F (149° to 260°C). Preheating to these temperatures can usually be done by playing the flame of the oxyacetylene torch over the work. For large or complicated parts, the preheating is done in a furnace.

It is very important that the preheating temperature does not exceed 500°F (260°C). If the temperature goes beyond this point, the alloy may be weakened or parts of the aluminum may collapse under its own weight.

The correct preheating temperature may be determined by the following three methods:

1. If a mark is made on the metal with a carpenter's blue chalk, it will turn white.

2. If a pine stick is rubbed on the metal, a char mark will be left on it.

3. If the metal is struck with a hammer, no metallic ring will be heard.

Selecting the correct tip size. Since aluminum has such a high thermal conductivity, it is necessary to use a tip slightly larger than the one ordinarily used for steel of the same thickness. Table 28-1 shows the recommended sizes and gas pressures to use for welding aluminum of varying thicknesses.

Adjusting the flame. Many operators use hydrogen instead of acetylene for welding aluminum, and this in many cases is more desirable, especially for welding light gage material. In either case, the torch should be adjusted so it will have a neutral flame. Some authorities rec-

ommend a slightly reducing flame, but usually a neutral flame will be found to be very satisfactory in producing a clean, sound weld. Whether using acetylene or hydrogen, the flame should be adjusted to a low gas velocity to permit a soft and not a blowy flame.

Welding Procedure[1]

After the pieces to be welded have been properly prepared and fluxed, the flame is passed in ever smaller circles over the starting point until the flux melts. Then the rod should be scraped over the surface at about 3 or 4 second intervals permitting the rod to come clear of the flame each time, otherwise the rod will melt before the parent metal and it will be hard to note when the welding should start. The scraping action will reveal when welding can be started without over-heating the aluminum.

The same cycle should be maintained throughout the course of welding except for allowing the rod to remain under the flame long enough to melt the amount of metal needed. The movement of the rod can easily be mastered with little practice. One of the difficulties sometimes experienced in aluminum welding is failure of the deposited metal to adhere. This is generally caused by attempts to deposit the weld metal on cold base metal. After the flux melts, the base metal must be melted before the rod is applied.

---

1. Reproduced through the courtesy of Reynolds Metals Co.

Or the metal and the filler metal may be brought to the molten state simultaneously.

Forehand welding is generally considered best for the welding of aluminum since the flame points away from the completed weld and thus preheats the edges to be welded. Too rapid melting is also prevented. The torch should be held at a low angle (less than 30° above horizontal) when welding thin material. For thicknesses ³/₁₆″ and above, the angle of the torch should be increased to nearer the vertical. Changing the angle of the torch to accord with the thickness minimizes the likelihood of burning through the sheet during welding.

In welding aluminum, there is little need to impart any motion to the torch other than moving it forward. On flanged material, care should be taken to break the oxide film as the flange melts down. This task may be accomplished by stirring the melted flange with a puddling rod.

With aluminum above ³/₁₆″ thick, the torch should be given a uniform (but not unduly) lateral motion in order to distribute the weld metal over the entire width of the weld. A slight back and forth motion will assist the flux in its removal of oxides. The rod should be dipped into the weld puddle periodically and withdrawn from the puddle with a forward motion. This method of withdrawal closes the puddle, prevents porosity, and assists the flux to remove the oxide film.

Aluminum welds should be made in a single pass as far as possible. This is especially true of alloys other than 1100 and 3003.

It must be remembered that the angle of the torch has much to do with welding speed. Instead of lifting the flame from time to time in order to avoid melting holes in the metal, it will be found advantageous to hold the welding torch at a flatter angle, thus increasing the speed. The speed of welding should also be increased as the edge of the sheet is approached.

CAUTION: The flame should never be permitted to come in contact with the molten metal. Also hold it no greater distance away from the material than 1 inch. See Fig. 28-7 for appearance of a good aluminum weld.

Welding Aluminum-Alloy Castings

In general, the welding of aluminum-alloy castings requires techniques similar to those used on aluminum sheet and other wrought sections. However, the susceptibility of many castings to thermal strains and cracks, because of their intricate design and varying section thickness, should be carefully considered. In addition, many castings in highly stressed structures depend upon heat treatment for their strength, and welding tends to destroy the effect of such initial heat treatments. If satisfactory facilities for reheat treatment are not available, the welding of such heat-treated castings is not recommended.

When a broken aluminum casting is to be welded, it is first cleaned carefully with a wire brush and gasoline to remove every trace of oil, grease, and dirt. Unless the casting has a very heavy cross-section, it is not necessary to tool the crack or cut out a V, as this can be accomplished by means of the torch and puddling iron. (A puddling iron is a piece of low-carbon stainless steel rod with a flattened end.) It is necessary, however, that the stock surrounding the defect be completely melted or cut away before proceeding with the weld. If a piece is broken

Fig. 28-7. Appearance of a properly made aluminum weld.

out, hold it in the correct position by light iron bars and appropriate clamps. The clamps should be so attached that the casting will not be stressed during heating.

If the casting is large or one with intricate sections, it should be preheated slowly and uniformly in a suitable furnace prior to welding. If the casting is small, or if the weld is near the edge and in a thin-walled section, the casting may be preheated in the region of the weld by means of a torch flame. Cast aluminum should be heated slowly to avoid cracking in the section of the casting nearest the flame.

Broken pieces are tack welded into place as soon as the casting has been preheated. The actual welding of the piece should commence at the middle of the break, and should be continued toward the ends. The welding rod must be melted by the torch, as the heat of the molten metal is not sufficient to melt it. When the weld is finished, the excess molten metal is scraped off with a puddling iron, and the casting allowed to cool slowly.

Holes in castings are welded in much the same manner as are cracked and broken castings. But it is necessary to melt away or cut away the sides of the hole in order to remove all pockets and to permit proper manipulation of the torch.

For welding ordinary castings, an aluminum-silicon or aluminum-copper-silicon welding rod is necessary. A flux must also be used. Puddling alone will merely break up the oxide film and leave it incorporated in the weld, while fluxing will cause the oxide particles to rise to the surface, resulting in a clean, sound weld. It is important that the added metal be completely melted and the molten metal thoroughly turned with the end of the welding rod or with a puddling iron. Thus the flux and oxide are worked to the surface of the molten metal, and there is very little danger of the finished weld becoming contaminated with particles of flux or other foreign material.

Points to Remember

1. Always use a flux to weld aluminum.
2. Use a 1100, 4043, or 5356 rod for welding aluminum.
3. Preheat the work before welding.
4. Keep preheat below 500°F (260°C).
5. Rub a pine stick or blue chalk on the aluminum to determine the correct preheating temperature.
6. Use a slightly larger tip than for welding steel.
7. Use a neutral or slightly reducing flame for all aluminum welding.
8. Hold the torch at an angle of 30° or less when welding thin material.
9. For thicknesses ³/₁₆″ and above, increase the torch angle to nearer the vertical.
10. Clean the surfaces thoroughly before starting the welding operation.

QUESTIONS FOR STUDY AND DISCUSSION

1. What are some of the characteristics that must be taken into consideration when welding aluminum?
2. How can you determine when aluminum has reached its preheating temperature?
3. Why must flux be used to weld aluminum?
4. How is this flux applied?
5. What type of welding rod is recommended for welding aluminum?
6. At what angle is the torch inclined when welding aluminum? Why?
7. Why should the rod be held in the flame and not away from it as in welding steel?
8. Why are the edges of the joint notched?

special welding processes

CHAPTER 29 carbon arc

Although metallic arc or gas-shielded arc are the most common welding processes, carbon arc welding is sometimes used for certain jobs. Its greatest application today is in welding non-ferrous metals, bronze welding, brazing, and cutting. See Chapters 30 and 33. The techniques for performing some of the common carbon arc welding operations are described in this chapter.

WHAT IS CARBON-ARC WELDING?

Carbon-arc welding is a process in which fusion is achieved by heat produced from an electric arc between a carbon electrode and the work. The carbon arc serves only as a source of heat and does not function as a medium for transferring metal as in the metallic arc process. When extra weld metal is necessary for the joint a welding rod is introduced into the heat of the arc and melted as required.

Carbon Arc Welding-DC Machine

Any standard DC arc-welding machine will supply the current for carbon arc welding. Better results will be obtained if a high-voltage welding generator is used because a greater arc length is needed for the carbon electrodes.

The holder. The holder shown in Fig. 29-1 is designed for carbon arc welding. The regular

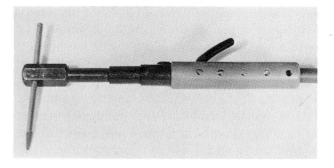

Fig. 29-1. A common type of carbon arc holder for DC machines.

metallic arc holder is not suitable since the carbon electrode becomes white hot and the intense heat soon ruins the ordinary holder.

Different sizes of clamps are available to accommodate electrodes of various diameters. A shield is often located near the handle to protect the operator's hand from the intense heat of the carbon electrode. The handle of the holder is made so air circulating around it keeps it cool. When the carbon arc is employed for continuous operations, especially on heavy work, the holder is often water-cooled.

Carbon electrodes. There are two kinds of electrodes for carbon arc welding, the *pure graphite* and the *baked carbon*. The pure graph-

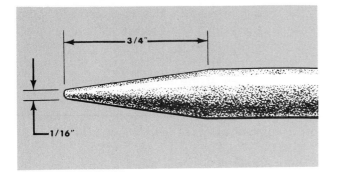

Fig. 29-2. Use a carbon electrode with a long, tapering point.

TABLE 29-1. APPROXIMATE CURRENT VALUES FOR CARBON ELECTRODES ON DC MACHINES

ELECTRODE dia (inches)	COMMON CARBON ELECTRODES (amperes)	SPECIAL GRAPHITE ELECTRODES (amperes)
1/8	15–30	15–35
3/16	25–55	25–60
1/4	50–85	50–90
5/16	75–115	80–125
3/8	100–150	110–165

Fig. 29-3. Shape the carbon electrode on an emery wheel.

ite type lasts longer under high currents but is more expensive. For most welding, baked carbon is satisfactory.

The electrode should have a long, tapering point as shown in Fig. 29-2, and should be about the same diameter as the thickness of the plates to be welded. The tip needs to be at least $1/16''$ in diameter with the taper extending back about $3/4''$. A carbon electrode with a blunt point burns off too quickly, leaving a broad face that makes the arc too difficult to control. The electrode is shaped by grinding it on an emery wheel as illustrated in Fig. 29-3.

Current values. The diameter of the electrode depends upon the current values you intend to work with. In turn, the current values will be governed by the thickness of the metal to be

welded. Thus, as in the metallic arc process, the thicker the metal, the greater the heat required and the higher the current setting. As a matter of fact, when it comes to selecting the proper current and electrode sizes, you may have to make a few tests. You will find that if the amperage is set too high, the carbon burns rapidly and overheats the metal. A good rule to remember is: *If the carbon burns cherry red more than $1^{1}/_{4}''$ from the tip, the amperage is too high.* Keep in mind, too, that if the amperage setting and the carbon size are correct, the flame will be mild and the arc quiet.

Table 29-1 gives the approximate current values for carbon electrodes to be used on DC machines.

Polarity. To carry out carbon arc welding operations with a DC welder, the machine must be set for straight polarity. If the welder does not have a polarity switch, be sure that the holder is connected to the negative terminal and the work to the positive terminal. Reversed polarity not only produces an unstable arc but, even worse, it causes greater quantities of vaporized carbon to enter the molten metal, producing brittleness in the weld area.

Filler rod. When the joint requires additional metal, a filler rod is necessary. The filler rod should be of the same composition as the base metal. For most copper and brass welding, a phosphor bronze rod is desirable. The diameter of the filler rod should be about the same size as the thickness of the base plates.

DC Carbon-Arc Welding Procedure for Non-Ferrous Metals

Prepare the joint. On light material of $1/8''$ in thickness or less, the edges need not be beveled. Plates over $1/8''$ should be beveled and if the thickness is $1/4''$ or more, a double V is necessary so both sides can be welded. Thin plates of 14 to 18 gage can often be joined without filler rod if the edges are shaped as shown in Fig. 29-4. The edges of these joints can be welded simply by fusing the two pieces together.

The use of steel or heavy copper backing bars is often advisable, particularly in welding nonferrous metals. The position of the backing bars for various joints is illustrated in Fig. 29-5.

For butt joints on light material, no spacing between the edges is necessary. Usually metal $3/16''$ or over in thickness should be spaced about $1/8''$ apart at the root.

The surfaces of the metal must be free from grease, oil, scale, or other foreign matter. If possible, clamp the pieces in position. If clamping is impractical, tack weld them at intervals of $8''$ to $10''$.

Position the joint. Place the piece to be welded in a flat position. Since carbon-arc welding is a puddling process, it is almost impossible to carry out the welding operation with the joint in a horizontal, vertical, or overhead position.

Adjust length of the electrode. Arrange the carbon so not more than $3''$ of the end extends

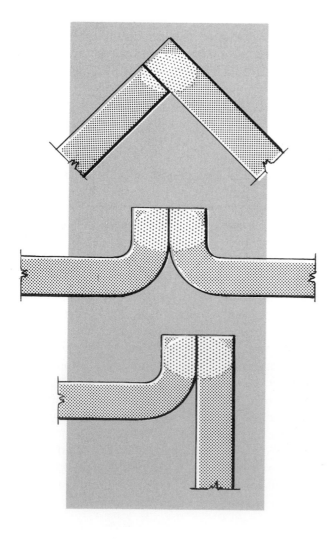

Fig. 29-4. These joints can be welded without using a filler rod.

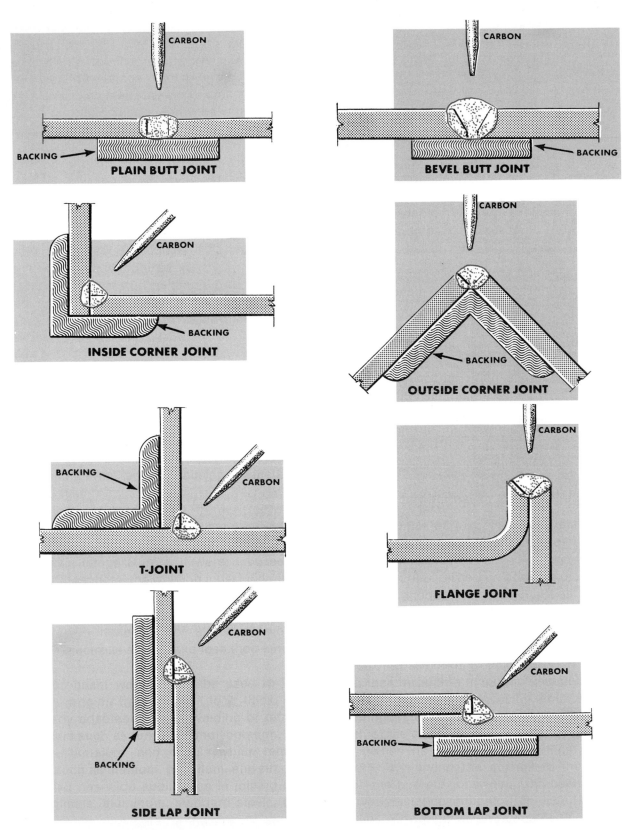

Fig. 29-5. Position of backing bars and electrodes for different joints.

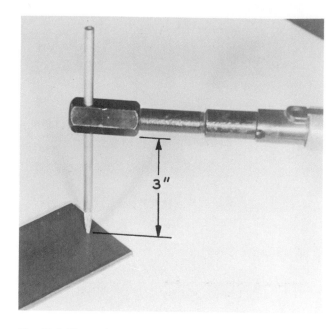

Fig. 29-6. The carbon electrode should not extend more than 3″ beyond the holder.

Fig. 29-7. Hold the filler rod in this manner.

beyond the grip of the holder as shown in Fig. 29-6. Be sure the carbon is shaped correctly. During the welding process, you may have to push the carbon through occasionally to compensate for burning off.

Adjust the machine. Make certain that the current is set for straight polarity and the amperage is correct for the electrode. If necessary, run a trial weld on a scrap piece before proceeding with the regular weld.

Strike the arc. To strike the arc, bring the carbon electrode in contact with the work and withdraw it to the proper arc length. A considerably longer arc is necessary than that used for metallic arc welding. As a rule, the arc length should be 3 to 4 times the thickness of the metal. This is necessary to prevent excessive transfer of carbon to the weld pool. For some metals, such as copper, an arc length up to 1″ will often produce better results.

If the arc is broken and needs to be re-started, do not strike the carbon directly over the hot metal weld. Bringing the tip of the carbon electrode directly in contact with the hot metal will cause a hard spot in the weld since some of the carbon will be absorbed by the metal. To re-start

the arc, strike it on the adjacent cold metal and then bring it over the weld area.

Add filler rod. Hold the filler rod almost parallel with the seam, with the end resting lightly on the work metal as shown in Fig. 29-7. Keep the electrode at right angles to the joint but inclined slightly away from the melting end of the filler rod. Play the arc on the filler rod and move the molten pool along the line of weld.

Carbon-Arc Welding with an AC Machine

The holder illustrated in Fig. 29-8 is especially designed for AC machines. This holder requires two carbon electrodes. In the single carbon electrode used on DC machines, the flame is maintained between the carbon electrode and the work, and as soon as the electrode is removed from the metal the flame goes out. With the AC holder, the flame is formed between the two electrodes and is kept intact even though the electrodes are withdrawn from the work.

The electrodes are inserted into the head of the holder and fastened by turning the thumb screws as shown in Fig. 29-9. Although the selection of carbons varies with each job, the following serves as a guide:

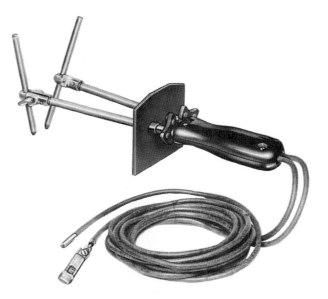

Fig. 29-8. This holder is used on AC machines. (The Lincoln Electric Co.)

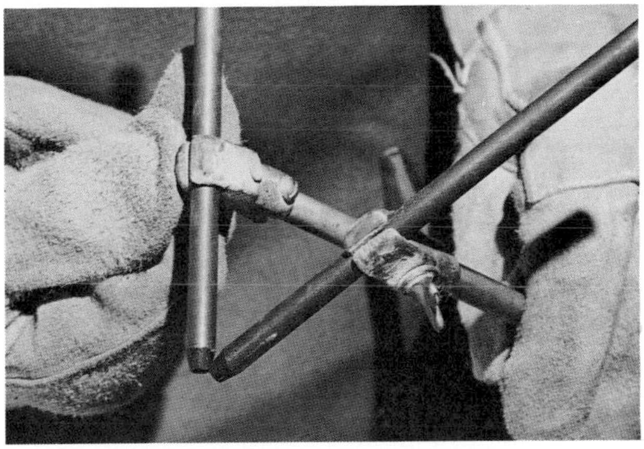

Fig. 29-9. The carbon electrodes in this holder are adjusted by turning thumb screws. (The Lincoln Electric Co.)

¹/₄″ dia carbon—20 to 50 amperes
⁵/₁₆″ dia carbon—30 to 70 amperes
³/₈″ dia carbon—40 to 90 amperes

The actual welding procedure is similar to that described for DC welding. The torch handle is gripped firmly with the thumb resting on the pushbutton. To establish the arc, push the button forward until the electrodes touch. Then release the pressure on the button just enough to permit the points of the electrodes to part and establish the arc. The control of the arc is maintained by pressure on this thumb button. When the correct distance between the two carbon electrodes is obtained, there will be a quiet, soft flame. As the carbon burns away, more pressure will have to be exerted on the button.

The size of the arc can be varied by raising or lowering the amperage within the recommended limits and by using large or small electrodes to fit the particular job. When the electrodes are burned away so the arc cannot be established, reset them in the head to the correct distance.

Points to Remember

1. Do not use the regular metallic arc holder for carbon-arc welding.

2. For most welding, use the baked-carbon type electrodes.

3. Shape the electrode so it has a long, tapering point.

4. Use a carbon electrode with diameter about equal the thickness of the metal to be welded.

5. When using a DC machine for carbon-arc welding, set the current for straight polarity.

6. Use a filler rod having the same composition as the base metal.

7. Whenever possible, use backing blocks when welding non-ferrous metals.

8. Either clamp or tack the pieces before welding.

9. Perform carbon arc welding with the pieces in a flat position.

10. Clamp the carbon electrode so not more than 3″ extends beyond the grip of the holder.

11. Maintain an arc length equal to 3 to 4 times the thickness of the plates.

12. Do not re-start the arc directly over the hot metal weld.

QUESTIONS FOR STUDY AND DISCUSSION

1. How does the carbon-arc process differ from the metallic arc process?

2. For what kind of welding is the carbon-arc process particularly useful?

3. Why must a special holder be used for carbon arc welding?

4. How does the holder used on a DC machine differ from the one commonly employed on an AC machine?

5. How should the carbon electrode be shaped?

6. What is the objection to using an electrode with a blunt point?

7. What determines the diameter of the electrode for welding with carbon arc?

8. How can you determine if the amperage is correct for the carbon electrode being used?

9. Why must the current be set for straight polarity for carbon arc welding?

10. What size and kind of filler rod should be used for carbon arc welding?

11. What consideration should be given to the preparation of the joint on pieces to be welded with the carbon arc?

12. How far should the carbon electrode project from the grip of the holder when used with a DC machine?

13. How long should the carbon arc be maintained on a DC machine?

14. With a DC machine, what is the procedure for restarting a carbon arc when welding on a non-ferrous metal?

15. Why shouldn't a carbon arc be restarted over the hot metal?

16. How should the filler rod be held during the carbon arc welding process?

17. How is the arc established with a dual carbon holder on an AC machine?

CHAPTER 30 brazing

According to the American Welding Society, *brazing* is defined as a group of welding processes wherein coalescence (forming together in one mass) is produced by heating the metal to suitable temperatures above 800°F (427°C) and by using a non-ferrous filler metal having a melting point below that of the base metals. The filler metal is distributed between the closely fitted surfaces of the joint by capillary attraction (power of a heated surface to draw and spread molten metal).

Most commercial metals can be brazed, such as copper and copper alloys, stainless steels, magnesium base metals, aluminum, low-carbon and low-alloy steels, cast iron, titanium, zirconium, and beryllium. Although most brazed joints have a relatively high tensile strength they do not possess the full strength properties of other conventional welding techniques.

The outstanding characteristic of brazing is that the mechanical properties of the base metal are not impaired since lower bonding temperatures are used than in fusion-welding. Brazing also has wide application in joining dissimilar metals. The one notable exception is that copper and copper alloys cannot be brazed directly to aluminum.

BRAZING

To achieve sound brazed joints the following are the preliminary requirements which must be recognized and followed a step at a time in order:

1. Joint design
2. Clean surfaces
3. Correct fluxes
4. Correct filler metals
5. Proper heating equipment

Joint Design. The two basic joints used for brazing are the butt and lap (T and corner are considered as butt welds). The lap joint probably is most common because it offers greatest strength. For maximum efficiency the overlap should equal or exceed three times the thickness of the thinnest member. The main drawback of the lap joint is that metal thickness at the joint is increased.

The butt joint does not provide the same degree of strength as the lap joint since its cross-sectional area is equal only to the cross-sectional area of the thinnest member. However, higher strengths can be achieved by scarfing the edge but this involves greater care in preparing the joint and keeping the pieces in alignment. The strength of a butt joint can also be improved by using a sleeve. See Fig. 30-1.

Another factor associated with joint design is joint clearance. If the surfaces are too tight together the plastic flow of the filler metal is hindered. Too great a distance will prevent the full effects of any capillary action, thus leaving voids and poor distribution of filler metal. Adequate joint clearance should fall in the range between 0.001″ and 0.10″.

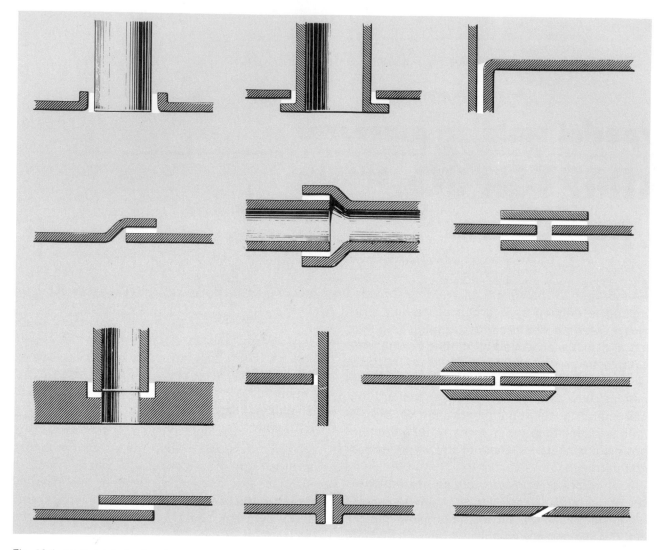

Fig. 30-1. Basic joints for brazing.

Surface preparation. Clean, oxide-free surfaces are absolutely necessary to make sound brazed joints. Uniform capillary action is possible only when surfaces are completely free of foreign substances such as dirt, oil, grease, and oxide. Dirt, oil, and grease can be removed by immersing the parts in some commercial cleaning solvent or salt bath, by pickling in acid (sulfuric, nitric, or hydrochloric) or using some type of vapor-degreasing unit. Surface oxides can be eliminated by sanding, grinding, filing, machining, blasting, or wire brushing. The method used will depend on the contaminants, the joint design, and type of metal to be brazed.

During any cleaning process care must be taken to avoid getting the faying (prepared adjoining) surfaces too smooth. Surfaces that are too smooth will prevent the filler metal from effectively wetting the joining areas. Such surfaces can be roughened slightly by rubbing with 30 or 40 grit (coarse) emery cloth. Brazing should be undertaken as soon as the metal is cleaned to prevent contamination from atmospheric exposure or handling.

Fluxes. The purpose of a flux is to prevent or inhibit the formation of oxide during the brazing process. It is not intended to remove oxides already formed, or dirt, grease, and oil. Metal

TABLE 30-1. FLUXES FOR BRAZING.

AWS BRAZING FLUX (Type No.)	RECOMMENDED BASE METALS	RECOMMENDED FILLER METALS	RECOMMENDED USEFUL RANGE (°F)	INGREDIENTS	FORMS SUPPLIED
1	All brazeable aluminum alloys	BAlSi	700–1190	Chlorides Fluorides	Powder
2	All brazeable magnesium alloys	BMg	900–1200	Chlorides Fluorides	Powder
3A	All except those listed under 1, 2 and 4	BCuP, BAg	1050–1600	Boric Acid Borates Fluorides Fluoroborates Wetting Agent	Powder Paste Liquid
3B	All except those listed under 1, 2 and 4	BCu, BCuP, BAg, BAu, RBCuZn, BNi	1350–2100	Boric Acid Borates Fluorides Fluoroborates Wetting Agent	Powder Paste Liquid
4	Aluminum bronze, aluminum brass and iron or nickel base alloys containing minor amounts of Al and/or Ti	BAg (all) BCuP (Copper base alloys only)	1050–1600	Chlorides Fluorides Borates Wetting Agent	Powder Paste
5	All except those listed under 1, 2 and 4	Same as 3B (excluding BAg-1 through -7)	1400–2200	Borax Boric Acid Borates Wetting Agent	Powder Paste Liquid

AWS

surfaces are easily contaminated in the atmosphere after they are cleaned. Some metals are more susceptable than others to attack. Moreover, any chemical reaction resulting from air exposure is accelerated as the temperature is raised during the brazing process. Hence a flux is needed to dissolve and remove oxides which may form during brazing.

An important requirement of a flux is its ability to readily dissolve and promote the fluidity of the filler metal. Equally important is its surface tension since this affects the wetability of the base metal and its flow in the joint. Finally a flux must last long enough to counteract any reactive effects developed during the brazing operation.

Fluxes are available in powder, paste, or liquid form. The main ingredients of fluxes are boric acid, borates, fluorides, fluoroborates, chlorides, and wetting agents. Fluxes must be selected to suit a particular metal. See Table 30-1 for matching of flux and metal.

A paste flux is the most common for brazing. It can be applied with good adherence to a joint before brazing. Powder flux is sprinkled on the joint or applied to the heated end of the filler rod by dipping it into the flux container. A liquid flux is used mostly in torch brazing where the fuel gas is passed through the liquid flux, carried along with it, and deposited wherever the flame is applied, as in Fig. 30-2, for flux application.

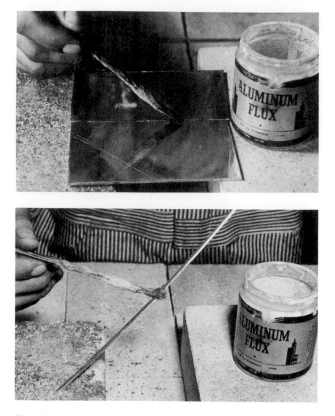

Fig. 30-2. A paste flux is readily applied by brushing.

fluxes is a time consuming task. Consequently controlled atmospheres are often used to prevent the formation of oxides during brazing. In a controlled atmosphere a gas is continuously supplied to a furnace and circulated within it at slightly higher than atmospheric pressures. Gas may consist of high-purity hydrogen, carbon dioxide, carbon monoxide, nitrogen, argon, ammonia, or some form of combusted fuel gas.

Filler metals.   The American Welding Society specifies that a brazing filler metal should have the following characteristics:

1. Ability to wet and make a strong bond on the base metal.

2. Suitable melting temperature to permit adequate distribution by capillary attraction.

3. Sufficient homogeneity and stability to minimize separation by liquation (separation of the solid and liquid portion) and not be excessively volatile.

4. Capable of producing a brazed joint to meet service requirements such as strength and corrosion resistance.

Brazing filler metals are available in wire, rod, strip, and powder forms. These various types are designed to braze different metals.

Flux removal.   Once a brazing operation is completed, all the flux residue must be removed, otherwise corrosion will set in. This residue can usually be recognized by its glass-like surface appearance.

Flux can be removed by washing the part in hot water. In some instances the joint can be immersed in cold water before it has completely cooled from the brazing temperature. The thermal shock of the cold water will usually crack off the residue. For heavy residue a chemical dip is sometimes used. Wire or fiber brushing, steam jet, or blast cleaning are also effective means of removing heavy residues or when large objects are involved. On some soft metals such as aluminum, mechanical removal of residues must be followed by fluid cleaning because small flux particles may have become embedded in the surfaces.

Controlled atmosphere.   Where mass production is involved and particularly when high quality joints are required the application of

TABLE 30-2.   BRAZING FILLER METALS.

AWS CLASSIFICATION OF BRAZING FILLER METALS	TYPES OF METALS TO BE BRAZED
BAlSi (aluminum-silicon)	Aluminum, aluminum alloys
BCuP (copper-phosphorus)	Copper, copper alloys
BAg (silver)	Ferrous and nonferrous metals except aluminum and magnesium
BAu (precious metals)	Iron, nickel and cobalt base metals
BCu (copper)	Ferrous and nonferrous metals
BNi (nickel)	Stainless steels, carbon steels, low-alloy steels, copper
BMg (magnesium)	Magnesium, magnesium alloys

AWS

Filler metals may be designated by commercial names or AWS classification symbols. The AWS classification consist of the letter B, which identifies it as a brazing filler metal, followed by the chemical symbols of the metallic elements included in the filler metal. If dash digits are shown at the end of the chemical symbols they simply specify some characteristic of the filler metal. Table 30-2 shows the basic AWS designated filler metals for brazing different metals.

Filler metal application. Brazing filler metals are applied either manually after the work is heated or are preplaced in a suitable position before the work is heated. Rod and wire are generally used for manual face-feeding. Preplaced filler metals are in the form of rings, washers, formed wire, shims, and powder. These are located in strategic places near the joint to assure a uniform flow of filler metal into the joining surfaces. Although preplaced filler metals can be used in manual brazing, their greatest application is in furnace, induction, or dip brazing. (See Fig. 30-4.)

Manual heating methods. The required heat for brazing purposes may be applied in various ways. For most manual brazing operations a gas torch is considered the most practical. See Fig. 30-3. The gas mixture may be oxy-acetylene, air-gas, gas-oxygen, oxy-hydrogen, or Mapp-oxygen. To a large extent the type of gas mixture used depends on the thermal conductivity, type, and thickness of the metal to be brazed.

Oxy-acetylene or Mapp-oxygen is generally more versatile because of its wide range of heat control. With this gas mixture a slightly reducing flame is required. A single or multiflame tip may be used. In either case, only the outer envelope of the flame and not the inner cone should be applied to the work.

The air-gas torch provides the lowest heat and has greater applications in brazing thin sections. The air-gas mixture may consist of air at atmospheric pressure and city gas or air and acetylene.

The gas-oxygen torch uses oxygen with city gas, bottled gas, propane, or butane. This mixture produces a high flame temperature and is effective where greater brazing heat is required.

The oxy-hydrogen torch is particularly adaptable for brazing aluminum and other nonferrous metals due to its low heat. The low temperature prevents overheating the metal and the hydrogen provides additional cleaning action and shielding during the brazing process.

Filler metal heat. A brazing filler metal must be completely molten before it flows into a joint. The melting temperature of filler metals will vary, depending on the type of filler used which in turn is governed by the kind of base metal to be brazed. In any event, the melting temperature (*liquidus*) of the filler metal must be lower than the *solidus* of the base metal. Solidus temperature is the highest temperature that the base metal can reach and still remain in a solid state. See Table 30-1. The lowest effective brazing temperatures are preferred to minimize the effects of heat on the base metal such as grain growth, warpage, and hardness reduction.

Production heating methods. Although torch brazing can be mechanized for production purposes, high production rates usually are better accomplished by means of furnace, induction, resistance, or dipping techniques. With these methods accurate heat control can be achieved, thereby insuring high quality brazed joints.

Furnace heating involves positioning parts to be brazed on trays and then placed in a gas, electric, or oil-fired furnace. See Fig. 30-4.

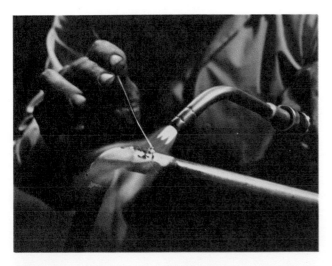

Fig. 30-3. Brazing with a multiflame gas torch. (Airco)

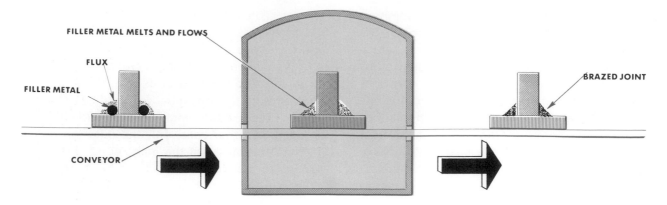

FILLER METAL MELTS AND FLOWS

FLUX

FILLER METAL

BRAZED JOINT

CONVEYOR

Fig. 30-4. Production brazing is frequently done in a furnace.

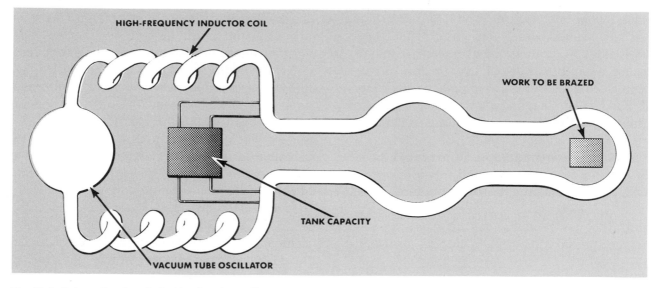

HIGH-FREQUENCY INDUCTOR COIL

WORK TO BE BRAZED

TANK CAPACITY

VACUUM TUBE OSCILLATOR

Fig. 30-5. Schematic of an induction heating coil.

Induction brazing consists of placing the work near an induction coil. As the current flows through this coil, resistance of the object to the flow of current causes instant heat to occur. See Fig. 30-5 for example.

Resistance brazing is very much like spot welding where heat is generated by the passage of a low voltage current through carbon electrodes that are clamped over the work. See Fig. 30-6 for electrode placement (C).

Dip-brazing consists of immersing parts in a bath of molten brazing metal. The brazing material is contained in a crucible which is externally

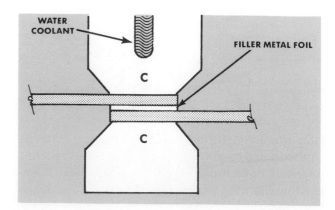

WATER COOLANT

FILLER METAL FOIL

C

C

Fig. 30-6. Resistance brazing. (AWS)

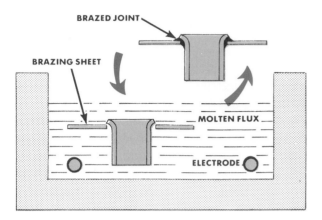

Fig. 30-7. Dip brazing. (AWS)

Fig. 30-8. Clamp pieces to be brazed.

heated. This method is limited to brazing small assemblies such as wire connections or metal strips when they can be easily held in fixtures. Another dip-brazing method involves the placement of parts in a bath of molten salt. The salt bath is heated either by passing an electrical current through the bath or heating the outside of the container. See Fig. 30-7.

Manual Brazing Procedure

The following are the basic steps which generally apply for brazing most metals:

1. Determine the appropriate type joint most suitable for the work to be brazed.

2. Remove all dirt, grease, oil, and oxides from the surfaces to be brazed.

3. Select the correct flux and apply it to both the work and filler metal by brushing, dipping, sprinkling, or spraying.

4. Assemble the pieces and keep them in alignment by means of clamps, fixtures, or jigs. See Fig. 30-8. Avoid too much pressure since enough clearance between mating surfaces must exist to allow a free flow of filler metal.

5. Preheat the entire work by playing the torch over the surfaces to bring them up to a uniform brazing temperature.

6. As soon as the flux is completely fluid, touch the filler metal to the joint, Keep applying filler metal until it flows completely through the joint. Do not apply the inner cone of the flame directly to the filler metal or work and be sure the flame is slightly reducing.

7. Clean the brazed work to remove all the flux residue or debris.

BRAZE OR BRONZE WELDING

Braze or bronze welding as it is sometimes called, is a little different from regular brazing. Whereas conventional brazing processes involve the joining of two surfaces by a thin bond of brazing material, bronze welding is carried out much as in fusion welding except that the base metal is not melted. The base metal is simply brought up to what is known as a *tinning temperature* (dull red color) and a bead deposited over the seam with a bronze filler rod. Although the base metal is not melted, the unique characteristics of the bond formed by the bronze rod are such that the results are often comparable to those secured through fusion-welding.

The bronzing operation is generally performed with an oxy-acetylene torch, although a carbon arc may be used.

Bronze welding is particularly adaptable for joining or repairing such metals as cast iron,

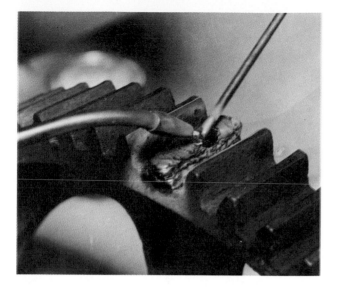

Fig. 30-9. Building up a missing gear tooth with a bronze weld. (Linde Co.)

malleable iron, copper, brass, and various dissimilar metals (cast iron and steel, etc.). See Fig. 30-9. *The one precaution that must be considered in bronze welding is not to weld a metal that will be subjected to a high temperature later,* since bronze loses its strength when heated to 500°F (260°C) or more. Also, bronze welding should not be used on steel parts that must withstand unusually high stresses.

Advantages. Since the metal does not have to be heated to a molten condition in bronze welding, there is less possibility of destroying the main characteristics of the base metal. Thus in repairing malleable castings the danger of affecting its ductility is minimized.

Equally important is the elimination of stored-up stresses which are often present in fusion welding. This is especially critical in repairing castings. The low degree of heat in bronze welding reduces to a minimum these expansion and contraction forces.

With bronze welding there is less need for extensive preheating. On thick sections where some preheating may be desirable the temperature is only brought up to a black heat.

Filler rod. The main elements of a bronze filler rod used in bronze welding are copper and zinc, which produce high tensile strength and ductility. In addition, it contains small quantities of tin, iron, manganese, and silicon. These elements help to de-oxidize the weld metal, decrease the tendency to fume, and increase the free-flowing action of the molten metal.

Flux. A clean metal surface is essential for bronze welding. If the bronze is to provide a strong bond, it must flow smoothly and evenly over the entire weld area. Adhesion of the molten bronze to the base metal will take place only if the surface is chemically clean. The flow action of the bronze over a surface is much like the flow of water over a piece of glass. If the glass is perfectly clean, the water will spread out into a thin, even film, whereas if the glass is dirty the water tends to gather into tiny drops and roll off.

Even after a surface has been thoroughly cleaned by mechanical means, certain oxides may still be present on the metal surfaces. These oxides can only be removed by means of a good flux. There are many satisfactory fluxes on the market.

The flux is applied by dipping the heated rod into the powdered flux. The flux adheres to the surface of the rod and thus can be transferred to the weld. Another method is to dissolve the flux in boiling water and brush it on the rod before welding is started.

Procedure:

1. Clean the surfaces thoroughly with a stiff wire brush. Remove all scale, dirt, or grease; otherwise the bronze will not stick. If a surface has oil or grease on it, remove these substances by heating the area to a bright red color and thus burn them off.

2. On thick sections, especially in repairing castings, bevel the edges to form a 90° V-groove. This can be done by chipping, machining, filing, or grinding.

3. Arrange the work so the weld travels upward on an incline. In this position the molten bronze cannot flow ahead of the heated welding area and the surface in front of the weld is left open to heating. See Fig. 30-10.

4. Adjust the flame so it is slightly oxidizing. Then gently heat the surfaces of the weld area.

5. Heat the bronzing rod and dip it in the flux.

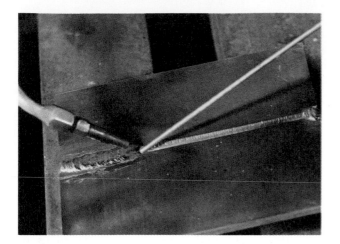

Fig. 30-10. Arrange the work so the weld travels upward on an incline.

Fig. 30-11. Here is an example of a properly deposited layer of bronze. (Linde Co.)

Fig. 30-12. When making a bronze weld in a vertical or semivertical position, first build up a shelf at the bottom. (Linde Co.)

(This step is not necessary if the rods have been prefluxed.) In heating the rod, do not apply the inner cone of the flame directly to the rod.

6. Concentrate the flame on the starting end until the metal begins to turn red. Melt a little bronze rod onto the surface and allow it to spread along the entire seam. The flow of this thin film of bronze is known as the *tinning* operation. Unless the surfaces are tinned properly the bronzing procedure to follow cannot be carried out successfully. You will find that if the base metal is too hot, the bronze will tend to bubble or run around like drops of water on a warm stove. If the bronze forms into balls which tend to roll off just as water would if placed on a greasy surface, then the base metal is not hot enough. When the metal is at the proper temperature the bronze spreads out evenly over the metal.

7. Once the base metal is tinned sufficiently, start depositing the proper size beads over the seam. Use a slight circular torch motion and run the beads as in regular fusion welding with a filler rod. Keep dipping the rod in the flux as the weld progresses forward. Be sure that the base metal is never permitted to get too hot. See Fig. 30-11.

8. If the pieces to be welded are grooved, use several passes to fill the V. On the first pass make certain that the tinning action takes place along the entire bottom surface of the V and about half way up on each side. The number of passes to be made will depend on the depth of the V. When depositing several layers of beads be sure that each layer is fused into the previous one.

9. When making a bronze weld with the work in a vertical position, first build up a slight shelf at the bottom. The shelf then acts as a support for further bronze. As the weld is carried upward, swing the flame from side to side to maintain uniform tinning and to produce even beads. See Fig. 30-12.

SOLDERING

Soldering is a form of brazing in which nonferrous filler metals having temperatures below 800°F (427°C) are used. The filler metal is called solder and is distributed between surfaces by capillary action.

There are two classifications of soldering: soft soldering and hard soldering. Soft soldering has greater applications in sheetmetal and plumbing industries. It uses filler metals composed of tin and lead and produces joints with relatively low tensile strength. Hard soldering is actually a regular brazing process very much like those previously described where filler metals use melting temperatures above 800°F (427°C). This soldering process is often referred to as silver soldering since the main element of the filler metal is silver. Silver soldering is used more in the electrical field, jewelry making, arts and crafts, and where higher strength joints are required than are obtained with soft soldering. The actual silver soldering is carried out like any ordinary brazing process.

Soldering Conditions:

1. Parts to be soldered must fit perfectly so the solder can travel by capillary action between the two surfaces. Solder will cease to flow where there is a gap between the two pieces.

2. Parts to be soldered must be absolutely clean because the solder will not stick to a dirty, oily or oxide-coated surface. Dirt and grease can be removed with a cleaning solvent. Steel wool or some form of abrasive cloth is used to eliminate the oxide. Application of a flux completes the cleaning process and keeps the metal free from oxide during the heating and soldering operation.

3. Parts must be held together during soldering so there is no movement. Any movement during the heating will cause the pieces to be misaligned and the slightest disturbance of the solder will cause it to solidify without forming a bond. The result is a weak joint.

4. Parts to be soldered must have a suitable joint design to withstand the necessary load imposed on it. A lap joint is the most satisfactory for most purposes. If greater strength is needed, some type of mechanical joint should be made before soldering. Typical joints for soldering are shown in Fig. 30-13.

5. Parts must be washed in hot water after soldering is completed to stop the corrosive action.

Soft solders. Tin-lead alloy solders have a melting range from about 370°F (188°C), for a mixture of 70% tin and 30% lead, to about 590°F (310°C) for a 5% tin and 95% lead mixture. The

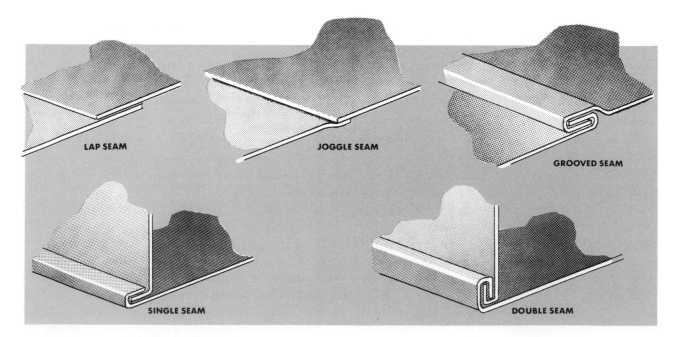

Fig. 30-13. Types of joints for soldering.

most common general-purpose solder is known as half-and-half or 50-50 solder. It contains 50% lead and 50% tin and melts at approximately 471°F (244°C).

Alloys with a low tin content have higher melting points and do not flow as readily as the high tin alloys. Solders with a high amount of tin have better wetting properties and produce less cracking.

Special solders are also available for specific purposes. Thus a tin-antimony solder is designed to solder food-handling vessels where lead contamination must be avoided. Tin-zinc solders are intended primarily for joining aluminum. Lead-silver solders are used where strength at elevated temperatures are required.

Solders are available in bar, cake, solid wire, flux-core wire, ribbon and paste forms. Flux-core wire solder has an acid or rosin flux in the center of the wire. With these solders no additional flux is needed.

Most metals such as steel, galvanized sheet steel, tin plate, stainless steel, copper, brass, and bronze can be joined with soft solder.

Fluxes. Just as in high temperature brazing, a flux is required for soldering. The flux prevents the formation of oxides during the soldering operation and increases the wetting action so the solder can flow more freely.

There are many excellent commercial fluxes available in paste, liquid, powder, and cake form. Some are general-purpose fluxes usable on most metals. Others are special fluxes such as those for soldering aluminum.

All fluxes are classified as corrosive or noncorrosive. Although the corrosive types are most effective, they must be washed away from the metal after soldering. They should never be used for electrical or electronics work. Rosin is the most common noncorrosive flux. Zinc chloride is the most frequently used corrosive flux.

Zinc chloride is prepared by adding small pieces of zinc to muriatic (commercial hydrochloric) acid until the zinc will no longer dissolve. The cut or killed acid is then diluted with an equal quantity of water.

CAUTION: In diluting the acid, the acid must always be added to the water. ALWAYS ADD ACID! Pouring water into the acid may result in a violent and dangerous action.

CAUTION: It is also important to remember that when zinc is dissolved in muriatic acid, injurious chlorine fumes are given off! Therefore the preparation must always be carried out near an open window or under a ventilation hood. Uncut or raw acid (straight or diluted) is preferred for galvanized iron but cut acid may be used and is safer to handle.

Heating devices. In any soldering operation both pieces of metal to be joined must be hot enough to melt the solder. A strong bond is achieved only if the molten solder spreads evenly over the surface. A number of devices are available for heating purposes. The type used depends on the size and configuration of the assembly.

Soldering coppers. A soldering copper consists of a piece of copper fastened to an iron rod with a wooden handle. These coppers vary in size with heads forged in several shapes. Generally a lightweight copper is used for soldering light-gage metal and a heavyweight copper for soldering heavy-gage metal. A lightweight copper on heavy metal does not hold enough heat to heat the metal or allow the solder to flow smoothly. Soldering coppers are heated in a furnace or with a blowtorch.

Tinning a soldering copper. The point of a soldering copper must be covered with a thin coat of solder to operate properly. Overheating or failure to keep the copper clean causes the point to become covered with oxide. The process of replacing this coat of solder is called *tinning*. To tin a copper:

1. File each side of the point until all oxide and pits are removed.

2. Heat the soldering copper until it is hot enough to melt solder.

3. Rub the point on a block of sal-ammoniac and apply a little solder as you continue to rub. Sal-ammoniac, which is ammonium chloride, helps to clean the point of the copper. An alternate way is to dip the point in a liquid or paste flux and then apply the solder.

4. Remove the excess solder by wiping the point with a clean cloth.

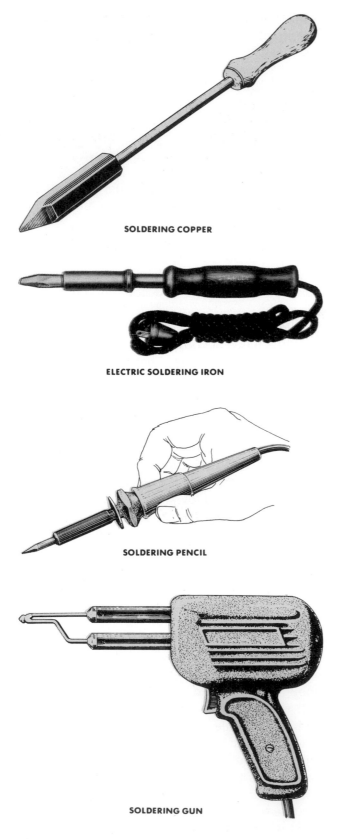

SOLDERING COPPER

ELECTRIC SOLDERING IRON

SOLDERING PENCIL

SOLDERING GUN

Fig. 30-14. Types of heating units for soldering.

Electric soldering irons and pencils. These devices are often more convenient than soldering coppers because they maintain a uniform heat. See Fig. 30-14. They vary in size from 25 watts to 550 watts. Light weight, low-voltage irons with replaceable heating elements and tips are called soldering pencils and are preferred for electric and electronic work. Electric soldering guns produce instant heat at the tip of a long small point when the trigger is pulled. On some guns the trigger also turns on a light—which focuses at the point. For these reasons the soldering gun is very popular for electronics soldering work.

Flame-burning devices. Some soldering operations are impossible or very difficult to perform with a soldering copper or iron. For such tasks a flame is used as the source of heat. See Fig. 30-15. The flame is produced either with an ordinary Bunsen burner or a gas torch depending on the nature of the job. The most efficient, safe, and versatile gas torch is one that burns natural gas and compressed air. These torches are equipped with changeable tips which can produce a wide range of flame sizes. The gas-air torch has two needle valves, one valve controls the pressure of gas and the other valve the compressed air. See Fig. 30-15. To light this torch the gas-needle valve is opened slightly and ignited with a spark lighter. Then the air-valve is turned on and adjusted until a blue flame results. The length of the flame is controlled by the amount of gas and air allowed to flow to the tip.

Bottled-gas torches are also used for soldering especially when the work is not at a fixed station where a gas-air torch is available. The bottled-gas torch must be operated with care, therefore the manufacturer's instructions should always be followed carefully. See Fig. 30-16.

Soldering Techniques

The two manually performed soldering operations are known as seam-soldering and sweat-soldering. See Fig. 30-17.

Seam-soldering. Seam-soldering involves running a layer of solder along the outside edge of the joint. To solder a seam directly, place the fluxed pieces together and tack the seam in

Fig. 30-15. A gas-air torch is very effective for soldering.

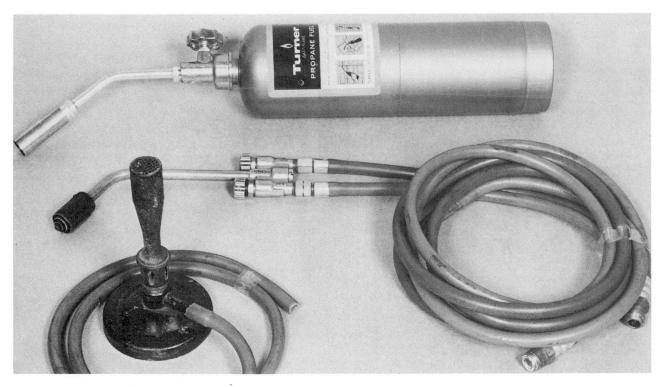

Fig. 30-16. Flame-burning devices for soldering.

several places. This is done by holding the copper on the metal until the flux begins to sizzle. Then apply a small amount of solder directly in front of the point. Do not apply the solder to the point. The hot metal should melt the solder. Now start at one end of the seam. Heat the metal and

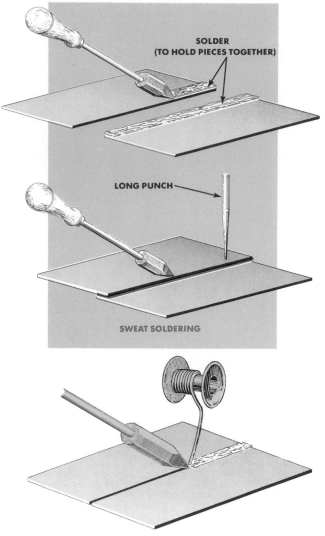

SOLDER
(TO HOLD PIECES TOGETHER)

LONG PUNCH

SWEAT SOLDERING

SEAM SOLDERING

Fig. 30-17. Soldering techniques.

apply solder as needed in front of the point. If necessary press each newly soldered portion together using the tang of a file.

Sweat-soldering. Sweat-soldering is a process whereby two surfaces are soldered together without allowing the solder to be seen. To perform such an operation proceed as follows:

1. Coat the pieces to be soldered with flux after all dirt, oil, grease, and oxide have been removed.

2. Apply a uniform coating of solder to each of the surfaces to be joined.

3. Place the surfaces together with the soldered sides in contact.

4. Place the flat side of a heated copper on one end of the seam. To avoid smearing the exposed surfaces of the metal with solder, remove any excess solder on the copper by quickly wiping the point with a damp cloth before placing it on the joint.

5. As the solder between the two surfaces begins to melt and flow out from the edges, press down on the metal with a stick or file tang. Draw the copper slowly along the seam and follow with the hold-down stick. Do not move the copper any farther than the solder melts.

Points to Remember

Brazing:

1. Avoid using a butt joint for brazing. A lap joint is stronger.

2. Be sure surfaces to be brazed are completely free of oil, grease, dirt, and oxide.

3. Always use a flux that is prescribed for the metal to be brazed.

4. Remove all flux residue after the brazing operation is completed.

5. Use an appropriate filler metal for the work to be brazed.

6. When using a gas torch, heat the surfaces with the outer envelope of the flame and not the inner cone.

7. Use the lowest effective heat for brazing.

Bronze welding:

1. Do not bronze weld a metal that will be subjected to high temperatures or high stresses.

2. If some preheating is necessary, bring the temperature up only to black heat.

3. Use only regularly approved rods for bronze welding operations.

4. Use a special bronzing flux for all bronze welding jobs.

5. Clean surfaces thoroughly before applying the bronze.

6. Arrange the work so the weld progresses upward at a slight angle.

7. Use a slightly oxidizing flame.

8. Be sure the surfaces are properly tinned before depositing beads.

9. Do not melt the surfaces to be welded; heat them only to a dull red.

10. Use a circular torch motion.

Soldering:

1. Always use a good flux when soldering.

2. Never prepare zinc chloride in a confined space, without ventilation.

3. Make sure parts to be soldered are clean and their surfaces fit closely together.

4. During the soldering process, do not allow the parts to move while the solder is in a liquid state as the work will be off-center.

5. Be sure the soldering heat is adequate for the soldering job to be done, including the types of metal and the fluxes.

6. Wash the soldered work in hot water to stop later corrosion action.

QUESTIONS FOR STUDY AND DISCUSSION

1. How does brazing differ from fusion-welding?

2. What specific advantage does brazing have over regular fusion welding?

3. What is the main limiting factor of any brazing process?

4. Why is a lap joint better than a butt joint for brazing purposes?

5. Why is joint clearance an important factor in brazing?

6. What procedure should be used in cleaning surfaces to be brazed?

7. Why is a flux needed for brazing?

8. Why should all flux residue be removed after brazing is completed?

9. When is a controlled atmosphere used in place of a flux?

10. What do the AWS brazing filler-metal classification symbols represent?

11. How should the torch flame be applied to the work to carry out a brazing operation?

12. What is meant by *liquidus* and *solidus* temperatures?

13. What brazing techniques are used for high-rate production work?

14. What is the difference between bronze welding and brazing?

15. What are some of the advantages of bronze welding?

16. When should bronze welding not be used?

17. What kind of rod is needed for bronze welding?

18. What is the function of the flux in bronze welding?

19. How should the flux be applied?

20. Why should the piece to be bronze welded be placed on a slight incline?

21. What kind of flame is recommended for bronze welding?

22. What is meant by tinning?

23. How can you tell when the surface is hot enough for bronze welding?

24. How does soldering differ from high-temperature brazing?

25. How does the tin content of solder affect its flowing properties?

26. How is a zinc chloride flux prepared? With what materials?

27. During the soldering process, why should parts be held firmly in place?

28. Why is a light-weight copper not suitable for soldering heavy-gage metal?

29. What is meant by tinning a copper?

30. What are some of the basic requirements which contribute to effective soldering?

31. How does seam soldering differ from sweat soldering?

32. What type of heating devices can be used for soldering purposes?

CHAPTER 31 *surfacing*

Surfacing is a process of applying a hard, wear-resistant layer of metal to surfaces or edges of worn out parts. It is considered as one of the most economical methods of conserving and extending the life of machines, tools, and construction equipment. Thus, the process may involve building up worn shafts, gears, or cutting edges of tools. See Figs. 31-1 and 31-2.

The two main types of surfacing are known as *hardfacing* and *metallizing.* Hardfacing or hard-surfacing, as it is sometimes called, is a fusion technique whereby a hard, tough overlay of metal is actually fused with the worn unit. Metal-

Fig. 31-1. The edges of these plowshares have been repaired with abrasive-resistant hardfacing material. (The Lincoln Electric Co.)

Fig. 31-2. Here are other examples of parts which have been repaired with hardfacing material. (The Lincoln Electric Co. and the Wall Colmonoy Corp.)

lizing is a spray coating procedure where finely divided particles of metal are deposited on worn away surfaces.

Types of Wear

Parts in service are subjected to three main types of wear: impact, abrasion, and corrosion.

Impact refers to crushing forces which cause parts to chip or crack.

Abrasion is associated with grinding, rubbing or gouging actions.

Corrosion involves the destruction of a surface because of atmospheric contamination (rusting), chemicals, and oxidation or scaling at elevated temperatures.

An understanding of what caused the wear is important, since any surfacing application requires different techniques.

Properties of Parts to be Surfaced

An additional requirement of any hardfacing operation is a knowledge of the composition of the component to be serviced.

Metals of these parts can be grouped into two categories. In one group are the metals whose physical properties are not changed significantly or are subject to cracking when heated and cooled in a hardfacing operation. These metals include the low-range carbon and medium-carbon steels, the low-alloy steels, and the stainless steels. The second group includes metal parts made of steels whose physical characteristics are changed with the application of hardfacing materials. These metals usually have been hardened by some heat-treating process and any subsequent exposure to heat may jeopardize this hardness or produce cracks. Metals in such a group include the higher range medium-carbon steels, high-carbon steels, cast irons, and other alloy steels.

Metals in the first group can be hardfaced without any particular precautions since no harmful cracking will affect the hardness of adjacent welds. With metals in the second group special care must be taken to minimize the sudden shock of localized heat. This is done by reducing the hardness by annealing or through gradual and uniform preheating and post cooling. Preheating from 300° to 500°F (149° to 260°C) will usually prevent weld hardening in medium and high-carbon steels. High-carbon alloy steels and wear-resistance alloy steels require preheating to the same temperature. After the surfacing operation is completed, reheating to a temperature of 800° to 1300°F (about 425° to 700°C) and slow cooling should follow.

HARDFACING

There are many different types of hardfacing materials. Most of them have a base of iron, nickel, copper, or cobalt. Auxiliary elements may be carbon, chromium, molybdenum, tungsten, silicon, manganese, nitrogen, vanadium, and/or titanium.

The alloying elements form hard carbides which contribute to the crystalline properties of the hardfacing metals. Thus a high percentage of tungsten or chromium with a high carbon content will form high carbide crystals that are harder than quartz. Materials having a high chromium content provide excellent resistance to oxidation and scaling. Nickel, cobalt, and chromium are particularly effective for corrosion resistance when added as hardfacing.

Hardfacing metals are either martensitic, pearlitic, or austenitic. Martensite is the hardest and strongest. Pearlite is moderately tough and hard. Austenite is soft, tough and resists impact.

Electrodes for hardfacing. Electrodes for hardfacing are divided into three main groups: severe abrasion resistant, moderate abrasion and impact resistant, and severe impact and moderately severe abrasion resistant.

Hardfacing metals are available as rods for oxy-acetylene welding, electrodes for shielded metal-arc welding, and hard wires for automatic welding. Some hardfacing materials come in tubular rods which contain a mixture of powder metal, powder ferroalloys and fluxing ingredients. The same material is available in powder form for hardfacing with the carbon arc.

1. Severe abrasion-resistant. Electrodes in this group are of the tungsten carbide and chro-

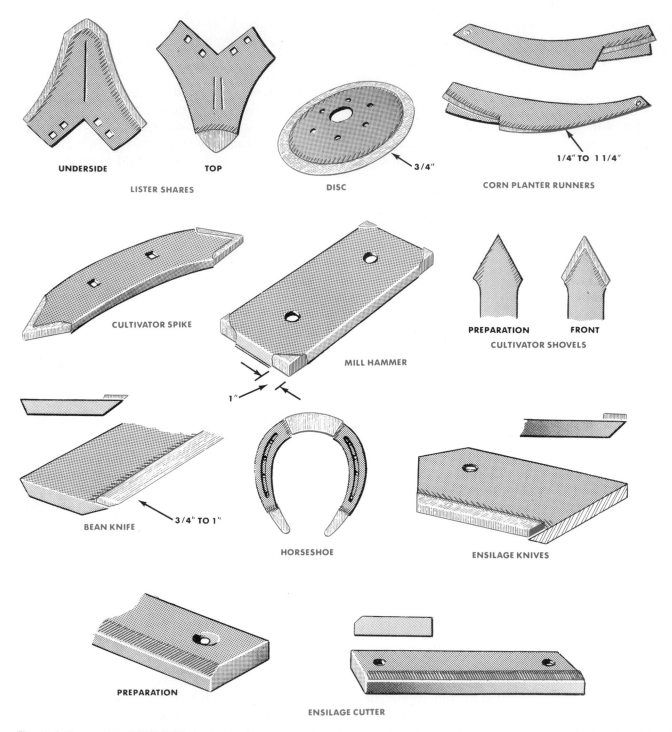

Fig. 31-3. Examples of farm tools that can be hardfaced with tungsten carbide electrodes to increase their useful life. (The Lincoln Electric Co.)

mium carbide types. These electrodes deposit a very hard abrasive-resistant material. They are not suitable for impact wear, since the deposit chips and cracks when subjected to shock.

Chromium and tungsten electrodes come either in coated tubular form or as regular coated cast alloy. The tube rods contain a mixture of powder metal, powder ferroalloys, and fluxing ingredients. The tubes are coated for arc stabilization and arc shielding. Both types of electrodes are used with the metallic arc. The same material is available in powder form. The powder alloy is used when hardsurfacing is performed with the carbon arc.

Tungsten electrodes have tiny crystals of tungsten carbide embedded in the steel alloy. When applied on a surface, the steel wears away, leaving toothlike particles of tungsten carbide exposed. Since tungsten carbide is very hard, the exposed particles make the edge of the part self-sharpening. This property is particularly desirable for earth digging equipment, scraping tools, plowshares, rotary digger blades, cultivator sweeps, and other similar machinery, as shown in Fig. 31-3.

Still another type of tungsten carbide electrode deposits fine particles of tungsten carbide that are so close that they form a smooth cutting edge. These electrodes are useful in repairing steel cutting edges such as lathe tool bits.

Chromium carbide electrodes are slightly less hard and less abrasion-resistant than the tungsten carbide type but are tougher. Most of them are not affected by heat treatment and are too hard to be machined. In addition to being hard, chromium carbide electrodes produce surfaces that provide better protection against oxidative corrosion.

2. Moderate abrasion and impact resistant. Electrodes in this group are of the high-carbon type and leave a very hard and tough deposit. They are excellent for repairing surfaces which must withstand both abrasion and impact forces, such as chisels, hammers, sprockets, gears, tractor lugs, bucket teeth on loaders, scraper blades, etc., as illustrated in Fig. 31-4. These electrodes are good for general purpose surfacing and cost considerably less than the tungsten carbide electrode.

Deposits from high-carbon electrodes can be heat treated to produce even harder surfaces, or they can be annealed to soften them for machining. The hardness of the deposit depends on the rate of cooling. The faster the part is cooled, the harder will be the deposit.

Fig. 31-4. Examples of tools surfaced with high-carbon electrodes. These tools must withstand both impact and abrasion forces.

Fig. 31-5. This dragline bucket was repaired with severe-impact hardfacing material. (Chemitron Corp.)

In this group are also the electrodes having a high percentage of manganese as an alloying element. Manganese electrodes are tougher but not abrasion-resistant like high-carbon types.

3. Severe impact and moderately severe abrasion-resistant. Deposits of these electrodes are tough but not so hard. They are highly resistant to impact and produce good resistance to abrasion. They are often referred to as self-hardening because the deposited surface hardens as it is pounded. While the outside surface is hard, the material underneath remains soft. This prevents cracking, even though there may be some deformation. These electrodes are especially adaptable for hardsurfacing rock crusher parts, chain hooks, scraper blades, pins, and links. See Fig. 31-5.

Stainless steel electrodes are often used for hardfacing parts that must resist impact forces without cracking. These electrodes offer the least resistance to abrasion in the "as deposited" condition; however, they will work-harden (cold working). In addition to their toughness they are very corrosion-resistant. Stainless steel electrodes are often used as base layers for other hardfacing electrodes.

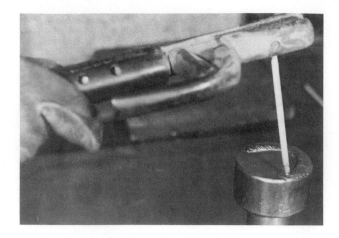

Fig. 31-6. Hardfacing with the shielded metal arc.

Hardfacing with Shielded Metal-Arc

Hardfacing with the shielded metal arc is probably used more extensively because of its high deposition rate. It also has wide application where large areas have to be surfaced or for heavy parts that normally would require excessive time to heat with the oxy-acetylene flame. This method of hardfacing is especially suitable for depositing overlays on manganese steel and other steel alloys where heat build-up must be restricted.

Either AC or DC current produces satisfactory hardfacing welds. The electrodes may be of the coated solid wire type or hollow tube containing alloy powder and flux.

Hardfacing electrodes are classified into three groups. Most of them are referred to by a manufacturer's trade name.

Procedure:

1. Clean the surface thoroughly of rust, scale, and all other foreign matter.

2. Use only enough amperage to provide sufficient heat to maintain the arc. This is very important to prevent dilution of the deposit by the base metal.

3. Arrange the work so it is in a flat position. Most hardfacing electrodes are designed to be run in the flat position only, as is illustrated in Fig. 31-6.

4. Maintain a medium long arc and do not allow the coating of the electrode to touch the base metal. In making the deposit, use either a

straight or weaving bead. A weaving bead is preferred when only a thin deposit is required. Do not extend the width of the weave over ³/₄".

5. Remove all the slag from the surface before depositing additional layers.

6. Manipulate the electrode carefully to secure adequate penetration into the adjoining beads. This can be done by holding the electrode a moment over the deposited bead to allow the heat to build up in the adjoining beads. Such a procedure will also minimize undercutting.

A whipping action is often used when surfacing an area along a thin edge. The arc is held over the heavy portion and then whipped out to the thin edge. In this manner a shallow deposit is made before the heat builds up enough in the base metal to burn through.

Hardfacing with the Carbon Arc

As a rule, the carbon arc is more satisfactory for hardfacing thin-edged parts requiring low heat. See Fig. 31-7. With the carbon arc the heat can be controlled more easily and therefore there is less danger of burning through the metal. For this process, the hardfacing material is in powder form and is applied as a paste as shown in Fig. 31-8.

Procedure:

1. Clean the surface and place the work in a flat position for welding.

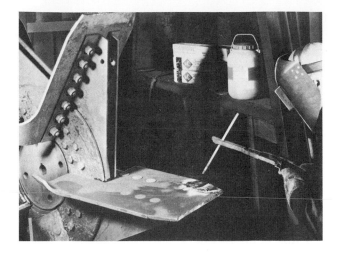

Fig. 31-7. The carbon arc is used here to hardsurface coal conveyor fan blades.

Fig. 31-9. Move the carbon in a circular motion, keeping the tip of the flame in contact with the surface.

Fig. 31-8. Spread hardfacing past evenly over the entire area. (Wall Colmonoy Corp.)

2. Spread hardfacing paste over the area to be faced. If the powder has not been mixed commercially to a paste, add water, alcohol, or other liquid as recommended by the manufacturer. Spread the paste evenly to a depth two to three times the desired thickness of the final deposit. Then allow the paste to dry momentarily.

3. Set the arc so just enough heat is provided to obtain a free-flowing puddle. If the amperage is too high, there will be an excessive dilution of the base metal and the hardness of the finished surface will be lowered.

4. Move the carbon electrode in a circular motion with the tip of the flame in contact with the surface. When the metal becomes red hot, the paste melts and fuses to the base metal as shown in Fig. 31-9.

5. If the hard-surfacing material is being applied to a section which tapers to a sharp edge or point, apply most of the heat to the heavier area and rotate the flame to the thin edge just long enough to carry the puddle to the edge or point, Fig. 31-10. Usually it is best to start the puddle at the point and work toward the heavier section.

6. Avoid heavy deposits. If after the first pass the surface is not sufficiently built up, add paste and repeat.

Hardfacing with Oxy-acetylene

The oxy-acetylene hardfacing technique is very useful in depositing overlays on small parts such as engine valves, plowshares, tools, and other similar items. With the oxy-acetylene flame, tiny areas can be surfaced and thin layers

Fig. 31-10. Hard surfacing the edges of a large screw. (Wall Colomony Corp.)

applied smoothly. Preheating and slow cooling are readily controlled minimizing cracking even with brittle wear-resistant surfacing materials. The principal limitation of the oxy-acetylene technique is its low deposition speed and need for heat control.

Metals for hardfacing with oxy-acetylene generally consist of low melting high-carbon filler rods. As a rule, a slightly reducing flame is recommended inasmuch as this will add carbon to the deposit.

The hardfacing operation is started by preheating the surface to produce a sweating condition. During the preheating cycle the tip of the hardfacing rod is held on the fringe of the flame. The rod is then moved into the center of the flame and melted. The actual deposition of the filler rod is carried out with a regular forehand welding technique using a slightly weaving motion.

Hardfacing with Gas-Shielded Arc

Both the gas tungsten-arc and gas metal-arc processes are ideal for hardfacing. In many cases, the gas shielded-arc processes are considered superior because of the ease with which an overlay can be made. The surfacing materials are readily deposited to form smooth, uniform, porosity-free surfaces.

Hardfacing with Tig is somewhat slower than with Mig but the overlays are of slightly higher quality. Tig is particularly effective in applying cobalt-base alloys. The process ordinarily requires very little preheating. Since the heat build-up is minimal there is less distortion and very little of the base metal is affected by the heat of the process.

The gas metal-arc process with its continuous wire is faster than Tig and produces excellent overlays. With both Tig and Mig the shielding gas provides an added feature when aluminum bronze surfacing materials are used. The shielding gas prevents oxidation and loss of alloying ingredients. A variety of special wires are available for practically every conceivable hardfacing operation.

Care must be taken in using Tig and Mig for surfacing to avoid dilution of the deposited weld metal. Helium and a mixture of helium-argon generally produces a higher arc voltage than pure argon and therefore increases the tendency for greater dilution, hence, argon or a mixture of argon and oxygen is recommended for hardfacing with the gas-shielded arc processes.

Hardfacing with Submerged Arc

The submerged-arc process is considered the most economical for hardfacing parts where heavy deposits are required and extensive areas are to be surfaced. Since the submerged arc utilizes high welding current its deposition rate is high and its deposits are of high quality. Smooth overlays can be made with little or no welding experience required of the operator.

The filler metal may be either solid or tubular and is especially suitable for surfacing that requires high compression strength. However, the relatively deep penetration of the submerged arc plus its protective flux covering usually develops more intensive heat in the welded area. Consequently, greater precautions must be taken to provide suitable preheat and post heat treatment for stress relief.

Very often the full strength of the hardfacing metal is attained only by depositing two or more layers. The initial layer frequently becomes diluted when fused into the base metal and therefore an additional layer is necessary to secure the required results.

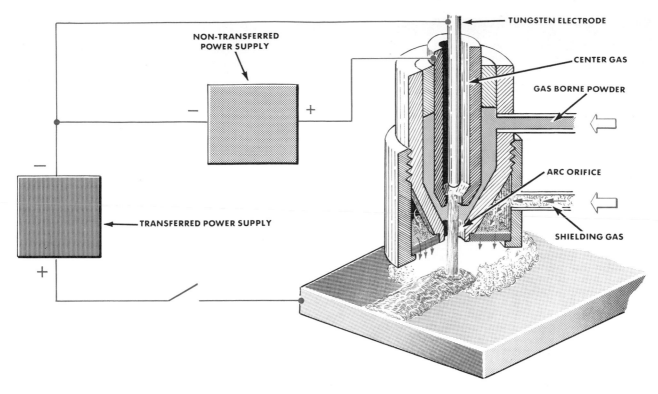

Fig. 31-11. Schematic view of hardfacing with plasma arc.

Hardfacing with Plasma Arc

Plasma arc surfacing is a mechanized tungsten-arc process that uses a metal powder as surfacing material. The metal powder is carried from a hopper to the electrode holder in an argon gas stream. See Fig. 31-11. From the torch the powder moves into the arc stream where it is melted and then fused to the base metal. The surfacing is an actual welding process and not a metal spray process. A wide variety of cobalt, nickel, and iron-base surfacing powders are available from manufacturers of welding supplies. These powders are fused materials and consequently are homogeneous in composition. They are classified as high-alloy materials having varying degrees of impact resistance qualities, abrasion resistance qualities, and corrosion resistance qualities. The type of application should be determined before selecting the powder to be used.

The power source consists of a conventional DC power supply unit with straight polarity. A second DC unit is connected between the tungsten electrode and the arc-constricting orifice to support a nontransferred arc. The second power supply supplements the heat of the transferred arc and serves as a pilot arc to start the transferred arc. Argon gas is used to form the plasma as well as the shield.

METALLIZING

Metallizing, sometimes referred to as metal spraying, is a process of depositing fine semi-molten metal particles or metal powder onto the surface of a metal to form an adherent coating. The powder or metal particles are shot through an intense heat and are spread on the surface where they form thin layers of metal. The powder or wire rod is fed into an oxygen-fuel gas flame and the small semi-molten droplets are driven onto the surface by a stream of high pressure air. The minute particles strike the surface at esti-

Fig. 31-12. Metallizing involves spraying tiny particles of metal onto the surface of a part to be rebuilt.

mated speeds of 250 to 500 ft per second, depending on the gun design. Cohesion is achieved by the mechanical interlocking and fusion of the tiny metallic particles and the bonding of the thin oxide film which forms on the particles while in motion.

Metallizing has made wide application in the machine field where worn surfaces need to be restored to their original sizes but where low tensile strength and porosity are not objectionable. It is a very functional process for jobs where welding or brazing heat is impractical or for applying deposits of dissimilar metals which otherwise are not possible. The process is unique since there is no limit to the size of the object or structure which can be coated. The added feature of metallizing is that no preheating or postheating is required. There is little or no distortion resulting from the spray and consequently no rigid sequence of operations will be found to be necessary. See Fig. 31-12.

Metallizing Process

The success of any spraying process depends on having a clean surface that has been properly roughened. All traces of oil, dirt, scale, rust, etc., must be removed. The roughing of the surface provides mechanical anchorages for the sprayed metal particles. Grit blasting is probably the most common method used for surface roughing. Blasting abrasives may be steel grit, hard sand, aluminum oxide, or silicon carbide. Sometimes after a surface is prepared for spraying, a thin layer of molybdenum is sprayed on. This produces a fusion bond which gives greater adherence to the subsequent spraying coats. Another method employed in roughing a surface is to run a chasing tool over the area. The chasing tool produces ragged threads.

The sprayed metal coatings are somewhat porous but in machine elements this is an advantage. The porous coating absorbs oil which provides more complete lubrication. Porosity becomes more critical in applications that are subject to severe attacks by acids and other corrosive materials.

To some extent porosity can be controlled by the adjustment of gas and air as well as the distance of the gun from the workpiece. However, too much effort to reduce porosity will usually result in hard, brittle and highly oxidized coatings which will fail in service.

Oxidation normally occurs in the melting flame and during the flight of the metal particles to the surface. As a rule, little oxidation will take place as the metal is melted unless the gas-fuel mixture is oxidizing. The greatest causes of oxidation are overheating of coating, excessive use of oxygen, and spraying too great a distance from the workpiece.

Wire spray guns. Metallizing is done with a special spray gun which weighs from 3 to 6 pounds and handles 20 gage to $^3/_{16}$" dia wire. Usually these guns will spray 4 to 12 pounds of metal per hour. Larger type guns are often mounted on a fixture and are designed for spraying large machine components.

The gun consists of two major parts: the power unit and the gas head. See Fig. 31-13. The power unit feeds the wire into the nozzle of the gun. The gas head controls the flow of oxygen, fuel gas, and the compressed air. The nozzle has a center orifice through which the wire is fed. Around this orifice are a number of gas jets which provides the flame and high velocity air stream. As the wire comes through the orifice, it is melted and atomized by the flame. The fine molten particles are picked up by the air stream and projected against the work. The most common gas for the oxy-fuel flame is acetylene which produces a temperature exceeding 5600°F (3094°C), although hydrogen or propane is sometimes used for lower melting metals.

Spraying operation. Any metal spraying operation should be carried out in a well ventilated area. Adequate ventilation is necessary to remove dust particles and fumes which are extremely hazardous to health. If positive ventilation is not possible, the operator should wear an effective respirator.

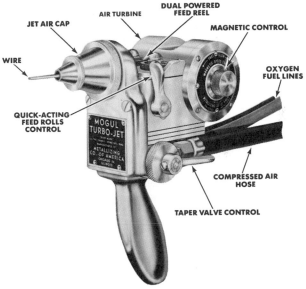

Fig. 31-13. Metallizing gun.

The wire speed, amount of spray, gas, and oxygen pressure must be regulated according to the recommendations established for the equipment to be used and the type of metallizing to be done. Air pressure is normally set for 60 psi. The use of a flowmeter will insure more accurate control of the gases. A slight increase in air pressure provides a finer coating and similarly, a decrease of air pressure produces a coarser coating.

The tip of the melting wire should project beyond the end of the air cap. This length depends to a large extent on the material being used. A recommended practice is to speed up the wire until chunks of it are being ejected.

Then the wire feed is reduced until the ejection of chunks discontinues.

Each coating should be kept as light as possible, somewhere around 0.003″ to 0.005″ in thickness. Too heavy a coat will produce an irregular and stratified surface. The actual movement of the gun is very similar to paint spraying. The nozzle should be kept approximately 4″ to 10″ away from the surface and moved with a uniform motion. If the gun is held too close to the work minute cracks will form in the coating. Too great a distance will produce a soft spongy deposit with low physical properties. The rate of gun travel is also important. When the travel is too rapid the coating develops a high oxide content.

In spraying a flat surface the gun is moved back and forth to allow a full uniform deposit. Spraying should begin beyond the edge of the area to be covered and continued beyond the end of the area. After the first layer the work or gun is often rotated 90° and this technique repeated for each subsequent coating until the required thickness is built up. On cylindrical pieces the work is generally fastened in a lathe with the gun mounted on the traveling carriage.

Electric-Arc Metallizing

The electro-spray unit shown in Fig. 31-14, deposits particles that are hotter and more fluid than those resulting from oxy-acetylene metalliz-ing equipment. Heat to melt the wire is generated by an electric arc instead of oxy-acetylene. The arc, which reaches a temperature of approximately 7000°F (3870°C), produces greater bond strength since the highly heated particles are able to create better fusion. Also the hotter particles will form coatings which will have a lower oxide content.

Metal Spraying Torch

A convenient metal spraying unit is the oxy-acetylene torch shown in Fig. 31-15. A hopper mounted on top of the torch body feeds powder metal alloy into the gas stream. The flow of the powder alloy is controlled by a thumb-operated lever. Alloy particles become molten as they through the flame and onto the workpiece. The same flame is used to pre-heat the work, to spray the part, and to fuse the deposited powder.

Fig. 31-15. An oxyacetylene torch metal spraying unit. (Wall Colmonoy Corp.)

Fig. 31-14. Electric-arc metallizing equipment. (Wall Colmonoy Corp.)

Points to Remember

1. Determine if the part must withstand impact or abrasion or both.

2. Use a tungsten carbide or chromium carbide surfacing electrode for parts that must withstand abrasion.

3. Use high-carbon or manganese electrodes for parts which will be subjected to moderate abrasion and impact.

4. Use special electrodes for hardfacing parts which must withstand severe impact loads.

5. Clean the surface thoroughly before hardfacing.

6. Place the work in a flat position.

7. Use the carbon arc when facing thin-edge materials.

8. Use a minimum amount of heat with both carbon-arc and metallic-arc.

9. Do not allow the coating of the electrode to contact the base metal.

10. With the metallic-arc, be sure to remove all slag before depositing all additional layers.

11. If the shielded metal-arc is employed to hardface thin-edge material, use a weaving or whipping action.

12. When using the carbon-arc, spread the paste evenly two to three times the thickness of the final deposit.

13. Move the carbon electrode in a circular motion and concentrate the heat on the thicker section.

QUESTIONS FOR STUDY AND DISCUSSION

1. What is hardfacing?

2. Of what value is hardfacing?

3. What types of wear do parts encounter in service?

4. Why must the correct type of electrode be used for hardfacing?

5. Why are tungsten carbide or chromium carbide electrodes used for hardfacing parts which must withstand heavy abrasion forces?

6. What are the advantages of chromium carbide electrodes over tungsten carbide?

7. For what type of surfacing are high-carbon electrodes used?

8. When are stainless steel electrodes used for hardfacing?

9. What are some factors in choosing the method of hardfacing?

10. Why should excessive heat be avoided in hardfacing?

11. Why should hardfacing be done in a flat position?

12. How should the shielded metal-arc be manipulated in hardfacing large objects where a high deposition rate is required?

13. When is a whipping action used in hardfacing with a shielded metal-arc?

14. How is hardfacing powder applied when using a carbon arc?

15. How thick should the hardfacing material be spread?

16. How should the carbon arc be manipulated in hardfacing?

17. What is the difference between hardfacing and metallizing?

18. When is the submerged-arc process used for hardfacing?

19. How does hardfacing with the plasma arc differ from that produced by the regular shielded metal-arc process?

20. When is it more advantageous to use oxyacetylene for hardfacing purposes.

CHAPTER 32 pipe welding

Pipe of all types and sizes is used a great deal today in transporting oil, gas, and water. It is used extensively for piping systems in buildings, refineries, and industrial plants. Furthermore, pipe has gained acceptance in construction, and often takes the place of beams, channels, angles, and other standard shapes. See Fig. 32-1.

Welding is the easiest and simplest method of joining sections of pipe together since it eliminates complicated threaded joint designs, permits free flow of liquids, and reduces installation costs.

Welding is also considered a practical and effective cost-cutting technique in joining non-critical low-pressure piping systems for refrigeration, air-conditioning or heating applications.

PIPE WELDING PROCESSES

Although some pipe sizes are occasionally welded with oxy-acetylene, most pipe welding is done with the shielded metal arc. However, a considerable amount of pipe is currently being welded with Mig, either manually, semi-automatic or fully automatic.

Pipe welding is recognized as a specialty in itself. See Fig. 32-2. Although many of the skills and practices are similar to other types of welding, pipe welders usually must develop certain techniques that are characteristic of pipe welding alone. Furthermore, since public health, environmental restrictions and safety are involved, especially in welding cross-country transmission

Fig. 32-1. Welding is used extensively in joining pipe. (Hobart Brothers Co.)

Fig. 32-2. Pipe welding is recognized as a trade in itself.

pipelines and high-pressure lines that are to convey steam, oil, air, and corrosive materials, pipe welders always have to pass certain tests to be certified.

It is impossible to list here any one set of specific certification standards because they will vary for different welding jobs and often are supplemented or modified by local specifications. Accordingly, anyone interested in qualifying as a pipe welder should be aware of certification requirements as they apply for the pipe welding job involved. Some general certification requirements for several classifications of welding are included in Chapter 40.

Roll and Position Welding

Pipe welding in the field is done in several ways. In one method two or more sections are lined up and tack welded. Special pipe clamps, as shown in Fig. 32-3, are used to hold the pipe in alignment until they are tacked. The weld is then completed in the flat position while helpers rotate the pipe. After the short sections are

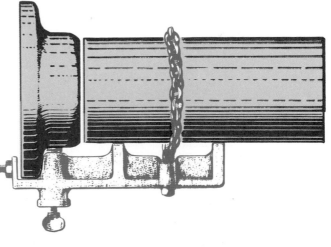

Fig. 32-3. Fast-action clamps hold the pipe in alignment while tack welds are made.

joined, the long pipe is placed in line with the connecting pipe and the weld made with the entire length in a stationary position. This operation is called *roll welding*. See Fig. 32-4.

Fig. 32-4. Roll pipe welding. (Hobart Brothers Co.)

Fig. 32-5. Tig welding is often used to join small diameter pipe. (Hobart Brothers Co.)

The *stove pipe* or position method consists of lining up each section, length by length, and welding each joint while the pipe remains stationary. Since the pipe is not revolved, the welding has to be done in various positions—flat, horizontal, vertical, and overhead.

As mentioned most pipe welding is done either with the shielded metal arc or gas metal arc (Mig). Gas tungsten arc (Tig) is occasionally used in shop welding of small diameter pipes. See Fig. 32-5. However, Tig is also used in certain pipe jobs to lay the root bead in large diameter pipes.

The advantage of Mig over stick welding as described in Chapter 19 is that no slag inclusions will occur in the weld. Furthermore, because of the excellent gas protection shield over the weld area, there is less danger of atmospheric contamination, thereby producing sound welds. Since no slag has to be removed in Mig welding less welding time is required.

Insofar as welding techniques and procedures are concerned there is no significant difference between the shielded metal arc and gas metal arc processes. Therefore, the general description of pipe welding techniques which follow are assumed to apply to both stick and Mig welding.

Pipe Joint Preparation

For welding most pipes a single-V joint is used. The beveling is usually done with a regular oxy-acetylene beveling machine, as indicated in Fig. 32-6.

The standard joint specifications for thick-wall and thin-wall pipe are shown in Fig. 32-7. Generally, pipes having wall thicknesses of 1/8" to 5/16" are classified as thin-wall pipe and pipes with wall thicknesses over 5/16" as thick-wall pipe regardless of diameter. Notice in Fig. 32-7 how the included angle, root face, and root opening will vary for both thin wall and thick wall pipe.

Small diameter pipes with wall thicknesses of less than 1/8" are normally welded without any edge preparation. The ends are simply butted together with a small separation to insure complete fusion. This classification of pipe is frequently welded by the Mig or Tig methods.

Fig. 32-6. A typical pipe beveling unit. (DND Corp.)

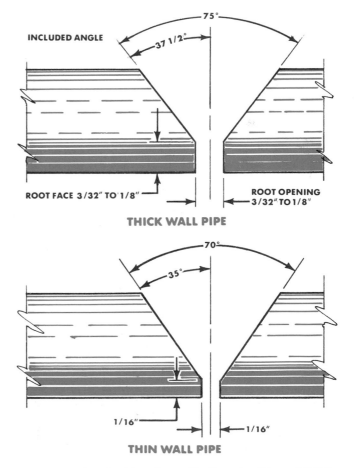

Fig. 32-7. Standard bevel specifications for thin and thick wall pipe.

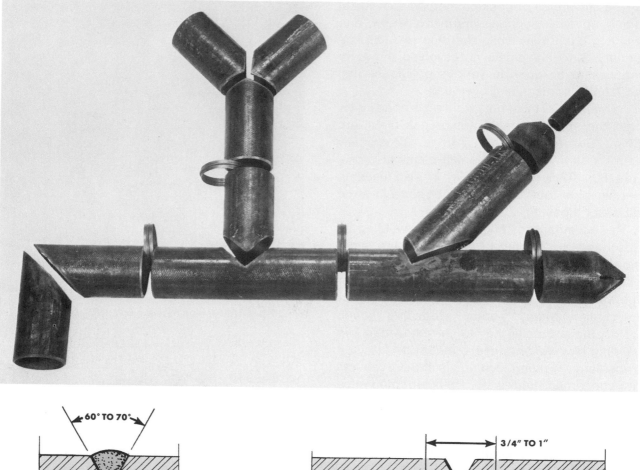

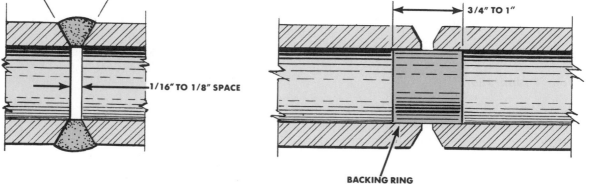

Fig. 32-8. Backing rings are sometimes fitted in small diameter pipe before welding. (Hobart Brothers Co.)

Occasionally, liners or backing rings are fitted into the pipe before welding. These rings assist the welder to secure penetration without burning through the surface as well as to prevent spatter and slag from entering the pipe at the joint. Backing rings also are useful to keep pipes in alignment and stop metal icicles from forming on the inside of the joint, Fig. 32-8.

Tack welding. Before welding, pipes are properly aligned and then tack welded. Special line-up clamps are used to insure correct alignment. See Fig. 32-9.

To maintain the required spacing between pipe sections a spacing tool is necessary. A wire of correct diameter placed between the pipes will provide the proper spacing.

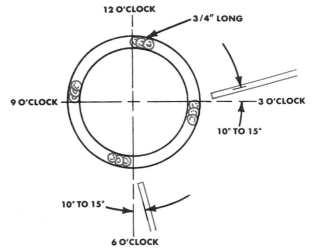

Fig. 32-10. Tack welding a pipe.

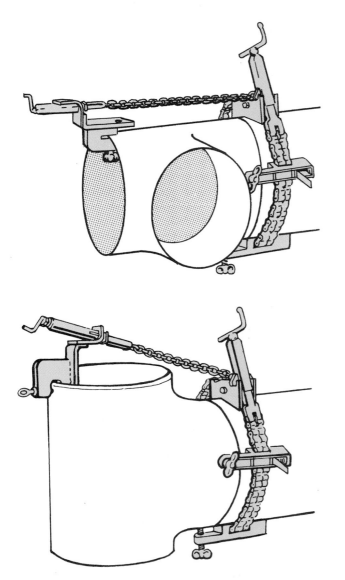

Fig. 32-9. Special clamps hold the pipe in alignment while tack welds are made. (CRC-Crose International, Inc.)

For most pipe welding, four tack welds are made. These tack welds are evenly spaced around the pipe and approximately ³/₄″ long. Tack welds should penetrate to the root of the weld since they become part of the root bead.

To make a tack weld, the electrode is inclined 10 to 15 degrees as shown in Fig. 32-10. The arc is struck in the joint slightly ahead of where the groove is to be made. Then the arc is quickly lengthened so as to stabilize it and give it time to form the protective gas shield. Now the rod is

pushed into the joint with a light pressure and a sliding motion started in the groove. If the electrode has a tendency to stick, it should be wiggled slightly but kept buried in the groove. When the tack weld is completed the electrode is pulled away. This procedure produces a strong and fully penetrating tack weld.

Welding Thin-Wall Pipe (Downhill Technique)

Most thin-wall pipe (¹/₈″–⁵/₁₆″) is welded by using the downhill technique. It is usually preferred for welding cross-country pipelines because it is faster.

The weld is started at the top or 12 o'clock position and carried downward to the bottom or 6 o'clock point of the pipe. After the 6 o'clock position is reached the same procedure is followed on the opposite side. See Fig. 32-11.

One of the problems in downhill welding is controlling the heat input. This is particularly true in welding small diameter pipe where the heat does not dissipate fast enough and excessive heat builds up in the weld zone. Generally heat input can be regulated by using a smaller diameter electrode and reducing the current setting.

Another problem in downhill welding is proper control of the puddle. The molten metal tends to flow downward in the same direction the arc is

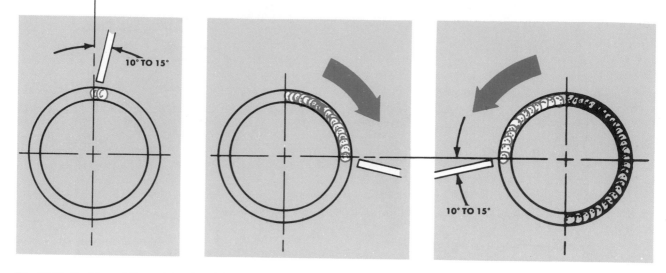

Fig. 32-11. Position of the rod for downhill welding.

moving. If this is not controlled penetration cannot be achieved and slag becomes entrapped in the molten metal, thereby producing slag inclusions in the weld. Slag inclusion, of course, is no problem in Mig welding. Control of metal flow is accomplished by keeping the arc ahead of the puddle. This can be done by using a fast travel speed and a high-current setting.

Welding procedure. After the pipes are securely tacked, a root bead is made completely around the joint. The electrode is held in approximately the same position as in making tack welds. The arc is struck slightly ahead of the weld to preheat the area where the weld bead is to be started. After the arc has stabilized, the electrode is lowered into the root opening and dragged along the edges of the groove. If the electrode has a tendency to stick and fails to glide smoothly because of the built-up heat, a slight side-to-side oscillating motion will usually correct the problem.

A properly deposited root bead should penetrate to the root and leave a solid bead below the surface with a slight crown not to exceed ¹/₁₆″. See Fig. 32-12. Some undercutting may occur on the faces of the groove but this is not objectionable since this defect will be eliminated by successive passes.

There will be times when the root opening will vary due to poor fit-up. If the root opening is

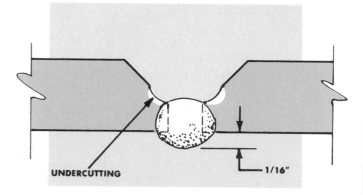

Fig. 32-12. Correct root bead penetration.

narrow the speed of travel and electrode angle should be reduced. Where a widened root opening exists the travel should be increased.

The success of a pipe weld depends on the correct penetration of the root bead because it forms the base upon which the successive layers are made. Without full penetration the final weld joint will not be sound.

Starting and stopping. Since there is a certain amount of starting and stopping a weld, due to changing electrodes or weld position, careful attention must be given to tying the ends of the weld together. To restart a weld the arc is struck about ¹/₂″ back of the bead and then moved forward with a long arc. As soon as the arc is stabilized the electrode is momentarily buried in the crater of the last beads to generate a pool of molten metal. The electrode is then raised slightly and the weld continued.

When a weld approaches the end and must be tied into the other deposited bead, the electrode is moved up the sloping sides of the previous bead and after the molten puddle blends smoothly between the two beads the direction of travel is briefly reversed. The arc is then withdrawn quickly by flicking the electrode downward and away from the center.

Successive passes. Upon completion of the root bead, additional layers of weld are deposited. The number of passes will depend on the thickness of the pipe. Usually the subsequent layers will consist of one hot pass, one or more filler passes and a final *cover* or *cap pass.* The specific function of the *hot pass* is to burn out the remaining particles of slag that may exist in the groove and to achieve a more complete fusion of the base metal and the root bead. This pass usually consists of only a light bead deposit and is made with a whipping motion as shown in Fig. 32-13. The electrode is moved down and up a distance of approximately 1½ electrode diam-

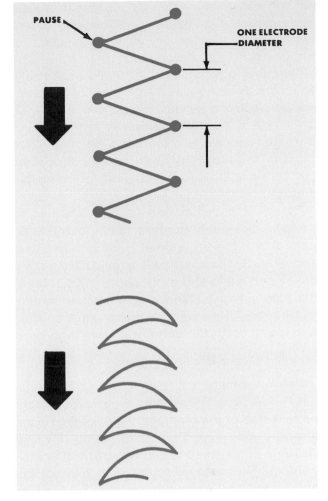

Fig. 32-14. Motion used in filler passes.

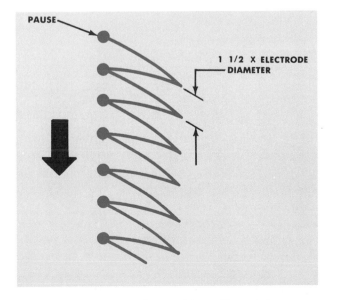

Fig. 32-13. Whipping motion used in making a hot pass.

eters, pausing momentarily on top of the up motion. The whipping action permits better control of the weld puddle and forces the metal to flow into the undercuts. The hot pass is made with the same diameter electrode used for the root bead but with slightly higher current.

The intermediate or filler passes are deposited with larger diameter electrodes and are intended to fill the weld joint. After each pass is completed the slag must be entirely removed if stick welding is used. Each layer should start and end at a different point to insure a uniform strong weld bond. Thus if the first pass is made at the 12 o'clock position the next pass should begin about one inch below this spot.

Filler passes are normally made with a slant motion. See Fig. 32-14 top. The side-to-side weave should travel downward a distance of about one electrode diameter per stroke. The electrode should pause at the end of each stroke to insure good fusion at each edge of the weld. As the electrode reaches the bottom of the weld or in the 6 o'clock position a semi-circle or horseshoe weave is often used. See Fig. 32-14 bottom. This motion permits better control of the puddle since it reduces the fluidity of the molten metal.

The final cover or cap pass is intended to provide maximum reinforcement to the weld joint and at the same time give the weld a neat appearance. The cover pass should have a slight crown extending about 1/16″ above the surface of the pipe. Either a slant or semi-circular motion can be used. However, the weave must be wide enough to cover the entire weld joint.

Welding Heavy-Wall Pipe (Uphill Technique)

Uphill welding is basically intended to join heavy-wall pipe. The welding progresses upward on one side of the pipe and then upward on the opposite side. See Fig. 32-15. As in downhill welding, a root bead is deposited first after tack welding. The weld is started just back of the bottom position or what is commonly known as

the 6:30 position. Actually the arc is struck ahead of this spot and a long arc maintained for a short period to preheat the surface, then it is brought back and the weld begun. While the root bead is being deposited, no electrode weaving motion is necessary. The electrode is simply advanced with a slow and uniform movement along the joint. As the electrode approaches the upper level, the molten metal begins to flow downward at a faster rate. When this happens a slight whipping action is desirable to achieve better puddle control.

After the root bead is completed, one or more filler layers are deposited followed by the final cover pass. Both fill and cover passes are made with a slant weave as described in downhill welding.

AUTOMATIC MIG PIPE WELDING

A considerable amount of large diameter (24″ and over) pipe is welded with automatic Mig welding equipment, Fig. 32-16. Automatic Mig welders not only speed up the welding process but produce welds without any danger of slag inclusions which constantly must be guarded against in regular stick welding.

Unlike the conventional pipe welding proce-

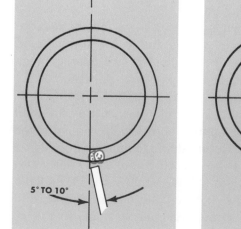 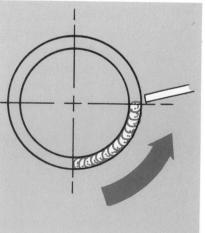 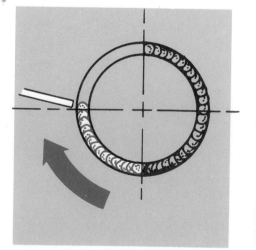

Fig. 32-15. Uphill welding.

Fig. 32-16. Automatic field welding. (Crutcher Resources Corp.)

dure where the root bead is deposited externally, in automatic welding the root bead is applied inside the pipe. A special bevel is made on the pipe for this purpose. See Fig. 32-17. Usually four welding heads mounted on an internal line-up clamp are used to make the internal root bead in a single pass.

The internal welding unit is self-propelled through the pipe and held by clamp shoes in place extending from the pipe. See Fig. 32-18. The welding heads are positioned precisely over the joints by means of special aligner blocks. Once the unit is correctly stationed, the next section of pipe is slipped over the reach rod of the unit. The joint is now properly spaced and another set of clamp shoes is actuated to hold the joint in place for welding. Then a button on the control box mounted on the handle of the reach rod is pushed and the welding commences.

Each welding head welds a 90° arc. All welds are made downhill with two heads moving clock-

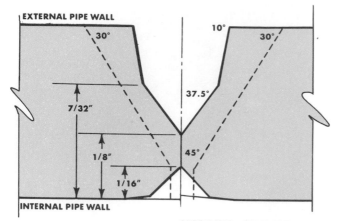

Fig. 32-17. Bevel for automatic versus conventional pipe welding. (Crutcher Resources Corp.)

Fig. 32-18. This internal welding unit is used to make the initial root bead. (Crutcher Resources Corp.)

Fig. 32-19. External welding bugs are used to make the hot fill-and-cover passes. (Crutcher Resources Corp.)

wise and the other two counterclockwise. Shielding gas for internal welding consists of 75 percent argon and 25 percent CO_2.

External welding. The external welding process includes the hot pass, fill pass, and cap pass. These are made with special welding units traveling externally around the pipe on prepositioned circumferential pipe bands. Two welding machines, sometimes referred to as bugs, move simultaneously on the pipe. One starts at 12 o'clock and travels downward to 6 o'clock. The other welding head starts at the 3 or 9 o'clock position and stops at 6 o'clock. The bug is then moved to the 12 o'clock position to complete the pass at the horizontal. See Fig. 32-19. All external welds are made with 100 percent CO_2, because of its higher rate of deposition and it produces better penetrating qualities.

Points to Remember

1. Most pipe welding jobs require that you be certified.
2. Usually small diameter pipe with wall thickness of less than $1/8''$ are not beveled.
3. Tack the pipes before welding them.
4. Use a down-hill technique to weld thin-wall pipe.
5. Be sure the root bead completely penetrates into the root of the joint.
6. Make certain that the ends of the weld are always tied together.
7. Use sufficient passes to fill the weld joint.
8. Finish a pipe weld with a final cap pass.
9. Use an uphill technique on heavy-wall pipe.
10. In uphill welding, start the weld just back of the 6:30 position.

QUESTIONS FOR STUDY AND DISCUSSION

1. Why do pipe welders usually have to meet stiff certification requirements?
2. How is thin-wall pipe distinguished from thick-wall pipe?
3. What is the function of a backing ring?
4. As a rule, how many tack welds are made on pipe?
5. When making a tack weld and the electrode sticks in the groove, what should be done?
6. Why is proper root opening very important in pipe welding?
7. What is meant when we say the weld should be started at the 6 o'clock position?
8. What is the difference between *uphill* and *downhill* welding?
9. What are some of the problems which may be encountered in downhill welding?
10. Downhill welding is used for welding what kind of pipe?
11. What is a root bead?
12. What is meant by tying the ends of a weld together?
13. When is the hot pass used and what is its function?
14. How many filler passes are deposited in a pipe weld?
15. What is the function of the cap pass?
16. Why is a whipping action of the electrode sometimes used in making a root bead?
17. Why should each layer start and stop at different points?
18. What rod motions are used in making fill passes?
19. At what angle should the electrode be held for downhill welding?
20. Why is Mig being used more and more in pipe welding?

CHAPTER 33 *cutting operations*

Metal can be cut with an oxy-acetylene flame, metallic-arc, carbon-arc, plasma-arc or arc air. The type of process used depends largely on the kind of metal to be severed or the economy of the operation. The cutting may be done manually or with mechanized equipment. In manual cutting, the operator manipulates a torch over the area to be cut. In machine cutting the torch is guided entirely by automatic controls.

Flame Cutting

Cutting metal by the oxy-acetylene process is widely used in many industrial fields. The cutting is done by means of a simple, hand cutting torch or by a more complicated, automatically controlled cutting machine as shown in Fig. 33-1.

Severing metal by this process is made possible by the fact that ferrous metals are subject to oxidation. When a piece of wrought iron or steel is left exposed to various atmospheric conditions, a reaction known as rusting begins to take place. Rust is simply the result of the oxygen in the air uniting with the metal, causing it gradually to change its nature and sometimes to wear away. Naturally this action is very slow. But if the metal is heated and permitted to cool, heavy rust scales form on the surfaces, showing that the iron oxidizes much faster when subjected to heat. Now, if a piece of steel were to be heated red hot and dropped in a vessel containing oxygen, a burning action would immediately take place, reducing the metal to an iron oxide commonly known as slag.

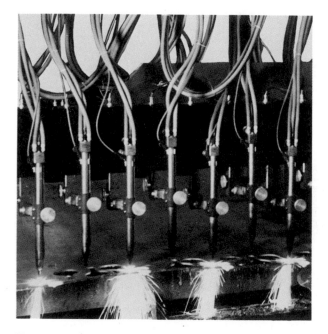

Fig. 33-1. An automatic flame-cutting machine. (Linde Co.)

In order to make possible the rapid cutting of metal, it is necessary to have an implement that will heat the iron or steel to a certain temperature and then throw a blast of oxygen on the heated section. The cutting torch, whether operated manually or by machine, functions in just such a manner. See Fig. 33-2.

The cutting torch. The cutting torch varies from the regular welding blowpipe in that it has an additional lever for the control of the oxygen used to burn the metal. It is possible to convert

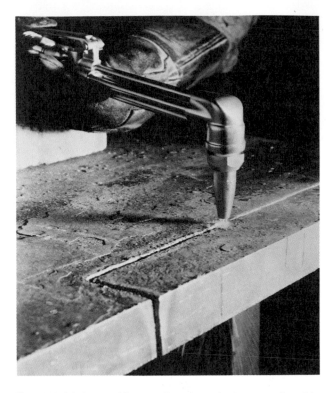

Fig. 33-2. Little metal is actually removed when severing this 1¹/₂″ plate by oxygen cutting. (Linde Co.)

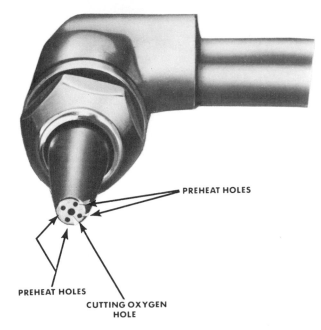

Fig. 33-4. The cutting tip for a flame-cutting torch. (Linde Co.)

the welding torch into a cutting torch by replacing the mixing head with a cutting attachment.

It will be noticed from Fig. 33-3 that the torch has the conventional oxygen and acetylene needle valves. These are used to control the passage of oxygen and acetylene when heating the metal. Many cutting torches have two oxygen needle valves for securing a finer adjustment of the neutral flame. The cutting tip is made with an orifice in the center surrounded by several smaller ones. The center opening permits the flow of the cutting oxygen and the smaller holes are for the heating flame. See Fig. 33-4.

A number of different tip sizes are provided for cutting metals or varying thicknesses. In addition, special tips are made for other purposes, such as for cleaning metal, cutting rusty, scaly or painted surfaces, rivet washing, etc.

Determining pressure. The pressure of oxygen and acetylene needed will depend upon the

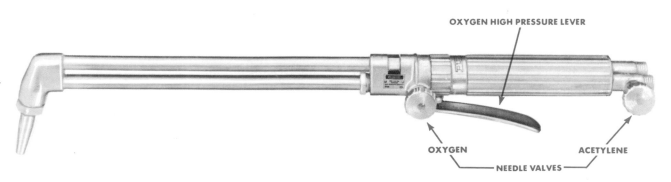

Fig. 33-3. A flame-cutting torch with one oxygen needle valve (left) and one acetylene needle valve (right). (Linde Co.)

TABLE 33-1.   CUTTING PRESSURE.

TIP NO.	THICKNESS (inches)	ACETYLENE PRESSURE (pounds)	OXYGEN PRESSURE (pounds)
0	1/4	3	30
1	3/8	3	30
1	1/2	3	40
2	3/4	3	40
2	1	3	50
3	1 1/2	3	45
4	2	3	50
5	3	4	45
5	4	4	60
6	5	5	50
6	6	5	55
7	8	6	60
7	10	6	70

Air Reduction Co.

Fig. 33-5. Starting the cut. (Linde Co.)

size tip used, which in turn is governed by the thickness of the metal to be cut. Table 33-1 points out the approximate pressure for various tip sizes for a particular type of cutting torch.

Always consult the manufacturer's recommendations for your particular torch as to the proper oxygen and acetylene pressure. The given oxygen pressure cannot always be strictly followed because, for example, steels may have an exceptionally heavy coating of rust or scale and may require a somewhat greater oxygen pressure to burn entirely through the metal.

Lighting the torch. Use this procedure to light the torch:

1. Turn on the acetylene needle valve and light the gas with a spark lighter as if for welding.

2. Turn on the oxygen valve and adjust it for a neutral flame. This flame is the one used to bring the metal to a kindling temperature which, for example in the case of plain carbon steel, is 1400°F to 1600°F (760° to 871°C).

3. Observe the nature of the cutting flame by pressing down the oxygen control lever. When the oxygen pressure lever is turned on it may be necessary to make an additional adjustment to keep the preheating cones burning with a neutral flame.

Cutting steel. It is well for the beginner to practice making straight cuts. To aid in keeping the cut in a horizontal line:

1. Rule a chalk line about 3/4" from one edge of the plate.

2. Place the plate so this line clears the edge of the welding bench as shown in Fig. 33-5.

3. With the torch adjusted to a neutral flame, grasp the blowpipe handle with the right hand in such a position as to permit instant access to the oxygen control lever. The valve is usually operated either with the thumb or forefinger.

4. The knack of making a clean, straight cut depends on how steady you hold the torch. Naturally, when the tip wavers from side to side, a wide kerf (cutting slit) will result, which means a rough cut, slower speed, and greater oxygen consumption. To help keep the blowpipe steady, hold the elbow or forearm on some convenient support.

5. Start the cut at the edge of the plate. Hold the torch with the tip vertical to the surface of the metal, with the inner cone of the heating flame approximately 1/16" above the chalk line. Keep the torch in this position until a spot in the metal has been heated to a bright red heat.

6. Gradually press down the oxygen pressure lever and move the torch forward slowly along the chalk line. See Fig. 33-6. The movement of the torch should be just rapid enough to insure a

Fig. 33-6. This is how a correct cut appears. The shower of sparks under the cut shows complete penetration. (Linde Co.)

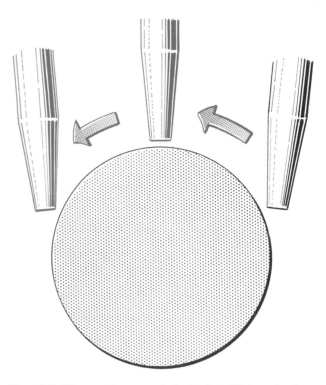

Fig. 33-7. When cutting round stock, start 90° from the top, and follow around the bar, as indicated by the arrow.

fast but continuous cut. A shower of sparks will be seen to fall from the underside, indicating that the penetration is complete and the cut is proceeding correctly.

7. If the cut does not seem to go through the metal, close the oxygen pressure lever and reheat the metal until it is a bright red again. If the edges of the cut appear to melt and have a very ragged appearance, the metal is not burning through and the torch is being moved too slowly.

8. When an exceptionally straight cut is desired, clamp a bar across the plate alongside the cutting line to act as a guide for the torch to follow.

9. At the start, the pieces may stick together even when the cut has penetrated through. This is due to the slag, produced by the cutting, flowing across the metal pieces. However, this is not serious, as the slag is quite brittle and a slight blow with a hammer will separate the two sections.

10. It may be necessary sometimes to start the cut in from the edge of the plate. In such a case, hold the preheating flame a little longer on the metal; then raise the cutting nozzle about 1/2″ and pull the oxygen lever. When a hole is cut through, lower the torch to its normal position and proceed with the cut in the usual manner.

Cutting round-bar steel. To sever round stock, start the cut about 90° from the top edge as shown in Fig. 33-7.

Keep the torch in a straight-up (perpendicular) position and gradually lift it to follow the circular outline of the bar. Maintain this position of the blowpipe while ascending as well as descending on the opposite side.

Beveling. To make a bevel cut on a steel plate, incline the head of the torch to the desired angle instead of holding it vertically. An even bevel may be made by resting the edge of the nozzle on the work as a support, as shown in Fig. 33-8, or be guided by means of a piece of angle iron clamped across the plate.

Piercing holes. Hold the torch over the spot where the hole is to be cut until the flame has heated a small, round spot. Gradually pull the oxygen lever and at the same time raise the nozzle slightly. In this manner a small, round hole can be pierced quickly through the metal.

When larger holes are required, trace the shape of the openings with a piece of chalk. If the holes are located away from the edge of the plate, first pierce a small hole, and then start the

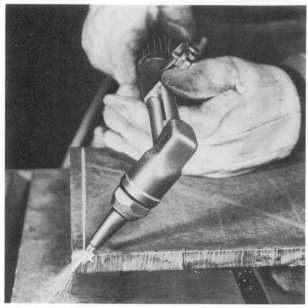

Fig. 33-8. Beveling with a cutting torch.

Fig. 33-9. Cutting holes (top) and large curves (bottom). (Linde Co.)

cut from this point, gradually working to the chalk line and continuing around the outline. For holes near the outer edge of the plate, proceed as shown in Fig. 33-9.

Cutting cast iron with oxy-acetylene.[1] To describe this process intelligently, we must first consider the nature of cast iron from two angles. Since it has such a wide range of use, it is natural that we may expect a wide difference in quality or chemical composition and a corresponding difference in ease of cutting. The better grades of castings are more easily cut, and those of random grades of scrap, such as counterweights, gratebars, floorplates, etc., present greater difficulty, requiring more gas, a wider kerf, and a corresponding slower rate of cutting speed.

Preparation. There is a rule for heavy steel cutting which certainly applies in every way to cast-iron cutting, and that is,

DO NOT START THE CUT UNTIL YOU ARE CERTAIN THAT YOU CAN COMPLETE IT.

It is obvious that, if the cut is stopped on a heavy section, it is extremely difficult to start again, and may doubly increase the cost as well as the annoyance. Considerably more heat, as well as sparks and slag, is generated in cast iron, so that proper protection to the body, face, and limbs is necessary. Asbestos gloves are essential, and a fire brick or suitable torch rest is desirable.

Cutting procedure:

1. For cutting the better grades of cast or gray iron, set the regulator to deliver the proper pressure as indicated in Table 33-2. The regulator adjustment, of course, is made with the high-pressure cutting valve open to compensate for the customary drop in pressure when the gas is released.

2. Next, light the torch and adjust the preheating flame so that it will show an excess of acetylene. This is done preferably with the high-pressure valve wide open to avoid any change in the character of the flame during the actual cutting operation. This is important!

The excess acetylene, as indicated by the

---

1. The Modern Engineering Co.

TABLE 33-2. PRESSURE TABLE FOR CUTTING CAST IRON.

TIP SIZE	THICKNESS (inches)	OXYGEN PRESSURE (pounds)	ACETYLENE PRESSURE (pounds)
	1/2	40	
	3/4	45	
L-3	1	50	7–8
	1 1/2	60	
	2	70	
	3	80	
	4	90	
L-4	6	110	8–10
	8	120	
	10	150	
	12	170	

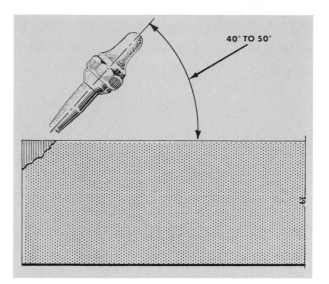

Fig. 33-11. Position the torch as shown here to start cutting cast iron.

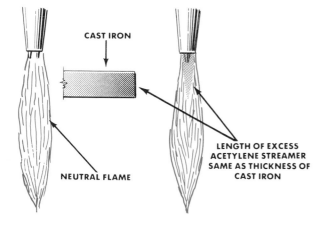

Fig. 33-10. For cutting cast iron use a flame with an excess of acetylene.

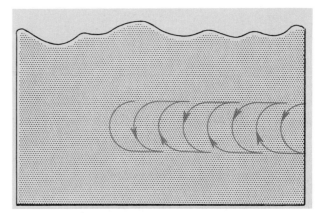

Fig. 33-12. Movement of the torch for cutting cast iron.

length of the white cone, must be varied to best suit the grade and thickness of the material cut. Experience is naturally the best guide on this point. However, it will generally vary from little or no excess of acetylene for the extremely light sections to an excess indicated by a one to two inch white cone for the heavier sections cut. See Fig. 33-10.

3. Bring the tip of the torch to the top or starting point as indicated in Fig. 33-11. Hold the torch on an angle of approximately 40° to 50° and

heat a spot about 1/2″ in diameter to a molten condition.

4. With the end of the shorter, preheating cone about 3/16″ from the metal, start to move the torch with a swinging motion, as shown in Fig. 33-12, and open the high-pressure cutting valve.

5. Gradually bring the torch along the line of the cut, continuing the swinging motion. As the cut progresses, gradually straighten the torch to an angle of 65° to 70°, as shown in Fig. 33-13, which will help to facilitate the penetration.

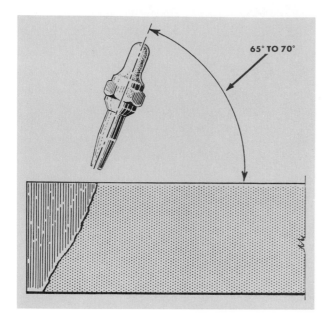

Fig. 33-13. Advanced position in cutting cast iron.

Fig. 33-14. Making a patterned cut with oxygen flame. (Cincinnati Milacron, Inc.)

6. The same swinging movement is continued throughout the entire length of the cut. As you gain experience and confidence, you will be able to reduce somewhat the length of the arc of this swinging motion, thereby reducing the width of the kerf and the amount of gas consumed.

7. On the heavier sections, sufficient heat is usually developed to allow the cut to proceed without interruption. On the lighter sections, however, more difficulty is experienced. The operator is frequently bothered with losing his cut; in other words, the surface of the metal cools too rapidly and only a slight groove is made in it by the flame.

Start over again by heating a small circle as previously described, but gradually raise the torch and incline it so as to cut away the lower portion of the section. Then proceed as before. As long as the exposed side of the cutting groove appears bright, continue the cut until it is completed.

Oxygen flame cutting is not limited to a hand operation. In Fig. 33-14 a skilled operator uses a mechanized oxygen flame to cut through a steel plate. The torch is automatically guided over the work from a pattern followed by the operator.

Powder Cutting

The regular oxygen cutting process is ineffective for cutting metals that form refractory oxides, such as aluminum, bronzes, and high-nickel alloys. To cut these metals an iron powder is fed into the oxygen stream. A mixture of iron powder and aluminum powder is sometimes used for cutting brass, copper, and high-nickel alloys. The aluminum releases more heat than just the iron powder alone. Iron powder. also produces a rapid cutting action on stainless steels and is very effective in getting smoother cuts in cast iron.

Powder cutting is done with a special powder-cutting torch as shown in Fig. 33-15. The torch is equipped with a powder tube, nozzle, and powder valve. The powder is stored in a dispenser and is carried to the powder valve by compressed air or nitrogen where it is fed to the flame. In operation the powder valve is opened first and then the oxygen valve.

Fig. 33-15. Powder-cutting torch. (Linde Co.)

Cutting with the Shielded Metal-Arc

Cutting with the arc does not produce a smooth, precision cut as with the gas cutting torch but it is quick and reasonably economical. The arc is particularly effective for cutting cast iron and steel for salvage purposes, small pieces, and areas which are hard to reach.

It is easy to understand why cutting with the arc is possible. Heat of the arc ranges from about 6500° to 10,000°F (about 3600° to 5500°C) whereas steel, for example, melts at about 2400°F (1315°C). Since cutting actually is a melting process, the excessive heat of the arc readily melts the metal.

Coated mild steel electrodes are used for cutting purposes with either AC or DC machines. Follow this procedure to carry out a cutting operation:

1. Set the machine for the same polarity as in welding for the electrode selected.

2. Use a mild steel coated electrode such as E-6010 or E-6011. The diameter of the electrode will depend on the thickness of the metal to be cut and the amperage capacity of the machine. For most general purpose cutting use $^3/_{32}$" electrodes for metals up to $^1/_8$" in thickness and $^5/_{32}$" electrodes for materials which exceed $^1/_4$" in thickness. Table 33-3 gives the approximate ampere setting for cutting.

Keep in mind that the amperage settings listed in Table 33-3 apply only to machines limited to 180 amperes. If you are using machines with

TABLE 33-3. SUGGESTED AMPERE SETTING FOR CUTTING.

THICKNESS (inches)	ELECTRODE dia (inches)	CURRENT RANGE (amps)
Up to 1/8	3/32	75–100
Up to 1/8	1/8	125–140
Up to 1/4	5/32	140–180

higher capacity, you can employ larger diameter electrodes with higher ampere settings.

3. Place the metal in a flat position. Start the cut at the bottom edge of the plate; that is, the edge in the lowest position. When the diameter of the electrode is larger than the thickness of the plate being cut, simply move the electrode in a straight line. Check Fig. 33-16 and note the position of the electrode.

4. When the metal is heavier than the electrode, use a weaving motion to make the cut. Move the electrode with a quick upward motion and then push the electrode downward as shown in Fig. 33-17. The downward movement helps to force the molten metal out of the slot.

5. Flat stock over $^1/_8$" in thickness is often easier to cut if placed in a vertical position.

6. To cut round stock, start the cutting at the outside edge so the metal can flow from the round bar. Carry the cut to the center of the

Fig. 33-16. To cut thin stock, move the electrode in a straight line. Use a guide to insure a straight cut.

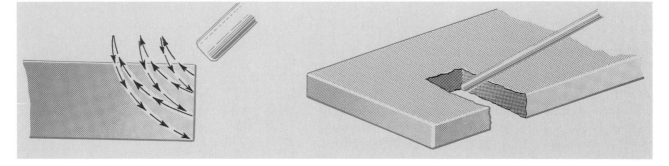

Fig. 33-17. Use a quick upward motion and then push the molten metal out with a downward motion.

Fig. 33-18. Cut round stock from each side. (The Lincoln Electric Co.)

piece and then start a new cut on the opposite side as shown in Fig. 33-18.

Piercing holes. To burn a small hole through thin stock, strike the arc and keep a long arc over the spot until the plate begins to sweat.

Next, bring the arc down into the molten pool of metal, moving the electrode in a circular motion as illustrated in Fig. 33-19. Continue the circular motion until the hole is pierced. It is a good idea to have a punch or bolt on hand so it can be driven into the hole while the metal is still hot. This will help true-up the hole.

To pierce holes in metal over $1/4"$ in thickness, place the metal so the hole is made in a vertical

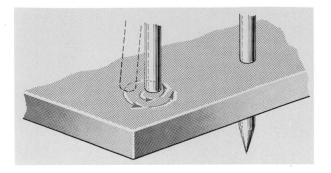

Fig. 33-19. To pierce a hole, bring the arc into the molten pool and move the electrode in a circular motion.

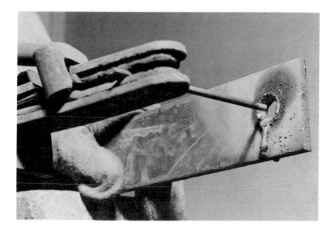

Fig. 33-20. To pierce holes in metal over ¹/₄″, place the metal on edge. (The Lincoln Electric Co.)

Fig. 33-21. For large holes, start the cut in the center and move the electrode around the marked edges.

position. This permits the molten metal to run out of the hole as shown in Fig. 33-20. The metal that adheres to the outside surface is easily removed with a chisel or by the heat of the arc.

Cutting large holes. For cutting large holes, follow this procedure:

1. With a soapstone or center punch lay out the hole to be cut.

2. Pierce a small hole in the center of the area to be cut as shown in Fig. 33-21.

3. Move the electrode from the center around the edge of the marked circle.

Cutting with the Carbon Arc

The carbon arc can also be used effectively for cutting. Just as with the metallic arc, the cutting is a melting process and as such does not produce smooth, even edges. Although the carbon arc will cut as well as the metallic arc, it is not generally as satisfactory for burning holes.

The actual cutting process is similar to that of the metallic arc. The current, as in the case of the metallic arc, must be set to a higher value than ordinarily used in welding material of comparable thickness.

Manipulate the carbon electrode so as to keep the lower side of the slot ahead of the upper edges as shown in Fig. 33-22. Move the arc from top to bottom of the cut to force the molten metal down. See Fig. 33-23.

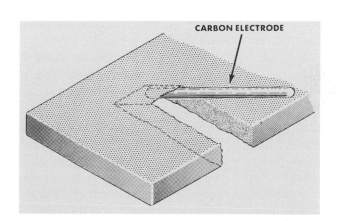

Fig. 33-22. Keep the lower side of the cut ahead of the upper edge.

Fig. 33-23. Cutting with the carbon arc.

Fig 33-24. Plasma arc cutting.

Plasma-Arc Cutting[1]

Plasma-arc cutting is regarded as one of the best processes for high-speed cutting of non-ferrous metals and stainless steels. It cuts carbon steel up to ten times faster than any oxyfuel, with equal quality and with greater economy. See Fig. 33-24.

Plasma is often considered the fourth state of matter. The other three are gas, liquid, and solid. Plasma results when a gas is heated to a high temperature, and changes into positive ions, neutral atoms and negative electrons. When matter passes from one state to another, latent heat is generated. Latent heat is required to change water into steam, and similarly, the plasma torch supplies energy to a gas to change it into plasma. When the plasma changes back to a gas, the heat is released.

In a Plasmarc torch, the tip of the electrode is located within the nozzle. The nozzle has a relatively small opening (orifice) which constricts the arc. The high-pressure gas must flow through the arc where it is heated to the plasma temperature range. Since the gas cannot expand, due to the construction of the nozzle, it is forced through the opening, and emerges in the form of a supersonic jet, hotter than any flame. This heat melts any known metal and its velocity

blasts the molten metal through the kerf. See Fig. 33-25.

Because maximum transfer of heat to work is essential in cutting, Plasmarc torches use a transferred arc (the workpiece itself becomes an electrode in the electrical circuit). The work is thus subjected to both plasma heat and arc heat. Direct current, straight polarity is used. Precise control of the plasma jet is feasible by controlling the variables—current, voltage, type of gas, gas velocity, and gas flow (cfh).

The power supply for cutting is a special rectifier type with an open-circuit rating of 400 volts. A control unit automatically controls the sequence of operations—pilot arc, gas flow, and carriage travel.

A water pressure input of 60 to 80 psi for gas cutting, and 100 psi for air cutting is necessary to keep the torch cool.

When cutting aluminum and stainless steel, best results are obtained with an argon-hydrogen, or nitrogen-hydrogen gas mixture. Carbon steels require an oxidizing gas. Air has proven to be the most efficient gas, however oxygen can also be used.

---

1. Linde Company

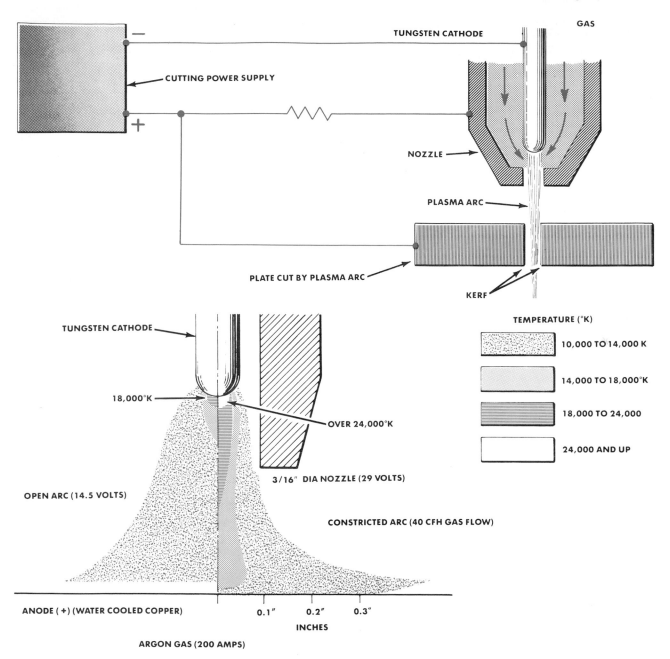

Fig. 33-25. In Plasmarc cutting the gas flowing through the torch produces a supersonic jet which is hotter that any gas produced flame. (Linde Co.)

The first steps in making a Plasmarc cut are— adjust the power supply and the gas flow to the appropriate settings. See Table 33-4. Then, when the operator pushes the START button on the remote-control panel, the control unit performs all ON-OFF and sequencing functions. The cooling water must also be turned on, or the waterflow interlock will block the starting circuit.

To make a mechanized cut, the operator locates the center of the torch about ¼″ above the surface of the plate to be cut and pushes the start button. Current flows from the high-

TABLE 33-4.   TYPICAL PLASMARC CUTTING CONDITIONS.

	THICKNESS (inches)	SPEED (lpm)	ORIFICE TYPE	INSERT (dia)	POWER (kw)	GAS FLOW (cfh)
STAINLESS STEEL	1/2	25	4 x 8	1/8	45	130 N₂
	1/2	70	4 x 8	1/8	60	130 N₂
	1 1/2	25	5 x 10	5/32	85	10 H₂ 175 N₂
	2 1/2	18	8 x 16	1/4	150	15 H₂ 175 N₂
	4	8	8 x 16	1/4	160	15 H₂
ALUMINUM	1/2	25	4 x 8	1/8	50	100*
	1/2	200	4 x 8	1/8	55	100*
	1 1/2	30	5 x 10	3/32	75	100*
	2 1/2	20	5 x 10	5/32	80	150*
	4	12	6 x 12	3/16	90	200*
CARBON STEEL	1/4	200	4 x 12M†	1/8	55	250
	1	50	5 x 14M†	5/32	70	300
	1 1/2	35	6 x 16M†	3/16	100	350
	2	25	6 x 16M†	3/16	100	350

*65 % Argon, 35% Hydrogen Mixture
†Multiport Orifice

frequency generator to establish the pilot arc between the electrode (workpiece) and the cathode in the nozzle. Gas starts to flow, and welding current flows from the power supply.

The pilot arc sets up an ionized path for the cutting arc. As soon as the cutting arc is established, the high-frequency current is shut off, and the carriage starts to move. See Fig. 33-26.

When the cutting operation is completed, the arc goes out because it has no ground, and the control stops the carriage, opens the main contactor, and shuts off the gas flow.

Air Carbon-Arc Cutting

Air carbon-arc cutting is a process wherein the severing of metals is accomplished by melting with the heat of an arc between a carbon electrode and the base metal. Power is supplied either with a regular AC or DC welding machine. However, the power requirements for a given diameter carbon electrode are higher than those for a comparable diameter shielded metal-arc welding. See Table 33-5.

The carbon-graphite electrode is held in a special holder as shown in Fig. 33-27. Plain or copper-clad carbon-graphite rods are used. Plain electrodes are less expensive, but copper clad electrodes last longer, carry higher currents and produce more uniform cuts.

As the metal melts, a jet of compressed air is directed at the point of arcing to blow the molten

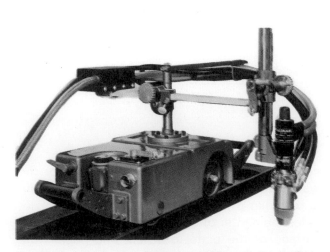

Fig. 33-26. A mechanical Plasmarc cutting unit. (Linde Co.)

TABLE 33-5.   SUGGESTED CURRENT RANGES FOR THE COMMONLY
USED ELECTRODE TYPES AND SIZES.

ELECTRODE DIA		DC ELECTRODE DCRP		AC ELECTRODE AC		AC ELECTRODE DCRP	
(inches)	(mm)	min/A	max/A	min/A	max/A	min/A	max/A
5/32	4.0	90	150	—	—	—	—
3/16	4.8	150	200	150	200	150	180
1/4	6.4	200	400	200	300	200	250
5/16	7.9	250	450	—	—	—	—
3/8	9.5	350	600	300	500	300	400
1/2	12.7	600	1000	400	600	400	500
5/8	15.9	800	1200	—	—	—	—
3/4	19.1	1200	1600	—	—	—	—
1	25.4	1800	2200	—	—	—	—

AWS

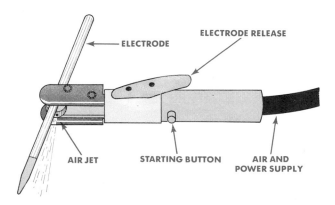

Fig. 33-27. Arc-air cutting electrode holder. (Arcair)

TABLE 33-6.   COMPRESSED AIR PRESSURE AND
CONSUMPTION RATE FOR VARIOUS
ELECTRODE SIZES.

ELECTRODE DIA		AIR PRESSURE	
(inches)	(mm)	(psi)	(kPa)
Under 1/4 (a)	6.4	40	280
Under 1/4 (a)	6.4	80	550
Under 3/8 (b)	9.5	80	550
Under 3/4 (c)	19.1	80	550
Under 5/8 (d)	15.9	80	550
Under 5/8 (e)	15.9	80	550
Under 5/8 (f)	15.9	·80	550

AWS

metal away. The compressed air line is fastened directly to the torch. The jet air stream is controlled by simply depressing a push-button on the holder. See Fig. 33-27.

Air is supplied by an ordinary compressor. In general pressure will range from 40 to 80 psi. See Table 33-6.

Gouging.   Grip the electrode so a maximum of 6″ extends from the electrode holder to the work. For aluminum alloys this distance should be reduced to 4″.

Hold the torch so the electrode slopes back from the direction of travel. The air blast should be behind the electrode. Maintain a short arc and move fast enough to keep up with metal removal. The arc must provide sufficient clearance so the air blast can sweep beneath the electrode and remove all molten metal. The depth and contour of the groove is controlled by the electrode angle and travel speed. For a narrow deep groove, a steep electrode angle and slow speed is used. A flat electrode angle and fast speed produces a wide shallow groove. The width of the groove is governed by the diameter of the electrode. In all instances, proper speed will produce a smooth, hissing sound.

FLAT POSITION

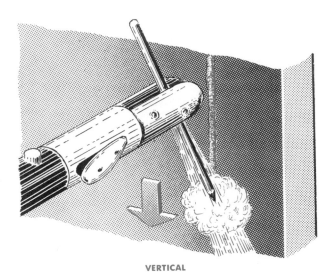

VERTICAL

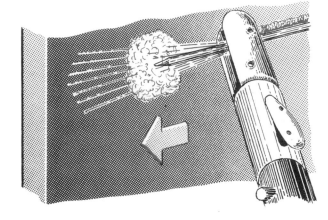

HORIZONTAL

Fig. 33-28. Position of the electrode for gouging. (Arcair)

For gouging in the vertical position, hold the electrode as in Fig. 33-28 and move downward. This permits gravity to assist in removing the molten metal.

Gouging in the horizontal position can be done by moving the electrode either right or left. When traveling to the right, hold the electrode as shown in Fig. 33-28. In gouging to the left, reverse the position of the electrode so the air jet is behind the electrode.

Cutting. The technique for cutting is the same as gouging except that the electrode is held at a steeper angle and is directed at a point that permits the tip of the electrode to pierce the metal being cut.

For cutting thick nonferrous metals, hold the electrode in a vertical position with a lag angle of 45° and, with the air jet above it, move the arc up and down through the metal with a sawing motion. Lag angle refers to the angle the electrode makes behind a line perpendicular to the cut axis at the point of cutting.

Washing. Washing is a process of removing metal from large areas, such as removal of surfacing and of riser pads on castings. In using the air carbon-arc for this purpose, weave the electrode from side to side in a forward direction to the depth desired. A lag angle of 55° is recommended with the air stream behind the electrode. The steadiness of the operator determines the smoothness of the surface produced.

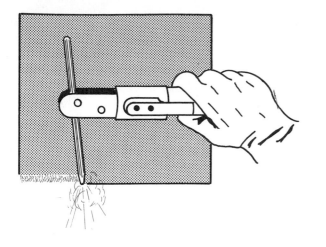 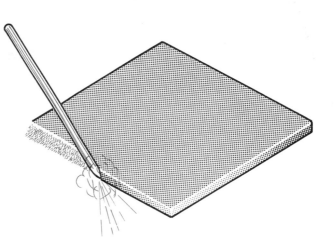

Fig. 33-29. Position of the electrode in beveling.

Beveling. For beveling purposes hold the electrode as shown in Fig. 33-29 left, with the air-blast between the electrode and the metal surface. Draw the electrode smoothly along the edge being beveled. When beveling light plates, hold the electrode as in Fig. 33-29 right with the air-blast behind the electrode in the direction of travel.

Safety Precautions

In any cutting operation, a large amount of metal always falls on the floor.

CAUTION: Be sure there are no combustible materials nearby.

Turn the cuffs of your trousers down over your shoes to prevent the possibility of molten metal getting inside the cuffs or shoes.

When an excessive amount of cutting is to be done, it is a good idea to sprinkle sand over the concrete floor. This prevents the molten metal from heating the concrete so that it cracks and causes particles to fly upward. Another method is to cut over a work bench drawer partially filled with sand. If the bench lacks a drawer, a sand-filled pan can be placed on the floor.

Points to Remember

1. In flame cutting, be sure to use the correct oxygen and acetylene pressure.

2. Keep the preheating cones burning with a neutral flame.

3. Hold the torch with the inner cone of the heating flame about 1/16″ above the metal until a spot is heated to a bright red heat.

4. Move the cutting torch just fast enough to make a fast but continuous cut.

5. If the cut does not go through the metal, start the cutting process over again, beginning at the original site.

6. In cutting cast iron, adjust the preheating flame so it is slightly carburizing.

7. Use an iron powder cutting process in cutting aluminum, bronzes, and the high-nickel alloys.

8. When cutting with the shielded metal-arc, use a mild steel electrode such as E-6010 or E-6011.

9. For high speed cutting of non-ferrous metals, plasma arc cutting is considered to be the most effective.

10. In plasma arc cutting, use direct current with straight polarity.

11. Use plain or copper-clad carbon-graphite rods when cutting metals with the air carbon-arc process.

12. When using any cutting process, be sure there are no combustible materials which might be set on fire.

QUESTIONS FOR STUDY AND DISCUSSION

1. What causes metal to rust?

2. What principle makes possible the cutting of metal by means of oxygen and acetylene?

3. Describe the process for lighting and adjusting the flame for a cutting torch.

4. What is the function of the high-pressure oxygen valve?

5. How does the cutting tip differ from the welding tip?

6. What governs the pressure of oxygen and acetylene that must be used for cutting.

7. As a general rule where should the cut be started? Why?

8. What aids may be used to facilitate an even cut?

9. How can one determine whether the cut is penetrating through the metal?

10. What happens when the cutting torch is moved too slowly?

11. What is the position of the torch when cutting round material?

12. How is it possible to make a bevel cut with a cutting torch?

13. Describe the operation followed in piercing small holes with a cutting torch.

14. What type of flame is used for cutting cast iron assuming a good grade of iron?

15. How is the torch held when cutting cast iron?

16. What torch motion is used for cutting cast iron?

17. Why is it possible to cut metal with a metallic arc?

18. What kind of coated electrodes can be used to cut metal?

19. What determines the diameter of electrodes to be used for cutting?

20. How does the amperage setting for cutting compare with the amperage for welding?

21. Why should the electrode be moved in a weaving motion when the electrode is smaller in diameter than the thickness of the metal?

22. How are holes pierced in metal with electrodes?

23. To pierce holes in metal over $1/4''$ thick, why is it better to do the cutting with the metal in a vertical plane?

24. What is meant by plasmarc cutting?

25. How does air carbon-arc cutting differ from metallic-arc cutting?

26. What are some of the precautions that should be observed before engaging in any cutting operation?

special welding processes

CHAPTER 34 *production welding*

Production welding refers to welding techniques used in the fabrication of goods on a mass production basis. Industries involved in manufacturing such products must rely on welding processes where hand manipulation is kept to a minimum and the joining of metal is performed rapidly and automatically. See Fig. 34-1.

Since production techniques depend on the nature of the goods made, the kind of welding equipment used will vary from one industry to another. Very often special welding machines are designed for a particular industry. Thus an aircraft company may need a spot-welding machine designed to join certain types of aluminum structures. An automotive manufacturer may require a resistance-type seam welder especially made to handle a body structure. Another concern may have to use a stud welding gun to fasten studs on some metal component. It is the purpose of this chapter to describe briefly some of the more common production welding techniques used in industry.

Fig. 34-1. An automatic welder is used in fabricating a missile. (Skiaky Bros., Inc.)

Resistance Welding

Of the many welding techniques applicable to production processes, resistance welding dominates the field. The fundamental principles upon which all resistance welding is based are these: (1) heat is generated by the resistance of the parts to be joined to the passage of a heavy electrical current, (2) this heat at the juncture of the two parts changes the metal to a plastic state, and (3) when combined with the correct amount of pressure, fusion takes place.

There is a close similarity in the construction of all resistance welding machines whether they are of simple design or are very complex and costly. The main difference is in the type of jaws or electrodes which hold the object to be welded. A standard resistance welder has four principal elements:

1. The frame is the main body of the machine which differs in size and shape for both stationary and portable types.

2. The electrical circuit consists of a step-down transformer which reduces the voltage and proportionally increases the amperage to provide the necessary heat at the point of the welding.

3. The electrodes include the mechanism for making and holding contact at the weld area.

4. The timing controls represent the switches which regulate the volume of current, length of current time, and the contact period.

The principal forms of resistance welding are classified as spot welding, seam welding, projection welding, flash welding, and butt welding.

Spot welding. This type of welding is probably the most commonly used type of resistance welding. The material to be joined is placed between two electrodes, pressure is applied, and a charge of electricity is sent from one electrode through the material to the other electrode.

There are three stages in making a spot weld. First the electrodes are brought together against the metal and pressure applied before the current is turned on. Next the current is turned on momentarily. This is followed by the third, or hold time, in which the current is turned off but the pressure continued. The hold time forges the metal while it is cooling.

Regular spot welding usually leaves slight depressions on the metal which are often undesirable on the "show side" of the finished product. These depressions are minimized by the use of larger-sized electrode tips on the show side.

Spot welders are made for both direct and alternating current. The amount of current used is very important. Too little current produces a light tack giving insufficient strength, too much causes burned welds.

To dissipate the heat and cool the weld as quickly as possible, the electrodes are water-cooled.

The two basic types of spot-welding machines are single spot and multiple spot. The single spot, Fig. 34-2, has two long horizontal horns, each holding a single electrode, with the upper arm providing the moving action.

Multiple-spot welders have a series of hydraulic or air-operated welding guns mounted in a framework or header but using a common or bar mandrel for the lower electrode. The guns are connected by flexible bands to individual transformers or to a common buss bar attached to the transformer. See Fig. 34-3. Two or four guns are often attached to a transformer.

Spot welders are utilized extensively for welding steel, and when equipped with an electronic timer, can be used for other commercial metals such as aluminum, copper, and stainless steel. They are also very effective for welding galvanized metal.

Although many spot welders are of the stationary design, there is an increased demand for the portable type. The portable, or spot-welding gun, as it is often called, consists of a welding head connected by flexible cables which run to the transformer. The jaws are operated either manually, pneumatically, or hydraulically. With this apparatus many spot welds may be made on irregular shaped objects as shown in Fig. 34-4.

A recent innovation in the spot welding field is the self-contained portable spot welder as shown in Fig. 34-5. The welding head contains a built-in timer, electrode contactors, and transformer which requires only a 115 volt power connection. It is especially suitable for sheet metal and repair shop auto body welding.

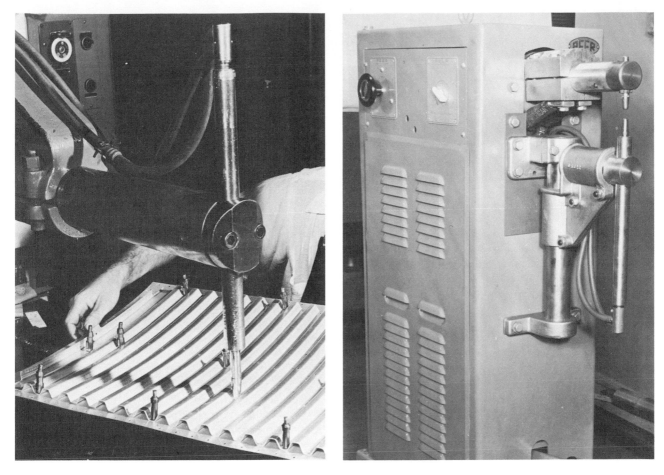

Fig. 34-2. Single-spot type of spot welding machine. (Peer Div., Landis Machine Co.)

Fig. 34-3. A multiple-spot welding machine. (The Taylor-Winfield Co.)

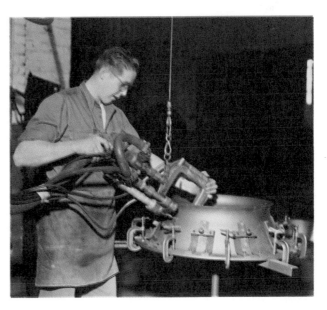

Fig. 34-4. A spot welding gun is used to weld the intake housing of a commercial fan. (Clarage Fan Co.)

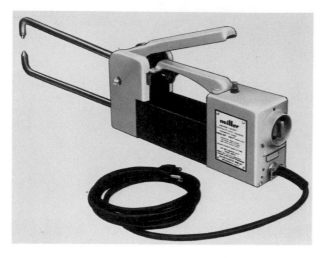

Fig. 34-5. A self-contained portable spot welder. (Miller Electric Manufacturing Co.)

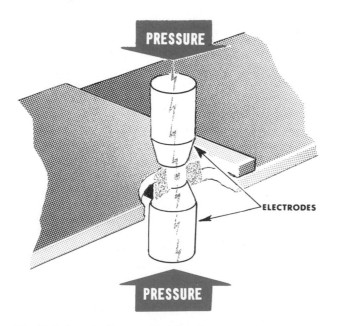

Fig. 34-6. In pulsation spot-welding the current flow is interrupted by precise electronic control. (Republic Steel Corp.)

On spot welders, the electrodes which conduct the current and apply the pressure are made of low resistance copper alloy and are usually hollow to facilitate water cooling. These electrodes must be kept clean and shaped correctly to produce good results. Thus if a 1/4" dia.

electrode face is allowed to increase to 3/8" by wear or mushrooming, the contact area is doubled with a corresponding decrease in current density. Unless this is compensated for by an increase in current setting, the result will be weak welds. Additional factors which cause poor welds are: misalignment of electrodes, improper electrode pressure, and convex or concave electrode surfaces.

Pulsation welding. This is a form of spot welding. In regular spot welding, interruption of the flow of welding current is controlled manually; with pulsation welding the current is regulated to go on and off a given number of times during the process of making one weld. This method permits spot welding thicker material as well as to increase the life of the electrode. The interrupted current helps to keep the electrodes cooler, thereby minimizing electrode distortion and reducing the tendency of the weld to spark. See Fig. 34-6.

Seam welding. Seam welding is like spot welding except that the spots overlap each other, making a continuous weld seam. In this process the metal pieces pass between roller type electrodes as shown in Fig. 34-7. As the electrodes revolve, the current is automatically turned on and off at intervals corresponding to the speed at which the parts are set to move. With proper control, it is possible to obtain airtight seams suitable for containers such as barrels, water heaters, and fuel tanks as shown in Fig. 34-8. When spots are not overlapped long enough to produce a continuous weld, as illustrated in Fig. 34-9, the process is sometimes referred to as *roller spot-welding* because of the intermittent current.

Because of the short current cycle, seam welding has several advantages. The rollers may be cooled to prevent overheating with consequent wheel dressing and replacement problems reduced to a minimum. Cooling is accomplished either by internally circulating water or by an external spray of water over the electrode rollers. Since the heat input is low, very little of the welded area is hardened and, therefore the yield point is not materially affected. The fact that very little grain growth takes place in the seam weld-

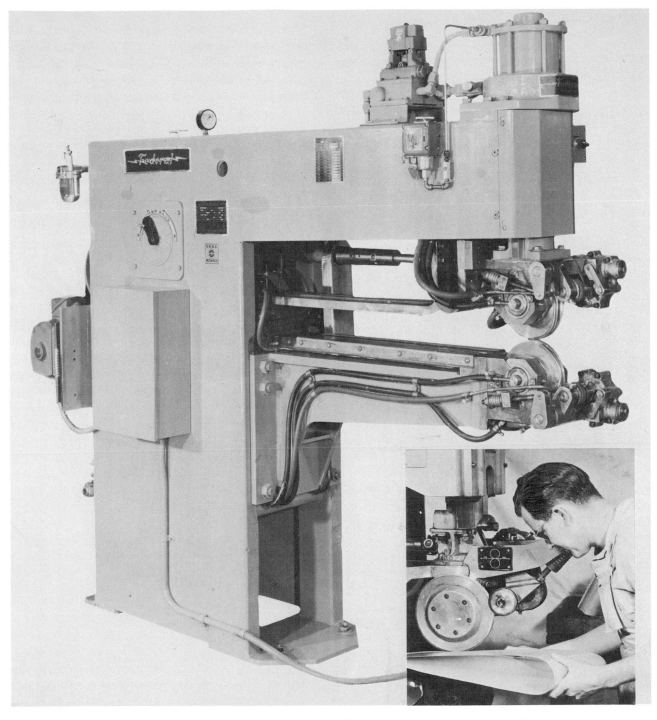

Fig. 34-7. Roller electrodes with automatic controls rapidly and effectively produce any desired length of weld along a seam.

ing process is also important from the standpoint of such corrosion-resistant alloys as stainless and other chromium alloy steels as their behavior is modified by grain growth.

An unusual seam welder is one having a combination of portable welding features with special longitudinal seam welding electrodes. This adaptation permits seam welding on large as-

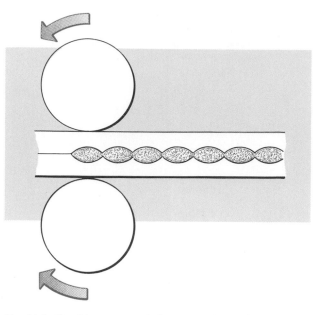

Fig. 34-8. If welds are spaced close enough, an air tight seam can be made.

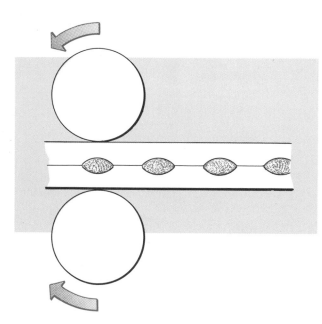

Fig. 34-9. With proper control setting, intermittent welds of a certain length are produced.

Fig. 34-10. A portable seam-welder is ideal where the structure to be welded is large and cannot conveniently be carried to the welding machine. (The Federal Machine & Welder Co.)

semblies where the welder must be brought to the work as shown in Fig. 34-10.

Projection welding. Projection welding involves the joining of parts by a resistance welding process which closely resembles spot weld-

ing. This type of welding is widely used in attaching fasteners to structural members.

The point where the welding is to be done has one or more projections which have been formed by embossing, stamping, casting, or machining. The projections serve to concentrate

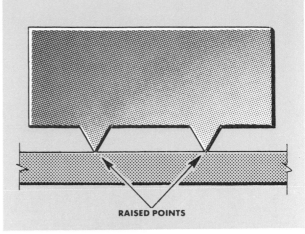

RAISED POINTS

BEFORE WELDING

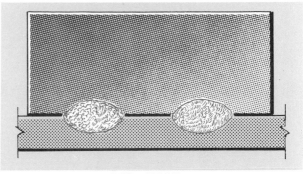

AFTER WELDING

Fig. 34-11. In projection welding, the weld area has been preformed with raised points.

the welding heat at these areas and cause fusion without employing a large current. The welding process consists of placing the projections in contact with the mating fixtures and aligning them between the electrodes as illustrated in Fig. 34-11. Either single or a multitude of projections can be welded simultaneously.

There are many variables involved in projection welding, such as stock thickness, kind of material, and number of projections, that make it impossible to predetermine the correct current setting and pressure required. Only by trial runs followed by careful inspection can proper control settings be established.

Not all metals can be projection welded. Brass and copper do not lend themselves to this method because the projections usually collapse under pressure. Aluminum projection welding is limited to extruded parts (shapes formed by forcing metal through a die). Galvanized iron and tin plate, as well as most other thin-gage steels, can be successfully projection welded.

Flash welding. In the flash welding process the two pieces of metal to be joined are clamped in dies which conduct the electric current to the work. The ends of the two metal pieces are moved together until an arc is established. The flashing action across the gap melts the metal, and as the two molten ends are forced together, fusion takes place as shown in Fig. 34-12. The current is cut off as soon as the forging action is completed, Fig. 34-13.

Flash welding is used to butt, or mitre-weld sheet, bar, rod, tubing, and extruded sections. It has almost unlimited application for both ferrous and nonferrous metals. It is not generally recommended for welding cast iron, lead, or zinc alloys for a variety of reasons.

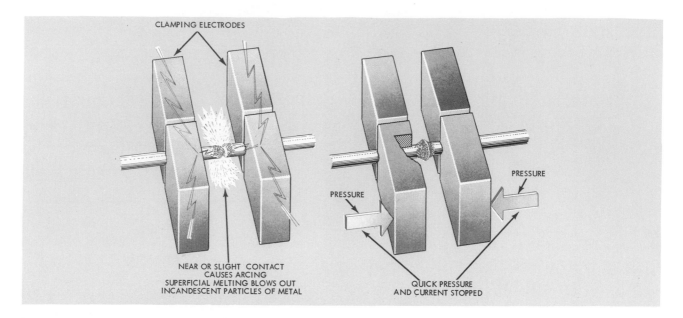

Fig. 34-12. In flash welding the two edges are brought together, causing intense arcing, which results in melting the metal. (Republic Steel Corp.)

Fig. 34-13. A steel ring is shown here being flash welded. (The Federal Machine & Welder Co.)

Parts to be welded are clamped by copper alloy dies shaped to fit each piece. For some operations the dies are water cooled to dissipate the heat from the welded area. The important factor in flash welding is the precision alignment of parts. Misalignment not only results in a poor joint but also produces uneven heat and telescoping of one piece over another.

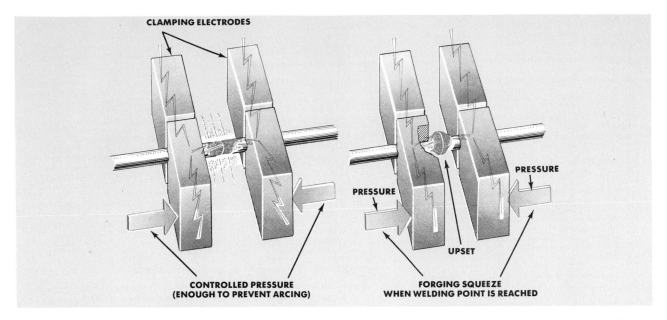

CLAMPING ELECTRODES

PRESSURE

PRESSURE

UPSET

CONTROLLED PRESSURE
(ENOUGH TO PREVENT ARCING)

FORGING SQUEEZE
WHEN WELDING POINT IS REACHED

Fig. 34-14. In butt welding the two ends are butted together and a high current passed through to melt the metal. Continuous pressure fueses the ends. (Republic Steel Corp.)

Fig. 34-15. An automatic butt-welding machine. (The Federal Machine & Welder Co.)

The only serious problem met in flash welding, if grain growth is not a problem, is the resulting bulge or increased size left at the point of weld. If the finish area of the weld is important, then it becomes necessary to grind or machine the joint to the proper size after welding.

Butt welding. In butt welding the metals to be welded are brought into contact under pressure, an electric current is passed through them, and the edges are softened and fused together as illustrated in Fig. 34-14. This process differs from flash welding in that constant pressure is applied during the heating process, which eliminates flashing. The heat generated at the point of contact results entirely from resistance. Although the operation and control of the butt welding process is almost identical to flash welding, the basic difference is in the use of less current and allowing more time for the weld to be completed. See Fig. 34-15.

Automatic Gas-Metal Arc Welding

Gas-metal arc welding has been acclaimed as the most economical and effective method of joining light gages of hard-to-weld metals such as nickel, stainless steel, aluminum, brass, copper, titanium, columbium, molybdenum, Inconel, Monel, and silver, as well as structural plates and beams. This process is performed manually, as in Chapter 19, or by automatic machines.

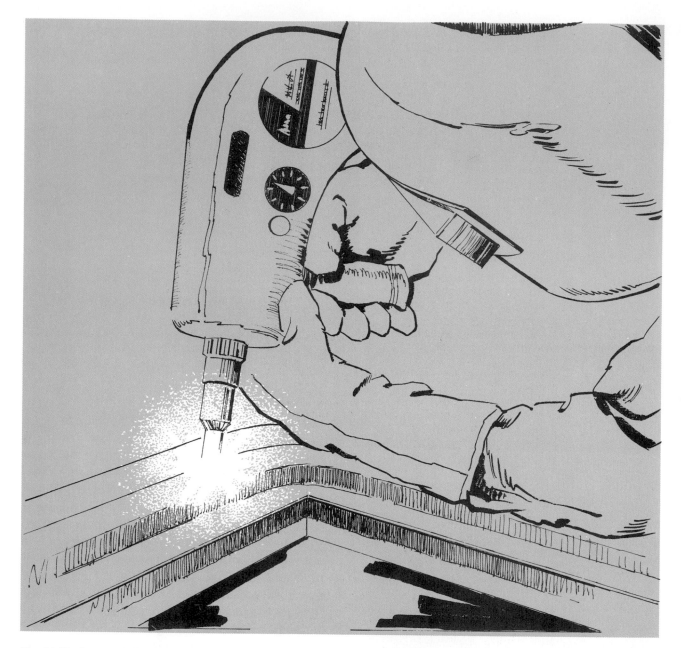

Fig. 34-16. A portable inert-gas-shielded arc gun being used to join a metal structure. (Airco, Inc.)

There are two types of automatic machines: the portable and the stationary. With the portable, the operator holds the unit as in Fig. 34-16. When the trigger is pressed, the gas, current, and wire automatically begin to flow. The operator simply has to concentrate on performing the weld in the designated area of the work piece, usually a single piece of equipment.

Portability of welding equipment is an increasing requirement in todays construction environment. Fig. 34-17 illustrates Airco's Super MIGet gun being used to join steel S-beam (formerly I-beam) sections high above ground level. This equipment can weld steel or aluminum and may be operated up to 50 feet from the power source. Standard spools of electrode wire are housed in

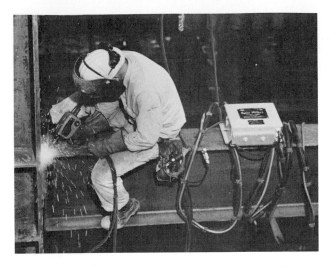

Fig. 34-17. Automatic portable welding guns are particularly adapable for many welding jobs. (Airco, Inc.)

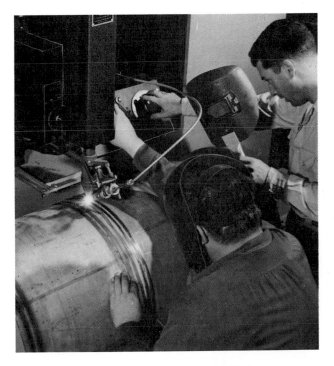

Fig. 34-18. The automatic inert-gas-shielded arc process produces dependable welds rapidly.

the cannister suspended above the operator and provide many hours of welding.

The stationary gas-metal arc unit differs from the gun type mainly in the welding head. The welding head is stationary instead of being port-

able. The head is either mounted on a carriage which travels over the work or it is in a fixed position and the structure to be welded is moved under the unit as shown in Fig. 34-18.

Stud Welding[1]

Stud welding is a form of electric arc welding. Two methods have been developed, each with a different principle of operation.

One of these is recognized by the use of a flux and a ceramic guide or ferrule. Equipment consists of a gun, a timing device which controls the DC welding current, the specially designed studs and ceramic ferrules. Studs are available in a wide variety of shapes, sizes and types to meet a variety of purposes. These studs have a recess in the welding end which contains the flux. This flux acts as an arc stabilizer and a deoxidizing agent. An individual porcelain ferrule is used with each stud when welding. It is a most vital part of the operation in that it concentrates the heat, acts (with the flux) to restrict the air from the molten weld, confines the molten metal to the weld area, shields the glare of the arc, and prevents charring of the material through which the stud is being welded.

In operation, a stud is loaded into the chuck of the gun and a ferrule positioned over the stud. When the trigger is depressed the current energizes a solenoid coil which lifts the stud away from the plate, causing an arc which melts the end of the stud and the area on the plate. A timing device shuts off the current at the proper time. The solenoid releases the stud and spring action plunges the stud into the molten pool and the weld is made.

Another method is characterized by a small cylindrical tip on the joining face of the stud. The diameter and length of this tip vary with the diameter of the stud and the material being welded. This method operates on alternating current, and a source of about 85 pounds air pressure is also required.

The gun is air-operated with a collet (to hold the stud) attached to the end of a piston rod. Constant air pressure holds the stud away from

---

1. Republic Steel Corp.

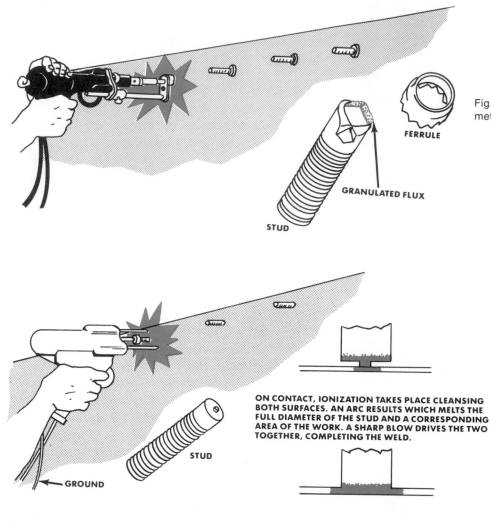

Fig. 34-19. Stud welding—Nelson method. (Republic Steel Corp.)

FERRULE

GRANULATED FLUX

STUD

Fig. 34-20. Stud welding—Graham method. (Republic Steel Corp.)

GROUND

STUD

ON CONTACT, IONIZATION TAKES PLACE CLEANSING BOTH SURFACES. AN ARC RESULTS WHICH MELTS THE FULL DIAMETER OF THE STUD AND A CORRESPONDING AREA OF THE WORK. A SHARP BLOW DRIVES THE TWO TOGETHER, COMPLETING THE WELD.

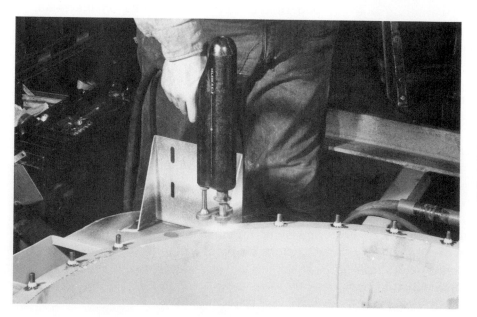

Fig. 34-21. Stud welding eliminates the necessity of drilling and tapping in the cover plate assembly of this tank. (Nelson Stud Welding Co.)

the metal until ready to make the weld, then air pressure drives the stud against the work. When the small tip touches the workpiece, a high-amperage, low-voltage discharge results, creating an arc which melts the entire area of the stud and the corresponding area of the work. Arcing time is about one mil second (0.001), thus a weld is completed with little heat penetration, no distortion and practically no fillet. The stud is driven at a velocity of about 31 inches per second and the explosive action as it meets the workpiece cleanses the area to be welded. A minimum thickness of the workpiece of 0.02 inches is desired, particularly if no marking on the reverse side is required.

Both methods of stud welding are adaptable to welding of most ferrous and non-ferrous metals, their alloys, and any combination thereof. See Figs. 34-19, 34-20, and 34-21.

Electron-Beam Welding

Electron-beam welding is essentially a fusion welding process. Fusion is achieved by focusing a high power-density beam of electrons on the area to be joined. Upon striking the metal, the kinetic energy of the high-velocity electrons changes to thermal energy, causing the metal to melt and fuse.

The electrons are emitted from a tungsten filament heated to approximately 3630°F (2000°C). Since the filament would quickly oxidize at this temperature if it were exposed to normal atmosphere, the welding must be done in a vacuum chamber. Therefore a vacuum chamber is necessary to prevent the electrons from colliding with molecules of air which would make the electrons scatter and lose their kinetic energy.

Electron-beam welding can be used to join materials ranging from thin foil to 2″ thick. It is particularly adaptable to the welding of refractory metals such as tungsten, molybdenum, columbium, tantalum, and metals which readily oxidize, such as titanium, beryllium, and zirconium. It also has wide application in joining dissimilar metals, aluminum, standard steels, and ceramics.

Advantages: Electron beam welding has several distinct advantages. It facilitates welding with a low total-energy input. Workpiece distortion and effects on the material properties of the workpiece are reduced to a minimum. The weld size and location can be controlled relative to the energy input. The process is chemically clean and facilitates welding without contamination of the workpiece.

Electron beam welding is often associated with the joining of difficult to weld metals. It is used in Aerospace fabrication where new metals require more exacting joining characteristics, however, adaptation of the process to commercial applications is increasing. There is every indication that this growth will continue.

Limitations. One of the major limitations of electron-beam welding is that the work must be done in a vacuum chamber. Consequently, the piece must be small enough to fit into the chamber. This limitation is being reduced to some extent because larger chambers are now manufactured to accommodate a vast variety of products. Another limitation is that when the workpiece is in the chamber in a vacuum it becomes inaccessible. It must be manipulated by some special device. However, engineers believe that electron beam welding, with all of its advantages, can be done without a vacuum chamber and applied to any normal atmospheric requirement.

Major units of the system. Electron beam welding equipment usually includes the following basic modules:

Electron gun. The gun consists of a filament, cathode, anode, and focusing coil. See Fig. 34-22. The electrons emitted from the heated filament carry a negative charge and are repelled by the cathode and attracted by the anode. The electrons pass through an aperture in the anode and then through a magnetic field generated by the electromagnetic focusing coil. An optical viewing system provides a line of sight down the path of the electron beam centerline to the weld area when the beam is OFF. See Figs. 34-23 and 34-24. By varying the current to the focusing coil, the operator can focus the beam for gun-to-work distances ranging from a half inch to 25 inches. The electron beam can be controlled with a focusing coil to produce a spot diameter of less than 0.005″.

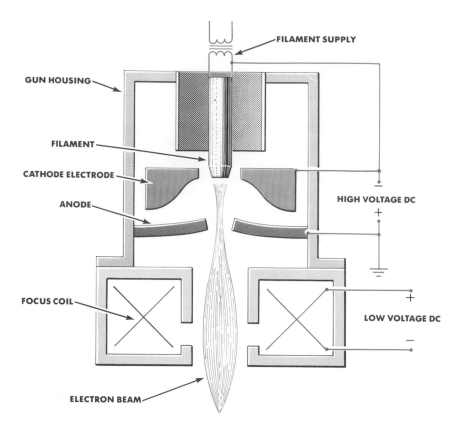

GUN HOUSING

FILAMENT SUPPLY

FILAMENT

CATHODE ELECTRODE

ANODE

HIGH VOLTAGE DC

FOCUS COIL

LOW VOLTAGE DC

ELECTRON BEAM

Fig. 34-22. Typical elements of an electron beam gun. (Skiaky Bros., Inc.)

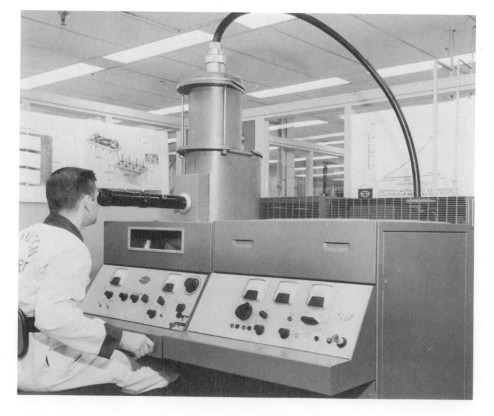

Fig. 34-23. Electron beam welding machine. (Hamilton Standard-United Aircraft Corp.)

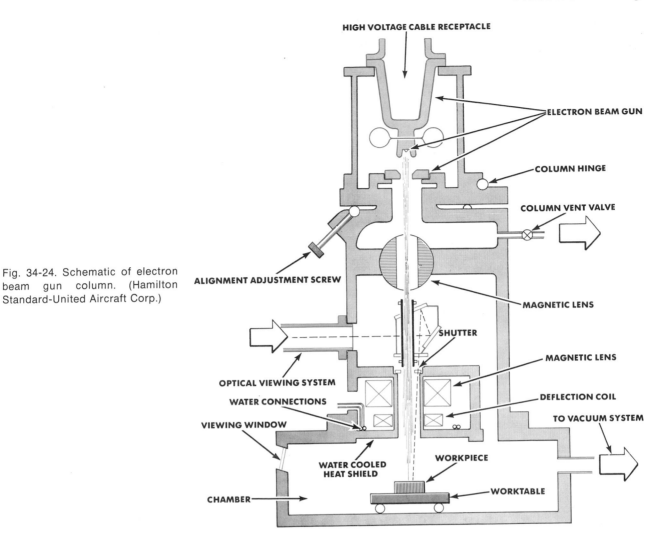

HIGH VOLTAGE CABLE RECEPTACLE

ELECTRON BEAM GUN

COLUMN HINGE

COLUMN VENT VALVE

ALIGNMENT ADJUSTMENT SCREW

MAGNETIC LENS

SHUTTER

MAGNETIC LENS

OPTICAL VIEWING SYSTEM

WATER CONNECTIONS

DEFLECTION COIL

TO VACUUM SYSTEM

VIEWING WINDOW

WATER COOLED HEAT SHIELD

WORKPIECE

WORKTABLE

CHAMBER

Fig. 34-24. Schematic of electron beam gun column. (Hamilton Standard-United Aircraft Corp.)

Vacuum chamber. The chamber is usually rectangular in shape, and has heavy glass windows to permit viewing the work. A work table in the chamber is arranged so it can be operated either manually or electrically in the X and Y directions. T-slots are provided on the table to attach fixtures or workpieces for welding as shown in Fig. 34-25.

Vacuum pumping system. The system is designed to provide a clean, dry vacuum chamber in a relatively short time. The capacity of the pump required is governed by the volume and area of the chamber and the time required to evacuate the chamber. The pumping equipment is usually completely automatic once the set-up has been completed.

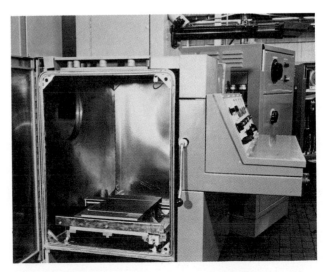

Fig. 34-25. Vacuum chamber where welding workpiece is positioned. (Hamilton Standard-United Aircraft Corp.)

Electrical controls. These include set-up controls and operating controls. The set-up controls include instruments required for the initial set-up of the welding operation, such as meters for beam voltage, beam current, focusing current, and filament current.

The operating controls consist of stop-and-start sequence, high voltage adjustment, focusing adjustment, filament activation, and work table motion. These controls are mounted so they are easily reached by the operator while he is observing the welding action through the viewing window.

Power unit. The power unit furnishes the main high voltage supply up to 150 kv and a low filament-power up to 6 volts.

Operating the equipment: *Set-up procedure.* The workpiece is positioned on the work carriage in the chamber. The set-up steps are begun. The beam gun, and work-to-gun distance are aligned manually, and visually by use of the optical system. Work travel or gun travel (depending on the type of welding facility used) are checked and adjusted for alignment.

The vacuum chamber is then closed. Vacuum-controls are started and the chamber is pumped down to the required vacuum (usually prescribed in a weld schedule).

Beam voltage, beam current, filament current and focusing current controls are then set from the weld schedule. This schedule is usually made by a welding engineer. When these control settings have been checked, the beam current may be switched ON and OFF for an instant for a weld spot alignment check. By viewing the weld spot through the optical system, determination is made with regard to operation for the actual welding.

The weld or weld area is viewed by opening the shutter only when the beam current is turned off. If the shutter is opened when beam current is on, the result will be severe damage to the optical system.

Weld procedure. After everything has been checked and all switching made operative, the welding is begun by switching the sequence start switch to the ON position. The weld is made automatically.

It has been mentioned that the growth of applications for electron beam welding no longer limit it exclusively to exotic metal welding. It may also be assumed that electron beam welding will never entirely replace other welding processes. All welding processes in their proper application will continue to serve if useful in the wide spectrum of welding functions.

Inertia Welding

Inertia or friction welding is a process where stored kinetic energy is used to generate the required heat for fusion. The two workpieces to be joined are aligned end to end. One is held stationary by means of a chuck or fixture, and the other is clamped in a rotating spindle.

The rotating member is brought up to a certain speed so as to develop sufficient energy. Then the drive source is disconnected and the pieces brought into contact under a computed thrust load. At this point the kinetic energy contained in the rotating mass converts to frictional heat. The metal at and immediately behind the interface is softened, permitting the workpieces to be forged together. See Figs. 34-26 and 34-27.

Inertia welding has several advantages over conventional flash or butt welding. It produces improved welds at higher speed and lower cost, less electrical current is required, and costly copper fixtures for hold parts are eliminated. With inertia welding there is less shortening of the components, which often results in flash or butt welding. Also the heat-affected zone near the weld is confined to a narrow band and therefore does not draw the temper of the surrounding area.

The inertia-welding process is applicable for welding many dissimilar or exotic metals as well as similar metals. Weld strength is normally equal to that of the original metals.

Laser Welding

Laser welding is like welding with a white-hot needle. Fusion is achieved by directing a highly concentrated beam to a spot about the diameter of a human hair. The highly concentrated beam generates a power intensity of one billion or more watts per square centimeter at its point of

ROTATING SPINDLE AND FLYWHEEL

1. PIECES ARE ALIGNED AND CLAMPED

2. FLYWHEEL IS ROTATED BY AN EXTERNAL ENERGY SOURCE

3. MEMBERS BROUGHT INTO CONTACT

Fig. 34-26. In inertia welding, heat resulting from stored kinetic energy is used to forge the pieces together. (Caterpiller Tractor Co.)

focus. Because of its excellent control of heat input, the laser can fuse metal next to glass or weld near varnish coated wires without damaging the insulating properties of the varnish.

Since the heat input to the workpiece is extremely small in comparison to other welding processes, the size of the heat affected zone and the thermal damage to the adjacent parts of the weld are minimized. Thus it is possible to weld heat-treated alloys without affecting their heat-treated condition. As a matter of fact the weldment can be held in the hand immediately after the weld is completed.

The laser can be used to join dissimilar metals such as copper, nickel, tungsten, aluminum, stainless steel, titanium and columbium. Furthermore, the laser beam can pass through transparent substances without affecting them,

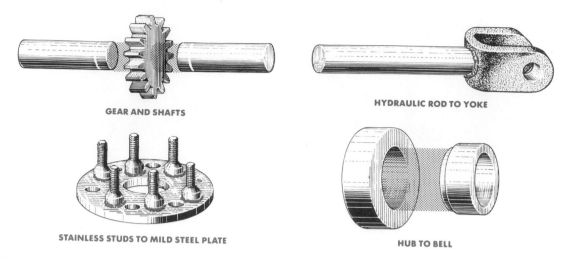

GEAR AND SHAFTS

HYDRAULIC ROD TO YOKE

STAINLESS STUDS TO MILD STEEL PLATE

HUB TO BELL

Fig. 34-27. Typical examples of parts welded by inertia welding. (Caterpillar Tractor Co.)

thereby making it possible to weld metals that are sealed in glass or plastic. Because of the fact that the heat source is a light beam, atmospheric contamination on the weld joint is not a problem.

The current application of laser welding is largely in aerospace and electronic industries where extreme control in weldments is required. Its major limitation is the shallow penetration. Present day equipment restricts it to metals not over 0.020″ thick.

The duration of the beam is usually about 0.002 seconds, with a pulse rate of one to ten times per second. As each point of the beam hits the metal, a spot is melted but solidifies in microseconds. The line of weld thus consists of a series of round, solid puddles each overlapping the other. The workpiece is either moved beneath the beam or the energy source is moved across the line of weld.

Focusing the beam onto the workpiece is accomplished with an optical system and the actual control of the welding energy by means of a switch. See Fig. 34-28.

Theory of the laser beam[2]. Atoms have been made to generate energy by exciting them in such common devices as fluorescent lights and television tubes. Fluorescence refers to the abili-ty of certain atoms to emit light when they are exposed to external radiation of shorter wave lengths.

In the laser welder, the atoms that are excited to produce the laser light beam are produced in a man-made ruby rod ³⁄₈″ in diameter. See Fig. 34-29. The ruby is identical to a natural ruby but has a more perfect crystal structure. About 0.05 percent of its weight is chromium oxide.

The chromium atoms give the ruby its red color because they absorb green light from external light sources. When the atoms absorb this light energy, some of their electrons are excited. Thus, green light is said to pump the chromium atoms to a higher energy state.

The atoms eventually return to their original state. In doing so, they give up a portion of the extra energy they previously absorbed (as green light) in the form of red fluorescent light.

When the red light emitted by one excited atom hits another excited atom, the second atom gives off red light which is in phase with the colliding red light wave. In other words, the red light from the first atom is amplified because more red light exactly like it is produced.

By using a very intense green light to excite the chromium atoms in the ruby rod, a larger number of its atoms can be excited and the chances of collisions are increased. To further enhance this effect, the parallel ends and the

2. Linde Division, Union Carbide Corp.

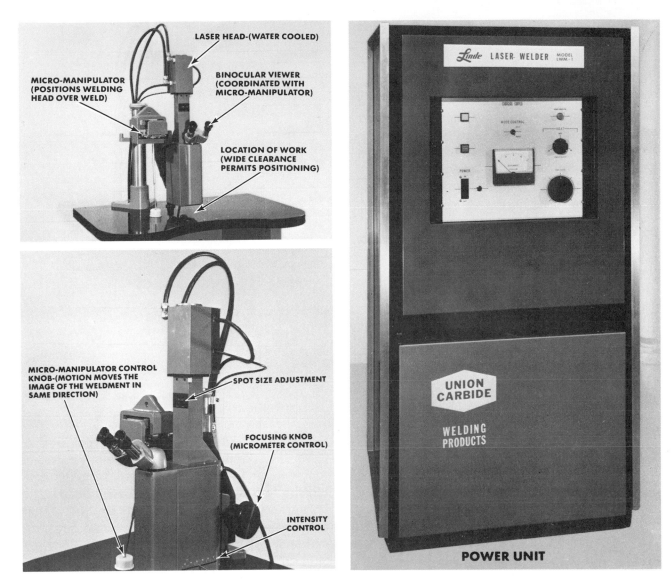

LASER HEAD-(WATER COOLED)

MICRO-MANIPULATOR
(POSITIONS WELDING
HEAD OVER WELD)

BINOCULAR VIEWER
(COORDINATED WITH
MICRO-MANIPULATOR)

LOCATION OF WORK
(WIDE CLEARANCE
PERMITS POSITIONING)

MICRO-MANIPULATOR CONTROL
KNOB-(MOTION MOVES THE
IMAGE OF THE WELDMENT IN
SAME DIRECTION)

SPOT SIZE ADJUSTMENT

FOCUSING KNOB
(MICROMETER CONTROL)

INTENSITY
CONTROL

Linde LASER-WELDER MODEL LWM-1

UNION CARBIDE

WELDING
PRODUCTS

POWER UNIT

Fig. 34-28. Laser welding unit. (Linde Co.)

Fig. 34-29. Schematic diagram of laser welder. (Linde Co.)

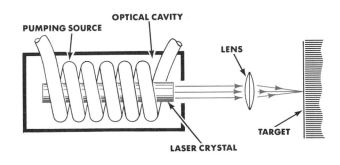

PUMPING SOURCE

OPTICAL CAVITY

LENS

TARGET

LASER CRYSTAL

sides of the rod are mirrored to bounce the red light back and forth within the rod. When a certain critical intensity of pumping is reached (the so-called threshold energy), the chain reaction collisions become numerous enough to cause a burst of red light. The mirror at the front end of the rod is only a partial reflector, allowing the burst of light to escape through it.

Plasma Welding

Plasma welding is a process which utilizes a central core of extreme temperature surrounded by a sheath of cool gas. The required heat for fusion is generated by an electric arc which has been highly intensified by the injection of a gas into the arc stream. The superheated columnar arc is concentrated into a narrow stream and when directed on metal makes possible butt welds up to one-half inch or more in thickness in a single pass without filler rods or edge preparation. See Fig. 34-30.

In some respects plasma welding may be considered as an extension of the conventional gas tungsten arc welding. The main difference is that in plasma welding the arc column is constricted and it is this constriction that produces the much higher heat transfer rate.

The arc plasma actually becomes a jet of high current-density. The arc gas upon striking the metal cuts or keyholes entirely through the piece producing a small hole which is carried along the weld seam. See Fig. 34-31. During this cut-

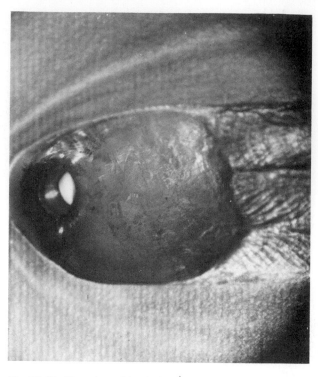

Fig. 34-31. Top view of keyhole with bead in ½'' thick plate of titanium. (Thermal Dynamics Corp.)

ting action, the melted metal in front of the arc flows around the arc column, then is drawn together immediately behind the hole by surface tension forces and reforms in a weld bead.

The specially designed torch, Fig. 34-32, for plasma welding can be hand held or mounted for stationary or mechanized applications. See Fig. 34-33. The process can be used to weld stainless steels, carbon steels, Monel, Inconel, titanium, aluminum, copper and brass alloys. See Fig. 34-34. Although for many fusion welds no filler rod is needed, a continuous filler wire can be added for various fillet types of weld joints.

Equipment. A regular heavy duty DC rectifier is used as the source of power for plasma welding. A special control console is required to provide the necessary operating controls. A water cooling pump is usually needed to assure a controlled flow of cooling water to the torch at a regulated pressure. Proper cooling prolongs the life of the electrode and nozzle. See Fig. 34-35.

Gas supply is either argon or helium. In some application, argon is used as the plasma gas and helium as the shielding gas. However in many

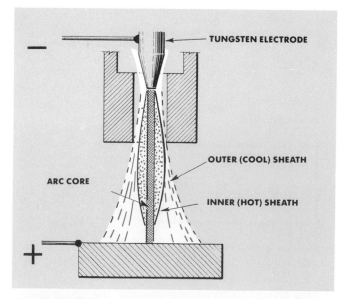

Fig. 34-30. Plasma welding uses a central core of extreme temperature surrounded by a sheath of cool gas. (Thermal Dynamics Corp.)

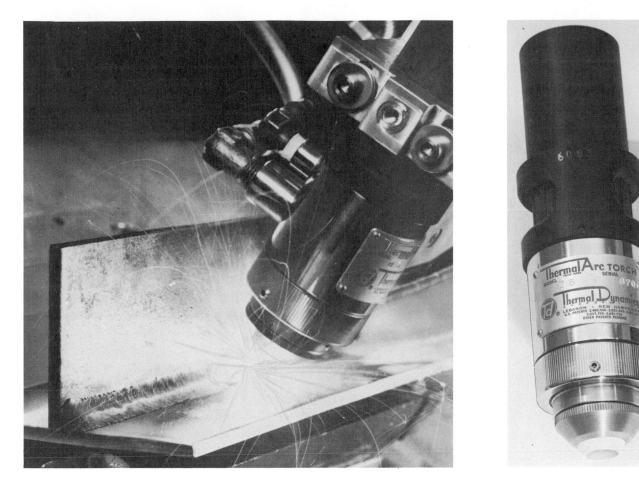

Fig. 34-32. Torch for plasma welding. (Thermal Dynamics Corp.)

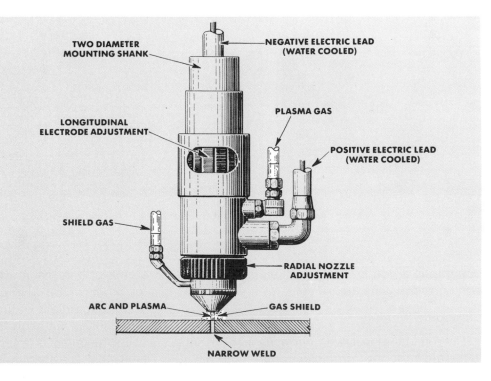

TWO DIAMETER
MOUNTING SHANK

NEGATIVE ELECTRIC LEAD
(WATER COOLED)

PLASMA GAS

LONGITUDINAL
ELECTRODE ADJUSTMENT

POSITIVE ELECTRIC LEAD
(WATER COOLED)

SHIELD GAS

RADIAL NOZZLE
ADJUSTMENT

ARC AND PLASMA

GAS SHIELD

NARROW WELD

Fig. 34-33. Plasma welding torch can be mounted for stationary or mechanized application. (Thermal Dynamics Corp.)

Fig. 34-34. Types of welding penetration on different types of metal produced with plasma welding. (Thermal Dynamics Corp.)

operations argon is used for shielding and generating the plasma arc.

Ultrasonic Welding

If two metal pieces with perfectly smooth surfaces are brought into close contact, the metal atoms of one piece will theoretically unite with the atoms of the other piece to form a permanent bond. Regardless of how smooth such surfaces are, a sound metallurgical bond normally will not occur because it is impossible to prepare surfaces that are absolutely smooth.

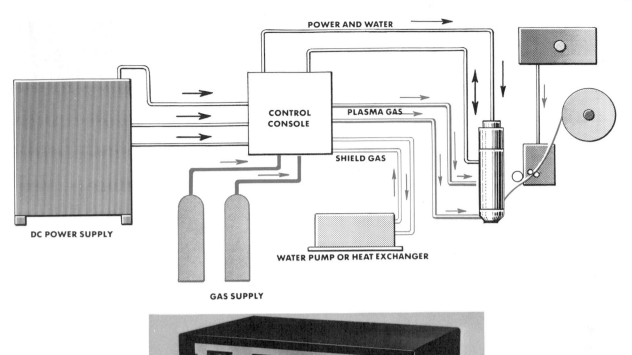

Fig. 34-35. Typical plasma welding installation and control console. (Thermal Dynamics Corp.)

No matter what means are used to smooth surfaces, they will still possess peaks and valleys as measured by a microscope. As a result only the peaks of two workpieces which come into close contact will unite, leaving the countless valleys without producing a bond. Furthermore, smooth surfaces are never actually clean. Oxygen molecules from the atmosphere react with the metal to form oxides. These oxides attract water vapor, forming a film of moisture on the oxidized metal surface. Both the moisture and oxide film also act as barriers to prevent intimate contact.

In the ultrasonic welding process, these three existing barriers are broken down by plastically deforming the interface between the workpiec-es. This is done by means of vibratory energy which disperses the moisture, oxide, and irregular surface to bring the areas of both pieces into close contact and form a solid bond. Vibratory energy is generated by a transducer. A schematic of a typical ultrasonic welding unit is shown in Fig. 34-36.

The welding equipment consists of two units: a power source or frequency converter, which converts 60 cycle line power into high-frequency electrical power; and a transducer, which changes the high-frequency electrical power into vibratory energy.

The components to be joined are simply clamped between a welding tip and supporting anvil with just enough pressure to hold them in

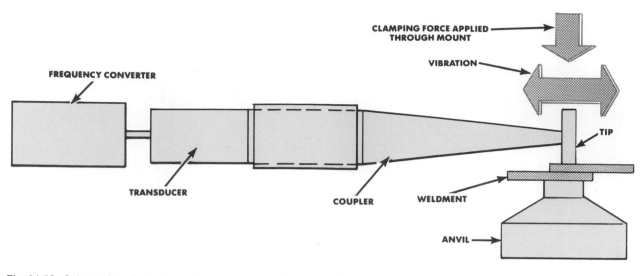

Fig. 34-36. Schematics of a typical ultrasonic welder. (Sonobond Corp.)

Fig. 34-37. Ultrasonic continuous-seam welder. (Sonobond Corp.)

close contact. The high-frequency vibratory energy is then transmitted to the joint for the required period of time. The bonding is accomplished without applying external heat, or adding filler rod or melting metal. Either spot-welds or continuous-seam welds can be made on a variety of metals ranging in thickness from 0.00017″ (aluminum foil) to 0.10″. Thicker sheet and plate can be welded if the machine is specifically designed for them. High-strength bonds are possible both in similar and dissimilar metal combinations.

Ultrasonic welding is particularly adaptable for joining electrical and electronic components, hermetic sealing of materials and devices, splicing metallic foil, welding aluminum wire and sheet, and fabricating nuclear fuel elements. See Fig. 34-37.

Welding variables such as power, clamping force, weld time for spot welds or welding rate for continuous-seam welds can be preset and the cycle completed automatically. A switch lowers the welding head, applies the clamping force and starts the flow of ultrasonic energy.

Successful ultrasonic welding depends on the proper relationship between these welding variables which is usually determined experimentally for a specific application. Thus clamping force may vary from a few grams for very light materials to several thousand pounds for heavy pieces. Weld time may range from 0.005 to 1.0 seconds for spot welding and a few feet per minute to 400 fpm for continuous-seam welding. The high-frequency electrical input to the transducer may vary from a fraction of a watt to several kilowatts.

Electro-Gas Welding

Electro-gas welding is a process which uses a gas-shielded metal arc and is designed for single-pass welding of vertical joints in steel plates ranging in thickness from $3/8''$ to $1^1/2''$. See Fig. 34-38.

The welding head is suspended from an elevator mechanism which provides automatic control of the vertical travel speed during welding. This mechanism raises the welding head automatically at the same rate as the advancing weld metal. The welding head is self-aligning and can follow any alignment irregularity in plate or joint.

Once the equipment is positioned on the joint, welding is completely automatic. Wire feed and current are constant. At the end of the weld the process automatically stops.

This welding technique is especially adaptable for ship building and fabrication of storage tanks and large diameter pipe.

Fig. 34-38. Electro-Gas welding. (Airco, Inc.)

Points to Remember

1. Spot welding is a form of resistance welding which has wide applications in industry.

2. Spot welders are available to produce single spot welds or multiple spot welds.

3. In pulsation welding, the current is regulated to go on and off a number of times during the welding process.

4. Seam welding produces a series of overlapping spot welds, thereby making a continuous weld seam.

5. Projection welding is widely used in attaching fasteners to structural members.

6. In flash welding, when the two metal pieces are brought together, an arc melts the edges and the molten surfaces are then forced together to form a permanent bond.

7. Butt welding is somewhat similar to flash welding except that constant pressure is applied during the heating process which eliminates flashing.

8. Electron beam welding is a fusion process where a high power-density beam of electrons are focused on the area to be joined.

9. In inertia welding heat, resulting from the parts being rotated together, is used to forge the pieces.

10. Laser welding is a fusion process where a high concentrated beam is focused on a tiny spot.

11. Plasma welding uses an electric arc which is highly intensified by the injection of gas into the arc stream and results in a jet of high current-density.

12. Ultrasonic welding is a process where vibratory energy disperses the moisture, oxide, and surface irregularities between the pieces, thereby bringing the surfaces into close contact to form a permanent bond.

QUESTIONS FOR STUDY AND DISCUSSION

1. Name several types of production welding processes?

2. What is the basic principle of resistance welding?

3. How does a portable spot welder differ from an ordinary stationary spot welder?

4. What is a seam welder?

5. What is roller spot-welding?

6. What is projection welding?

7. How does flash welding differ from butt welding?

8. What is meant by pulsation welding?

9. How does the stud-welding gun operate?

10. What are the two basic types of automatic Mig welding machines?

11. How is fusion obtained in electron beam welding?

12. What are some advantages, and limitations of electron-beam welding?

13. How does plasma welding differ from regular Tig welding?

14. What is the function of the transducer in ultrasonic welding?

15. How is fusion of metal accomplished in ultrasonic welding?

16. In laser-beam welding how is the high-intensity laser light beam generated?

17. What is the principle of inertia welding?

18. What is electro-gas welding and what is it designed for?

CHAPTER 35 underwater cutting and welding

In principle, underwater cutting and welding is much the same as conventional cutting and welding in air, except that environmental conditions impose additional limitations on the operator. Diving apparel (Fig. 35-1), water depth, adverse currents, low temperatures, poor visibility, and unstable footing, are all factors which make underwater cutting and welding difficult. Since the diver can work only a short time on the bottom, particularly at lower depths, the use of correct techniques and equipment is extremely important, *if not a matter of life or death.*

UNDERWATER CUTTING—PRINCIPLES OF OPERATION

Metals may be cut or pierced underwater by any of three methods. The two most common methods are known as oxy-hydrogen and arc-oxygen and they depend on the oxidation of metals for cutting action. In these two processes, heat is applied to a spot on the metal. When the metal has reached the kindling temperature, pure oxygen is directed at the heated spot. This causes the metal to burn very rapidly. Both the oxy-hydrogen and arc-oxygen methods are limited to cutting plain carbon and low alloy steels.

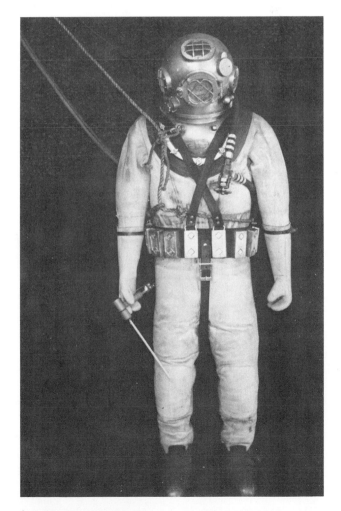

Fig. 35-1. Special equipment is used for underwater cutting and welding. (Craftsweld Equipment Corp.)

1. Courtesy Craftsweld Equipment Corp. and U.S. Navy Dept.

The third method of cutting uses the metallic arc, a melting rather than a burning operation. The metallic arc cutting process is better for cutting corrosion resisting and austenitic steels or other metals which do not corrode readily.

Oxygen-Hydrogen Cutting

The oxy-hydrogen process of flame cutting underwater involves the use of compressed oxygen, compressed hydrogen, and air under pressure. The technique does not differ radically from open-air cutting practice, since the torch performs in much the same way as a standard torch being operated in air.

One thing that sets underwater cutting apart from ordinary cutting is that underwater cutting requires the operator to become accustomed to working with relatively high gas pressures.

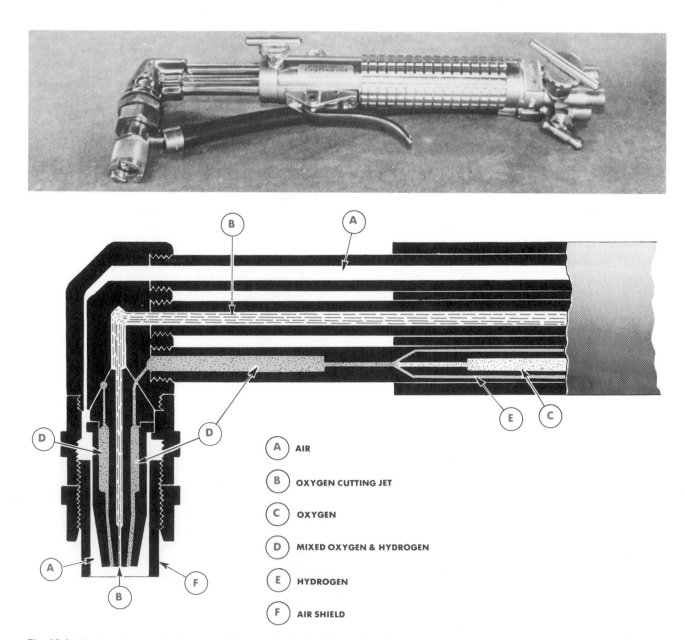

A AIR

B OXYGEN CUTTING JET

C OXYGEN

D MIXED OXYGEN & HYDROGEN

E HYDROGEN

F AIR SHIELD

Fig. 35-2. Underwater oxy-hydrogen cutting torch. (United States Navy)

These pressures must increase with the water depth at which the work is being performed. Mechanically, underwater cutting is accomplished by the same means used topside, except that an additional hose is used to deliver compressed air to an air shield, or *skirt,* that surrounds the cutting tip and sheathes it with a bubble of air.

This shield is not necessary to keep the flame alight in that the fuel gas and oxygen support combustion even when the flame is immersed in water. The purpose of the shield is to stabilize the flame and hold the water away from the area of metal being heated. The higher pressure in

underwater work is necessary to provide the required increase in flame intensity and to overcome the water pressure at whatever depth the torch is being operated. Hydrogen therefore, is the fuel gas used almost exclusively in this type of cutting, because it is generally unsafe to use acetylene at pressures higher than 15 pounds.

Equipment. The *underwater cutting torch* is made to operate with one hand, leaving the other hand free or for steadying the progress of the torch. It is generally manufactured with a 90° head, but may be especially ordered with straight head, 75° or 45° specified. See Fig. 35-2. An *air shield,* or jacket, screws over the cutting

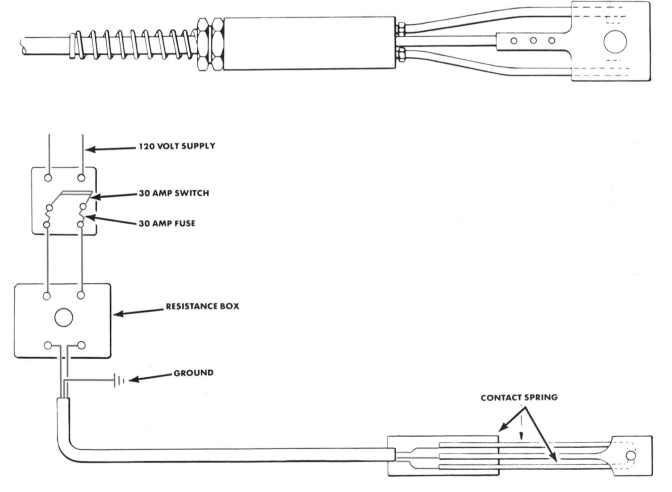

DIAGRAM OF CONNECTIONS

Fig. 35-3. Underwater torch lighter. (United States Navy)

tip with the front end extending about ¹/₈″ beyond the end of the tip.

The *oxygen* and *hydrogen regulators* are of the heavy-duty type equipped with adequate safety blow-off devices.

A special *electrical lighter* has been developed to ignite the torch underwater. This lighter is not designed for operation in the air. When used above water, the tips burn and have to be renewed. The lighter operates on 110–120 volts, and is controlled from the surface. It consists of an insulated handle with two spring copper jaws terminating in carbon tips. See Fig. 35-3. The current is always turned off when the lighter is not in use.

Lighting the torch in air.   The following procedure is to be used when lighting the oxy-hydrogen cutting torch in the air, following the normal cautions for handling combustable gases.

1. Open the hydrogen cylinder valve and adjust the pressure regulator to 25 to 30 lb, plus hose friction and additional pressure to compen-sate for water depth where cutting is to be done. Added pressure should be approximately 50 lb per 100 feet of water depth and 5 lb per 100 feet for hose friction. See Table 35-1.

2. Open the oxygen cylinder valve and adjust pressure regulator. Oxygen pressure should be proportional to surface cutting, depending upon the thickness of metal, plus additional pressure to compensate for depth of water and hose friction.

3. Adjust the compressed air control valve to provide about 25 lb of pressure plus additional pressure calculated to compensate for depth and hose friction. The compressed air is not a critical factor, except if the pressure is too great it may affect vision because of the excessive flow of air bubbles.

4. Open the oxygen and hydrogen valves to clear the hose, and then shut them.

5. Now proceed to light the torch and adjust it with both the oxygen and hydrogen preheating valves as wide open as possible and still have long sharp cones in the flame. If you cannot see

TABLE 35-1.   RECOMMENDED PRESSURES (GAGE PRESSURE AT SURFACE) FOR OXY-HYDROGEN UNDERWATER CUTTING TORCH.

WORKING DEPTH (ft)	WATER PRESSURE (lb)	LENGTH HOSE (ft)	AIR (lb)	HYDROGEN (lb)	OXYGEN (lb)
10	4	100	55	55	75
20	9	100	60	60	80
30	13	100	65	65	85
40	17	150	75	75	95
50	22	150	80	80	100
60	26	200	90	90	110
70	30	200	95	95	115
80	35	250	100	100	120
90	39	250	105	105	125
100	43	300	115	115	135
125	54	300	125	125	145
150	65	300	140	140	160
175	76	400	155	155	175
200	87	450	170	170	190
225	97	450	185	185	200

U.S. Navy

the cones in the sunlight, hold the torch in the shade against a dark object.

Testing flame. Before the torch is lowered to the diver, the flame should be tested for underwater stability. This may be done as follows:

1. Hold the hose at a point 6' or so back from the torch. Drop or throw the lighted torch into the water so that the flame is three feet or more underwater.

2. If close to the water level, the torch may be held in the hand and moved back and forth with the flame well underwater. The properly adjusted torch will not go out under these conditions.

3. The torch may be tested for cutting by making a short cut on a piece of scrap steel. A torch that will not cut in air can not be expected to cut underwater. Care should be taken not to foul the tip.

Lowering the torch. A lighted torch may be lowered underwater in one of three ways:

1. In moderate or shallow depths and easily accessible locations, the lighted torch may be carried below by hand. In such case, the diver should be prepared to make adjustments to the flame to compensate for increased pressure during descent.

2. In shallow depths and locations where it is absolutely certain that the flame can be kept clear of all hoses, lines, diver's suit, and helmet, etc., the lighted torch may be lowered directly to a point within reach of the diver. Never lower a lighted torch until it is certain that the diver is ready and watching for it and is in the clear in such a position that the torch flame cannot possibly strike his helmet, suit or air line.

3. The torch may be carried below by the diver or bent to a shackle and sent down on a finder-line.

Lighting torch underwater. The following procedure is to be used when lighting the oxy-hydrogen torch underwater:

1. Open the air valve until the bubble coming from the tip is 3" long.

2. Open the hydrogen valve until the bubble from the torch tip is about 3" long. Note the setting and close the valve.

3. Open the preheating oxygen valve until the bubble is about 2½" long.

4. Reopen the air and hydrogen valves to the settings determined in steps 1 and 2.

5. Signal the tender to turn on the igniter.

6. Hold the torch horizontally with the tip pointing away from the lines, hose, and person.

7. Hold the igniter so that the gas from the torch tip blows through the hole in the igniter guard and past the igniter points.

8. Squeeze the igniter contacts together and then release. The spark formed as the igniter points spring apart will ignite the preheat flame, if adjustment of the torch valves is correct.

9. If the torch fails to ignite, open the hydrogen valve a little more and try again.

10. If the torch still fails to ignite, enrich the mixture by turning on a little more oxygen.

11. Adjustment may be continued, turning on first more hydrogen and then more oxygen, until it is certain that adjustment is right.

Cutting operation. Successful underwater cutting means the ability to adjust the torch for preheating in both fresh and salt water. Also necessary is the skill to recognize the action of the torch when it begins to cut and when the cutting stops. With proper flame, instantaneous preheat results and cutting is rapid and continuous in any position, even under poor visibility when guiding the torch by sound, glow, and feel.

The lighted torch underwater gives off a characteristic rumbling or bubbling sound. By listening to this sound, you can get an indication of the condition of the flame.

Here are some general conditions:

1. Compressed air too high. Cools metal too rapidly and excessive force is required to hold the torch head against the work.

2. Compressed air too low. Fails to displace water, causing the flame to flicker and sputter.

3. Hydrogen too high. This will blast or flare and may melt the torch tip.

4. Hydrogen too low. Will not maintain flame or may not light or may backfire into the tip. Flame will not cut.

5. Oxygen too high. Torch sputters and flares. Excessive force is required to hold torch to the work because of pressure.

6. Oxygen too low. Preheat flame will not light or will pop out.

On the bottom, place yourself in a comfortable position with your eyes below the level of cutting whenever possible. Hold the preheating flame against the metal. If the flame is right, there should be a bright yellow glow. If not, readjust the valves until you get it, first the hydrogen and then the oxygen. Move the valves very little at a time and watch the results.

When the proper preheat is reached, squeeze the high-pressure oxygen lever. Pull the trigger gradually rather than suddenly to avoid chilling the preheated spot. Hold the end of the air nozzle about ³/₈″ away from the metal and point it slightly in the direction of the cut. Some divers hold the edge of the nozzle against the metal as a gage. Others use nozzles with prongs, and rest the prongs lightly on the metal. However, the skillful operator usually is able to hold the torch a uniform distance without any steadying devices. Occasionally, even the skillful operator will use prongs when cutting in muddy waters or in difficult positions.

Advance the cut by moving the torch at a steady, even rate of speed, along the line of cut, fast enough to keep the cut going, and slow enough to cut completely through the metal.

If the torch is advanced too slowly, the steel will cool below the ignition temperature and the cutting action will stop. You then will have lost the cut and will have to start over.

If the torch is advanced too rapidly, the metal will not be completely cut in two. The skipped spots will be very hard to cut out.

If the cut has been lost, it may be restarted by going back ¹/₂ to ³/₄″ and starting on the side of the old cut. The purpose of this is to be sure that the cut is complete and no *holiday* left to give trouble.

After the cut is completed, close tightly all valves on the torch before returning to the surface to prevent clogging of passages. At the end of the day, blow out the torch, dry it and put it away with all valves open.

Arc-Oxygen Cutting

The Arc-Oxygen cutting process is in wide use throughout the world for both surface and underwater cutting because of the many advantages it offers. The principles of arc-oxygen cutting are very simple. An electric current is applied through a tubular steel electrode, creating an arc that heats the metal to be cut. Oxygen, delivered through the hollow core of the electrode directly to the point of reaction, serves as the cutting agent. The arc has sufficient heat to start melting the metal instantly, even underwater.

In cutting mild steel for example, the application of oxygen to the molten metal results in the oxidation or combustion of the steel. The additional heat liberated by this oxidation superheats the arc and, as the cut proceeds, the steel is actually consumed by a self-sustaining process—not just melted away.

In application, arc-oxygen cutting is fast, easy, safe and economical. No pre-heating is required, and no previous skill or experience is needed. It can be used to cut metals of almost any thickness. An unskilled operator will cut a 1″ mild steel plate with arc-oxygen in less than half the time required for oxy-acetylene on the surface. The heat affected area of the cut metal is about the same as in arc welding, and since the heat penetration is quite shallow, there is practically no warpage of the cut metal, and therefore no attendant stresses.

For underwater, the process has all but replaced oxy-hydrogen because, with the drag technique, arc-oxygen cutting can be accomplished in zero visibility at any practical depth, while oxy-hydrogen involves considerable skill, pre-heating, and the difficulty of maintaining proper flame, advance of cut, etc.

Equipment. Underwater cutting with arc-oxygen requires an oxygen supply, a 300 ampere welding machine (preferably DC) and a fully insulated torch. See Fig. 35-4. A double-pole single-throw knife switch must be installed along the torch and ground welding cables at a point that is within easy reach of the diver's tender. See Fig. 35-5. The switch is turned on only when the diver is actually making a cut. At all other times, the switch must be kept in the OFF position.

Additional equipment includes a ground clamp designed for underwater service, a diver's flip-type eyeshield, diving outfit equipped with

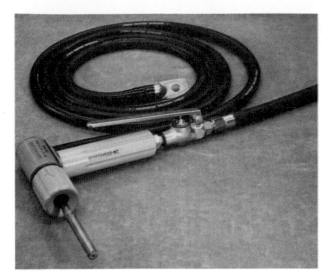

Fig. 35-4. Arc-oxygen underwater cutting torch. (Craftsweld Equipment Corp.)

Fig. 35-5. 300-ampere shut-off switch shown installed along ground and torch cables. (Craftsweld Equipment Corp.)

telephone, oxygen cylinder with pressure regulator, oxygen hose, and welding cables.

Cutting operation. DC current with straight polarity is recommended for underwater cutting. AC current is less desirable because it increases the electrical hazards to the diver and causes greater electrolytic damage to underwater

equipment. The cutting arc is produced by special high-speed $5/16'' \times 14''$ tubular steel waterproofed electrodes.

Arc-oxygen cutting electrodes are manufactured to the highest commercial and military standards. They are recommended for use with a 300-ampere DC welding machine, but are almost equally as efficient with AC. These electrodes have a special extruded flux covering that makes some very important contributions to the cutting process. First, the flux burns at a rate slower than the consumable hollow steel core of the electrode, and thereby maintains an automatic arc gap at the cutting point. This makes it unnecessary for the operator to hold an arc (as in conventional arc welding) and therefore the simple technique requires only that *the electrode be dragged across the work.* Second, the flux is an insulating medium, and eliminates all chance of arcing from the side of the electrode to any adjacent metal. Third, the flux performs the very special function of helping any non-oxidized metal or slag flow from the cut. In addition, the steel core of the electrode, by introducing molten steel into the reaction, makes it possible to cut "hard-to-cut" metals in a way much like the iron powder cutting process. (See Chapter 33) Thus, one electrode combines the qualities of flux-injection and iron powder cutting without the need for special additional equipment.

Fig. 35-6. Diver's flip-type eyeshield shown installed on helmet. (Craftsweld Equipment Corp.)

All cable and connections must be in perfect condition. All connections should be insulated and made watertight by careful taping, especially those that will be submerged during operations underwater.

The flip-type eyeshield should be attached to the outside of the diver's helmet faceplate (Fig. 35-6) with appropriate lenses as follows:

#4 lens for very muddy water

#6 lens for average conditions

For ease in handling, the power cable and oxygen hose should be taped together at intervals of 2 feet.

The diver should wear a watertight dress, dry rubber mittens and a beany that covers the ears. The helmet chin button is usually insulated with rubber tape or a baby bottle nipple.

The actual cutting operation should proceed in the following manner:

1. Start the welding machine and set for 300 amperes. Be sure the knife switch on the welding cable is open (OFF).

2. Set the oxygen regulator for the required pressure. See Table 35-2.

CAUTION: Cutting must be done when facing the ground connection to avoid serious electrolytic damage to the helmet. See Fig. 35-7.

4. Insert an electrode in the torch and lock it in position with a slight turn of the torch locknut. Close the eyeshield over the faceplate and crack open the oxygen valve on the torch by squeezing the lever slightly. Cracking the oxygen valve keeps the electrode from welding itself to the work when the arc is started.

5. Call to the surface for *Current On.* This tells the tender to close the knife switch on the welding cables.

6. Place the tip of the electrode at right angles to the work and strike the arc. As soon as the arc is started, squeeze the oxygen lever all the way and then drag the electrode along the line of cut.

7. Bear down on the torch so as to keep pressing the electrode against the work. Move forward as fast as possible while maintaining full penetration of cut. Lack of penetration will be evident by the amount of "back spray." Always keep the electrode against the work while making a cut.

TABLE 35-2. ARC-OXYGEN CUTTING DATA.

(Average conditions for mild steel)

THICKNESS OF STEEL	1/4″	1/2″	3/4″	1″
Tubular steel electrodes consumed	21 lb	29 lb	29 lb	33 lb
	or	or	or	or
per 100 ft of cut	70 pcs	100 pcs	100 pcs	100 pcs
Oxygen consumed	250 cu ft	300 cu ft	350 cu ft	400 cu ft
per 100 ft of cut	at	at	at	at
Oxygen pressure*	20 psi	30 psi	40 psi	50 psi

*Add 5 psi for each 10 ft depth of water, plus 5 to 10 psi for each 100 ft of hose used
*Use additional 10 to 20 psi for cutting cast iron, stainless steel and nonferrous metals

3. When submerged to the proper cutting position, attach the ground clamp to a clean area on the work that will provide a good electrical connection. Locate the ground clamp so you always face it when the current is on. Clear the area around you so that footing is secure, and no obstacles are in the way.

8. When the electrode has been consumed, call to the surface for *Current Off.* After the tender reports that the current is OFF, unscrew the torch locknut slightly and blow the electrode stub out of the torch by squeezing the oxygen lever. Then insert a new electrode and continue the cut, following the same procedures.

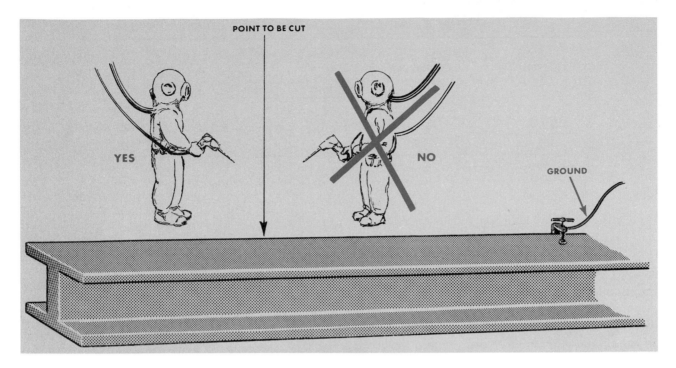

POINT TO BE CUT

YES

NO

GROUND

Fig. 35-7. Diver must always face ground connection when the current is on. He should do all cutting between himself and the ground. To avoid serious damage to helmet, the diver must never stand between himself and the ground. (Craftsweld Equipment Corp.)

Metallic-Arc Underwater Cutting

The process of arc cutting is purely a method of applying the heat of the electric arc to melt the metal along a desired line of cut. In underwater cutting, this method may be used in situations in which no oxygen is available. When ordinary covered welding electrodes are employed, it provides a means for cutting both ferrous and non-ferrous metals. This process, known as metallic arc cutting, is superior to those employing oxygen when cutting steel plate of thickness less than 1/4″ as well as for non-ferrous materials such as brass, copper, nickel-copper, and manganese bronze, regardless of thickness.

Equipment. The recommended power supply for metallic arc underwater cutting is a DC welding generator having a capacity of at least 400 amperes and connected for straight polarity. Satisfactory cutting can be done with a 300 ampere DC machine but the cutting rate will be slower. An alternating current power source is usually not recommended for underwater cutting because the heat produced is less and the hazard greater from a safety standpoint.

CAUTION: As in all underwater cutting operations involving an electrical circuit, it is important that a positive disconnecting switch be used in the circuit. This gives the diver full protection by having the current ON only when he is actually cutting.

The electrode holder must be completely insulated and designed specifically for underwater cutting.

Special waterproof electrodes are available for cutting purposes. See Table 35-3.

Operation. When employing 3/16″ electrodes with a power source of 300 amperes at about 40 volts, steel plate up to 1/4″ thickness can be cut by simply dragging the electrode along the desired line of cut. The thickest plate which can be cut using this drag technique is about 1/4″ where the capacity is 300 amperes, and about 3/8″ where the amperage can be increased to 400.

TABLE 35-3.   METALLIC ARC UNDERWATER CUTTING.

ELECTRODE SIZE	UNIT	NO. OF ELECTRODES APPROX	POWER SOURCE	STEEL PLATE CUT IN FT/BOX OF ELECTRODES		
				1/4″ THICK	1/2″ THICK	3/4″ THICK
3/16″	50 lb box	410	300 amps	185	102	—
3/16″		410	400 amps	307	135	58
1/4″		220	400 amps	176	77	44

Note: With 300 amperes available for underwater cutting, use 3/16″ diameter electrodes; 5/32″ may be used but the burnoff rate is very rapid. With 400 amperes available, use 3/16″ diameter or preferably 1/4″ diameter electrodes

To cut thicker plate, a slow, short stroke, sawing motion must be used to push the molten metal out of the far side of the cut. Skillful application of this sawing technique makes the metallic arc-cutting process practical over a wide range of thicknesses even where the large electrodes and heavy currents recommended for use with the drag technique are not available.

Where large electrodes and heavy currents can be had, the drag technique has the advantages of speed and simplicity of operation. Mastery of the cutting technique will be aided by understanding that the metal is merely melted, not oxidized or consumed in any way, and that if the molten metal does not run out of the cut by itself, it must be pushed out by manipulation of the electrode.

UNDERWATER WELDING

Underwater welding is done primarily on mild steel with special underwater welding electrodes. These electrodes will produce welds that develop 80 percent of the tensile strength and 50 percent of the ductility of similar welds made on the surface. The reduced ductility and strength is the result of the quenching action of the water.

If a weld to be made is of a critical nature, it is generally advisable to make a test weld at the same water depth and working conditions. This test should then be brought to the surface for inspection and testing.

Welding Process

Equipment.   The preferred power source for underwater welding is a 300 ampere DC generator connected for straight polarity. A positive operating safety switch must be installed in the welding circuit. See Fig. 35-8.

The electrode holder must be completely insulated, durable, and allow for easy changing of electrodes. Holders with metallic jaws as used for welding in air are not recommended even if they are fully insulated. A specially designed plastic holder for underwater welding is shown in Fig. 35-9.

Operation.   Welding is done either with 3/16″ or 5/32″ dia electrodes. No specifically designed electrodes have been developed for use in underwater welding. The two commercial brands of electrodes which have given satisfactory performance in underwater welding test are Westinghouse *Flexarc SW* and Lincoln *Fleetweld 37*. These are designated for all-position welding.

Since all electrode coatings deteriorate when immersed in water, it is a good idea to waterproof the covering by dipping the electrode in some waterproofing solution. Several solutions are commercially available. However, waterproofing is not entirely necessary if the electrodes are used within a short time after immersion. This can be achieved by sending the diver a few electrodes at a time. If the electrodes are waterproofed, care must be taken to clean off the tips so enough of the bare wire is exposed for easy starting of the arc.

The actual welding procedure underwater should be carried out as follows:

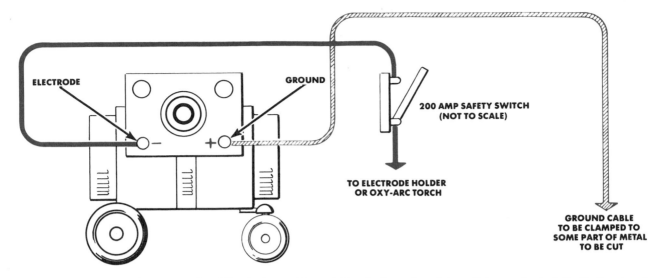

ELECTRODE

GROUND

200 AMP SAFETY SWITCH
(NOT TO SCALE)

TO ELECTRODE HOLDER
OR OXY-ARC TORCH

GROUND CABLE
TO BE CLAMPED TO
SOME PART OF METAL
TO BE CUT

Fig. 35-8. All underwater cutting and welding is done with straight polarity, and a safety switch must be installed in the circuit. (United States Navy)

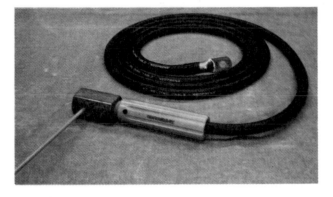

Fig. 35-9. Underwater welding electrode holder. (Craftsweld Equipment Corp.)

1. With 3/16" electrodes, set the welding machine for 225 to 280 amperes at water depths up to 50 feet (15.24 meters). The correct amount of current should result in consuming one electrode in 49 to 55 seconds. If the electrode is not consumed in this time, proper current is not reaching the work due to long cables, poor connections, or improper setting of the welding machine. The 5/32" electrodes require somewhat less current and are consumed in slightly shorter time than 3/16" electrodes.

2. Because of poor visibility under water, fillet welds should be used wherever possible, in order to provide an edge that the diver can use as a guide. The work must be clean of rust, paint and marine growth at all times.

3. Attach the ground clamp as close as possible to the section to be welded.

CAUTION: It must be located so you always face the ground clamp when the current is on. (See Fig. 35-7.)

4. In making a *horizontal weld,* place the end of the electrode against the left end of the work at an angle of about 15° to 45° (right-hand operator) and call for *Current On.* The arc should start as soon as the tender closes the knife switch. It may be necessary to scrape or tap the electrode against the work in order to start the arc if the metal is rust coated.

When the arc has started, press the electrode against the work so that it consumes itself, and move the electrode along the line of weld at an even speed. Generally, 10" of electrode will deposit about 8" of weld.

Do not hold or space the arc the way welding is done in air when welding underwater, just keep holding the electrode against the work at the same angle that was used when starting the weld. After the electrode has burned down to the stub, call for *Current Off.*

5. Before starting with another electrode, clean the end of the previous deposit and then

make a slight overlap when the weld is resumed. When making multiple passes, the entire weld must be thoroughly cleaned before the next pass is made.

6. Use the same technique for *vertical welds,* starting from the top down, with the electrode pointing upward.

7. *Overhead welds* require greater skill and more exact manipulation but can be made successfully. Increasing the angle of the electrode to the work up to 55° will help improve the contour of the weld.

SAFETY IN UNDERWATER CUTTING AND WELDING

In underwater cutting and welding, no matter what procedure is used, the importance of safety must be constantly emphasized. In underwater operations, the life and safety of the divers are ever dependent upon the strict observance of safety regulations.

CAUTION: The use of electric power in underwater cutting and welding can be very hazardous. This is especially true in sea water which is an excellent conductor of electricity.

With proper safeguards and reasonable care, underwater cutting and welding can be accomplished with comparative safety if the basic protective measures which follow in this section are strictly observed. Furthermore, personnel engaged in welding or cutting operations should exercise their imagination to insure that no hazard exists without all possible steps being taken to eliminate it or to provide emergency controls.

SAFETY CAUTIONS

1. Only a qualified diver, assisted by an experienced tender, should use underwater welding equipment.

2. The diver should practice cutting or welding above water before attempting underwater work.

3. Detailed operating instructions should be obtained from the equipment manufacturer and followed carefully.

4. The oxygen regulator used should be adequate for delivery of the required volume without freezing up. Be sure oxygen regulator, hose, fittings and torch are all clean and free of oil. *Never use oil, grease or any other combustible lubricant on equipment that uses oxygen.*

5. The diver must be attired so that his or her body is fully insulated from the work, torch, and water. Precautions must be taken to insulate the diver's head from the helmet by wearing a skull cap.

6. The current should be OFF at all times except when the diver is actually cutting or welding at the work.

7. After inserting the electrode in the holder, the diver should place the electrode against the work before signalling for *Current On.* The tender, after closing the switch, should immediately confirm this action by repeating *Current On.*

8. When the electrode is consumed, the diver should not attempt to remove the stub until signal has been made for *Current Off* and then the diver must wait until the request is confirmed from on top.

9. A careful examination must always be made before starting a cut to prevent the piece that is cut away from falling or rolling over, or striking another structure and pinning the diver down or fouling his lines.

10. The cable and torch insulation, and all joints in electrical circuit, should be checked for current leakage at frequent intervals. All underwater cable connections should be fully insulated and watertight.

11. No work of any kind should ever be permitted on the surface over the area in which the diver will be working within a radius of at least equal to the depth of underwater operations.

12. The diving gear should be in good condition, and equipped with a reliable loudspeaker telephone. The diver should always wear dry rubber mittens.

13. Before starting any cutting or welding, be sure there are no highly combustible or explosive materials (whether gases, liquids or solids) close to the point to be cut or welded, or within a radius of at least 50 feet. Sparks have been known to travel that far, especially upwards (and in other directions through existing channels such as ship bottoms or holds).

14. Because difficult footing and poor visibility generally prevail underwater, the diver should handle torch or welding holder with care, staying completely clear of all hoses, and avoiding too much slack in his lines. *Keep all hoses and lines away from the cutting or welding operation.*

The diver must also be careful not to get into any position where something might fall on the precious life lines.

15. The diver must not permit any part of his or her body or gear to become a part of the electric circuit.

16. If AC current must be used for cutting, the resulting shock can be more severe if the diver's body or gear accidentally enters the circuit. (AC CURRENT IS NOT RECOMMENDED FOR WELDING AT ANY TIME.)

17. The diver should always telephone the tender to shut OFF current before changing electrodes, and keep it OFF except when actually cutting or welding. Be sure to close eyeshield (with proper lens) before striking an arc. Always remove the electrode before taking torch or welding holder underwater or returning it to the surface.

18. The diver should inspect his or her helmet and all other metallic parts of the gear regularly for signs of deterioration resulting from electrolysis. The diver should make the ground connection at a point with reference to the welding or cutting position that will reduce electrolysis to a minimum.

CAUTION: The diver should never turn his or her back on the ground connection.

19. After each day's use, the torch or welding holder should be thoroughly rinsed in fresh water, and then dried, to help maintain proper operating efficiency.

20. All Arc-Oxygen Cutting Torches are equipped with a removable Flash Arrester Cartridge containing a monel screen. This screen is very inexpensive and easy to replace. If the screen gets clogged with dirt or slag, or burns out, it should be replaced. Operating the torch without flash arrester or screen could be a safety hazard. (Frequent clogging or burning out of the screen may occur if too little oxygen pressure reaches the electrode or if you attempt to burn the electrode down to the very last.)

Points to Remember

1. Be sure your attire fully insulates your body from the work, torch, and water.

2. Before lowering a lighted torch to the diver, test it for stability by dropping it into the water.

3. Never lower a lighted torch until the diver is ready for it.

4. In lighting a torch under water, hold the torch horizontally with the tip pointed away from the lines.

5. A lighted torch under water, if it is functioning properly, gives off a rumbling or bubbling sound.

6. Be sure the compressed air is of the right pressure—if too high the metal cools too rapidly making the cutting process difficult—if too low the flame flickers and sputters.

7. Maintain correct hydrogen and oxygen pressure, otherwise proper cutting is impossible.

8. Advance the torch at a steady, even rate.

9. Always check if the piece to be cut will fall free after the cut is completed.

10. Always check your equipment before going under water.

11. Watch your footing and stay clear of all hoses and lines.

12. When using the arc-oxygen cutting process, be sure the switch is turned on only when making the cut. At all other times the switch must be turned off.

13. In arc-oxygen cutting, make certain that all cables are in perfect condition and all connections are insulated and watertight.

14. Remove electrode stub only after the tender has been told to turn the current off.

15. Always face the ground connection when the current is on.

16. Do all cutting between yourself and the ground.

QUESTIONS FOR STUDY AND DISCUSSION

1. Why is hydrogen rather than acetylene used as a fuel in underwater cutting?

2. How does an underwater cutting torch differ from a cutting torch used in open air?

3. Approximately how much additional hydrogen pressure must be provided to compensate for water depth and hose friction?

4. Why is excessive compressed air pressure undesirable?

5. How should a flame be tested before lowering the torch to the diver?

6. In what ways can a lighted torch be lowered underwater?

7. How should a torch be lit underwater?

8. How is it possible to tell underwater if a torch flame is correctly adjusted?

9. What happens when the compressed air is too high or too low?

10. What is likely to occur if the hydrogen pressure is too high or too low?

11. What occurs when the oxygen pressure is too high or too low?

12. Why is the proper speed at which a torch is advanced so important in making a clean cut?

13. What is the basic principle by which the arc-oxygen cutting process operates?

14. In the arc-oxygen system, why must a switch be installed on the surface along the torch and ground welding cables?

15. Why is DC current preferable for underwater cutting?

16. What type of electrodes are used for underwater cutting in the arc-oxygen process?

17. What is meant by the *drag technique* in underwater cutting?

18. When cutting with arc-oxygen, why should the operator always be sure to face the ground connection?

19. When is the metallic arc process used in underwater cutting?

20. Why is an AC machine not recommended for underwater cutting?

21. What type of motion is used to cut light materials with the metallic arc?

22. Why do welds that have been made underwater have less ductility than those that have been made on the surface?

23. What type of electrodes are used for underwater cutting?

24. Why are fillet welds most often recommended for all underwater welding?

CHAPTER 36 *plastic welding*

In the fabrication of many plastic products a welding process is often used to fasten parts together. Joining plastic edges by welding is a common technique in assembling such products as storage tanks boxes, and other containers. Installation of pipe and duct work by welding has become common. The manufacture of many other custom products of plastic are possible by welding techniques. See Fig. 36-1.

Plastic welding is very much like welding metal where localized heat is used to produce fusion. The same material preparation such as fit-up, root gap, joint design and beveling are required in plastic welding as in metal welding. However, there is one significant difference. In metal welding a sharply defined melting point develops and both the metal and welding rod melt and flow together to form the weld joint. Plastics, on the other hand, are poor heat conductors and consequently do not melt and flow; they simply soften. To achieve a permanent bond, a soft (heated) filler rod has to be forced into the softened surface of the joint. Thus plastic welders have to develop a knack of working within narrower temperature ranges than those doing metal welding.

Some plastics are easier to weld than others and one group of plastics cannot be welded at all. The basic plastic welding techniques are hot-gas, heated tools, induction, and friction.

TYPES OF PLASTICS

Most plastics are known by trade names or by the principal substance from which they are made. The two main plastic families are *thermosetting* and *thermoplastics.*

The thermosetting plastics will soften only once when exposed to heat. After they are molded into a particular form and become hard no subsequent heating will soften them. These plastics are not weldable. Typical thermosetting plastics are ureas, phenolics, melamines, polyesters, silicones, epoxies, and urethanes.

Thermoplastics will repeatedly soften whenever heat is applied. These plastics can easily be welded. There are many kinds of thermoplastics, such as acrylics, polystyrenes, polyamides, polyfluorides, hydrocarbons, and vinyls. Generally the more common thermoplastics used where welding is involved are the polyethylenes, polyvinyl chlorides (PVC) and polypropylenes. Welding these plastics will produce seams that are as strong or stronger than the materials being bonded.

PLASTIC WELDING TECHNIQUES

Hot Gas Welding

Hot gas welding is accomplished with a special design gun containing an electrical heating

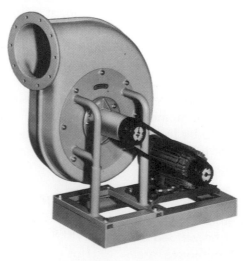

Fig. 36-1. Welding is a common practice in assembling many plastic products. (Industrial Plastic Fabricators, Inc.)

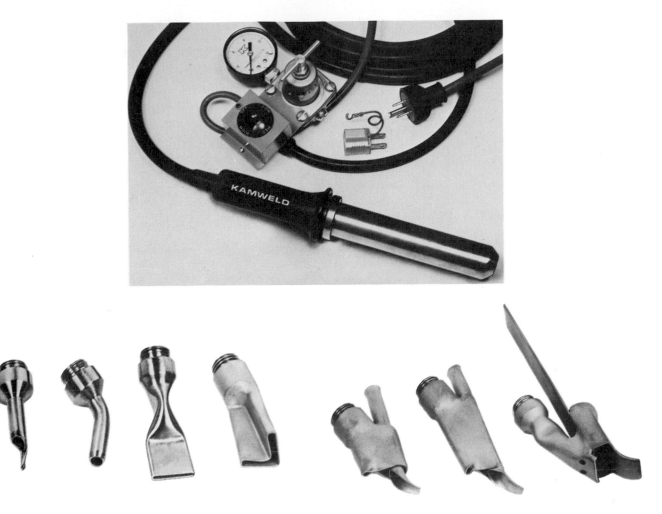

Fig. 36-2. Types of tips used in plastic welding.

unit. A stream of compressed air or inert gas (nitrogen) is directed over the heated element which then flows out of the nozzle onto the surface of the material being bonded. The gun permits the use of several tips for different welding operations. The basic tips are round for general purpose welding and welding in tight corners, flat for straight welding, and V-shaped for corner welding. A special tip is also available for high-speed welding. The increased speed is achieved by the design of the tip which holds the filler rod and applies the needed pressure as the weld is made. A tacker tip is sometimes used for tack welding. See Fig. 36-2.

Joint preparation: The type of joints used in plastic welding are the same as those in metal welding—butt, lap, T, corner, and edge. The edges of the joint should be beveled to have a sufficient area so a good bond can be formed. Notice in Fig. 36-3 that the beveled edges have an included angle of 60°, with a root gap between $1/64''$ to $1/16''$.

Welding procedure. The technique for welding plastic is very similar to oxy-acetylene welding of metals. The gun is held in one hand and the filler rod in the other. Welding goggles or helmet are not necessary. In general the following procedure is used:

Select the correct shape tip and insert in the gun. Normally guns should be able to supply a temperature varying from 400°F to 600°F (204 to 316°C) or more (up to 925°F or 496°C). Different materials and different thicknesses require different degrees of heat.

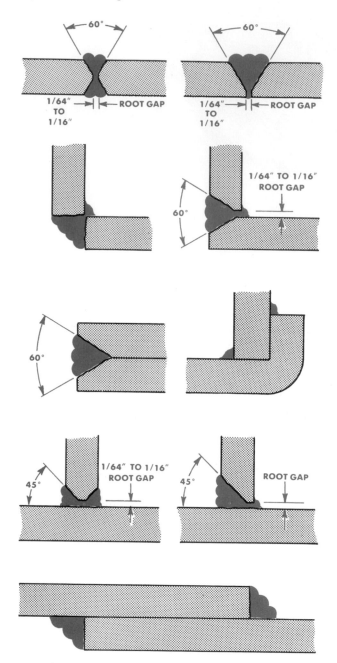

Fig. 36-3. Joints for hot-gas welding of thermoplastics.

TABLE 36-1.   AIR PRESSURE RECOMMENDATIONS.

ELEMENT* (watts)	AIR PRESSURE (pounds)	TEMPERATURE (°F at 3/16" from tip)
320	2–3	400
340*	2–3	410
350	2 1/2–3 1/2	430
450	3–4	540
460*	3–4	600
550	4–5	700
650	4 1/2–5 1/2	800
750	5–6	860
800*	5–6	900

*Note: Three-heat unit with a rotary heat selector switch: Low—340 w, Medium—460 w, High—800 w.

Compressed air is best for welding PVC and several other types of plastics. However, better results are obtained when welding oxygen-sensitive plastics such as polyethylene and polypropylene by using nitrogen gas. See Table 36-2. Both the gas and compressed air should be controlled by regulators to provide the correct pressure flow. With some installations a gas flow meter on the nitrogen tank is equipped with a Y or by-pass valve to permit shutting off the nitrogen and switching to compressed air. This allows the use of compressed air to keep the weld at the proper heat when welding is temporarily stopped, to conserve nitrogen.

2. *Select the correct filler rod and cut its end at a 60° angle.*   The rod should be of the same basic composition as the parent plastic. Either flat, round or triangular rods may be used. Triangular rods are particularly advantageous in V or fillet welds since the area can be filled with one pass. This reduces welding time and minimizes chances of porosity which may occur with multiple passes of round rods. See Fig. 36-4.

3. *Hold the tip of the gun about 3/16" to 1/2" away from where the weld is to be started and begin a fanning motion.*   Place the rod in a vertical position so the heat from the gun is directed both on the rod and base material. When the base material and rod become tacky, press the rod firmly into the joint and bend it

1. *Set the air or gas pressure according to the manufacturer's recommendations.*   Although the wattage of the heating element determines the range of heat, the air or gas pressure determines the actual amount of heat at the tip. See Table 36-1.

TABLE 36-2.   THERMOPLASTIC WELDING CHART.

THERMO-PLASTIC	PVC	POLY-ETHYLENE	POLY-PROPYLENE	PENTON	ABS	PLEXIGLASS
Welding temp	525	550	575	600	500	575
Welding gas	Air	Nitrogen	Nitrogen	Air	Nitrogen	Air

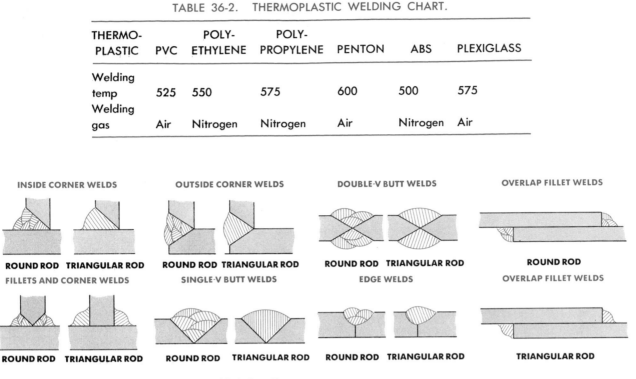

INSIDE CORNER WELDS   **OUTSIDE CORNER WELDS**   **DOUBLE-V BUTT WELDS**   **OVERLAP FILLET WELDS**

ROUND ROD TRIANGULAR ROD   **ROUND ROD TRIANGULAR ROD**   **ROUND ROD TRIANGULAR ROD**   **ROUND ROD**

FILLETS AND CORNER WELDS   **SINGLE-V BUTT WELDS**   **EDGE WELDS**   **OVERLAP FILLET WELDS**

ROUND ROD TRIANGULAR ROD   **ROUND ROD TRIANGULAR ROD**   **ROUND ROD TRIANGULAR ROD**   **TRIANGULAR ROD**

Fig. 36-4. Triangular rods produce better welds in less time. (Kamweld Products Co.)

back at a slant with the point away from the direction of welding. See Fig. 36-5. As the gun is moved along the seam continue to exert pressure on the rod to force it into the groove. Maintain a constant fanning motion at a 45° angle so both the rod and joint area are heated equally. On heavy gage sheet with light rod most of the heat should be directed on the joint.

During the welding cycle it is important not to exert too much pressure on the rod since it will cause the rod to stretch excessively. The length of the rod used should be no more nor less than the length of the weld. Equally important is to avoid overheating because it will cause the rod and base material to char and discolor. Overheating will weaken the weld and cause cracks to radiate into the sheet from the weld. Underheating is also objectionable since it produces a cold weld which has poor tensile strength.

4. *Check weld.*   Bend a test weld 90°. If the weld is made properly, the weld beads will not separate from the base material nor will it be possible to pry the rod out of the weld when cooled. Cutting a cross-section through a test weld will disclose whether or not there is correct penetration. See Fig. 36-6.

CAUTION: Some plastic materials produce obnoxious odors and poisonous fumes (polyvinyl chloride). Therefore, precautions must be taken to avoid inhaling these fumes. In any case plastic welding should always be done in a well ventilated area. It is a safe practice to follow the manufacturer's recommendations for the type of plastics to be welded.

Tack welding. Tack welding is simply a means of fusing pieces together prior to welding them in order to eliminate the use of clamps or fixtures. See Fig. 36-7. A tacker tip is used for this purpose which is quickly drawn at intervals along the joint. The gun is held at an angle of about 80° with the point of the tip touching the material. Tacks should be about 1/2" long. When a tack welding tip is not available, tacking can be done with a regular round tip and filler rod.

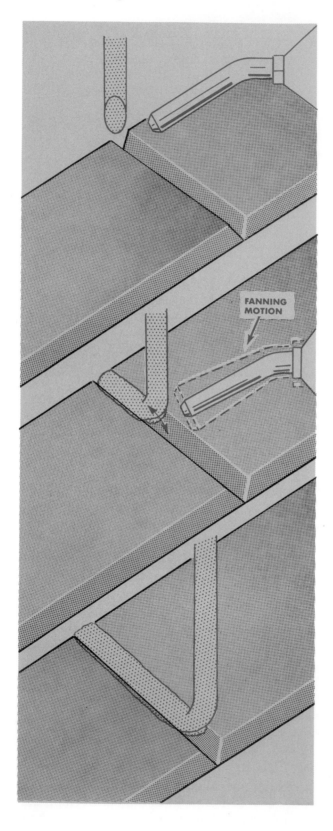

Fig. 36-5. Procedure for thermoplastic welding. (Seelye Plastic, Inc.)

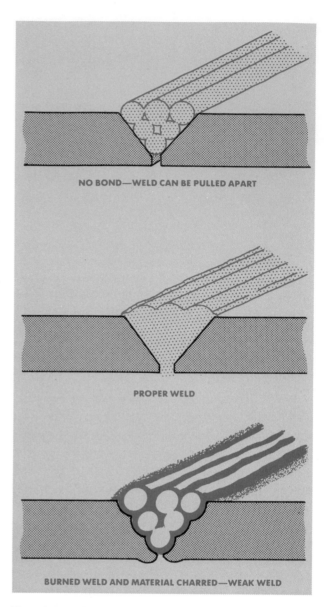

NO BOND—WELD CAN BE PULLED APART

PROPER WELD

BURNED WELD AND MATERIAL CHARRED—WEAK WELD

Fig. 36-6. Welding analysis of a test weld. (Seelye Plastic, Inc.)

High-speed welding. The speed of making welds can be substantially increased with the use of a special tip that holds the welding rod in the correct position. Flat strips are often used for this operation although round or triangular rods will also serve the same purpose. Since the rods are supplied in roll form they first must be cut into required lengths with one or two inches allowed for trimming. Welding procedure should be as follows:

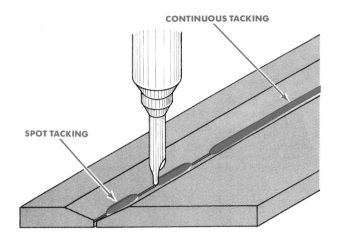

Fig. 36-7. Making a tack weld. (Kamweld Products Co.)

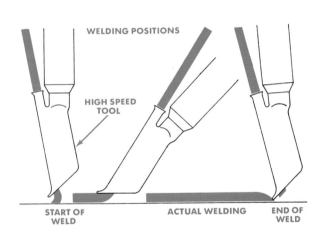

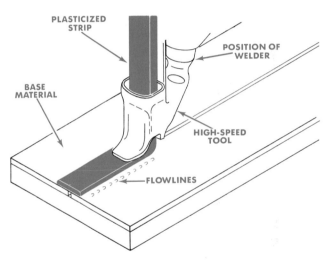

Fig. 36-8. High-speed welding. (Kamweld Products Co.)

1. *Insert the rod into the feeder tube.* Start the weld by tamping the broad shoe of the tip on the surface until the first inch of the rod adheres firmly to the base material. Hold the welding gun at a 90° angle to the work and press the end of the rod into the weld. Feed the weld rod manually until the weld bead is well started. See Fig. 36-8 as an example.

2. *Move the welder forward along the seam,* dropping the angle of the welder to 45°. Once the welding operation is well underway the rod will feed automatically into the pre-heated tube by simply exerting firm downward pressure on the gun with the wrist.

3. *Keep the gun moving at sufficient speed.* Correct speed can be observed by the formation of flow lines on both sides of the rod. Insufficient speed will cause the rod to stretch because of the built-up excessive heat. This condition can be corrected by a quick tamping motion of the shoe as used in starting the weld.

To stop the weld, lift the welder and allow the remaining rod to pull through the feeder tube. Then cut the rod with a special curved knife. Never allow the filler rod to remain in the feeder tube because it will clog the opening.

When a weld is to be made from one corner to another such as in a tank, start the weld in one corner and proceed only halfway. Then begin the weld from the other corner and overlap the scarfed end of the first weld.

Heated-Tool Welding

In the heated-tool process the edges to be joined are heated to a fusing temperature and then brought into contact and allowed to cool under pressure. The edges of the plastic sheet are softened with some heat producing unit such as an electrical strip or bar heater, hot plate, or resistance-coil heater. The heaters should be of the aluminum or nickel plate type since hot steel and copper units have a tendency to decompose plastic. This welding technique is frequently used in joining sections of pipe and tubing and in the assembly of many molded articles.

Heat is applied by holding the edges in contact with the heating unit until the surface is softened. When the material is sufficiently in a molten state it is removed from the heater and the edges quickly pressed together. The pressure between the pieces should be enough to force out air bubbles and form a solid contact. Normally, pressures of 5 to 15 psi will produce good bonded joints. Pressure can be applied by hand or in production work by jigs. The pressure must be maintained until the weld is cooled.

The most important factor in securing sound welds by the heated-tool technique, outside of proper softening of materials and firm contact, is the elapsed time between removing the pieces from the heating unit and joining them together. This interval should be as short as possible to prevent any degree of solidification before the edges come in contact.

Induction Welding

In induction welding, heat is generated by causing a high-frequency current to flow into a metal insert placed between the areas to be joined. Although induction welding is one of the fastest methods of joining plastic, its greatest limitation is that the metal insert must remain in the weld.

The metal inserts usually consist of metallic foil, a coil of wire, wire screen, metallic conducting particles or any other configurations of con-

ductive metal. They must be placed in the interface so they are not exposed to the air, otherwise rapid heating is induced which may cause the inserts to disintegrate. Fusion occurs only in an area immediately near the insert. When the edges become soft, uniform pressure is applied to bond them together. As a rule welds made by the induction process are not as strong as those obtained by other heating methods.

Friction Welding

Friction or spin-welding consists of rubbing the surfaces of the parts to be joined until sufficient heat is developed to bring them up to a fusing temperature. Pressure is then applied and maintained until the unit is cooled. Usually in friction welding one piece is held stationary and the other is rotated. When sufficient melt occurs the spinning is stopped and the pressure increased to squeeze out air bubbles and distribute the softened plastic uniformly between the surfaces.

The principal advantages of spin-welding are speed and simplicity of the process itself. However, this technique is limited to circular areas. Sometimes spin welding produces a flashing out of soft material beyond the weld area, but usually the excess flashing can be directed to the interior of the part if the weld is properly designed. Excess flashing can also be avoided by preventing the parts from overheating and by maintaining the proper pressure.

Spin welding has wide applications in fastening instrument knobs, pipe fittings, container caps, handles and other units where parts can readily be spun.

Points to Remember

1. Weld plastics only in a well ventilated area.
2. Bevel all edges to secure a proper weld joint.
3. Use a filler rod of the same composition as the parent material.
4. Set the air or gas pressure to provide the proper flow as prescribed by the manufacturer.
5. Use a fanning motion to insure uniform heat distribution over the rod and edges of the joint.

6. Avoid exerting too much pressure on the filler rod as it will snap.

7. Do not allow the surface to char or discolor.

8. Do not let the rod stretch.

QUESTIONS FOR STUDY AND DISCUSSION

1. What is the main difference between plastic welding and metal welding?

2. Why are thermosetting plastics not weldable?

3. At what range of temperatures are plastics generally welded?

4. What governs the degree of heat which is to be used in plastic welding?

5. What is the particular advantage of using a triangular shape filler rod over a round rod?

6. How far away from the surface should the gun be held when welding plastics?

7. Why is a fanning motion necessary in manipulating the gun over the weld joint?

8. Why should excessive pressure on the rod be avoided?

9. How can you tell if the heat is too great?

10. What happens if insufficient heat is used when a welder is making a plastic weld?

11. How can you check to see if a weld is made properly?

12. What precautions should be taken when welding plastics?

13. How does a high-speed plastic welding technique differ from the regular hot gas-welding technique?

14. When using a high-speed welding tip, why shouldn't the filler rod be allowed to remain in the feeder tube?

15. How is the heated-tool welding technique accomplished?

16. What is one of the main limitations of induction plastic welding?

17. How are plastic joints bonded by spin welding?

18. For what kind of application is spin welding best suited?

CHAPTER 37 testing welds

In the fabrication of any welded product, tests are generally employed to determine the soundness of welds. The nature of the test is governed to a great extent by the service requirements of the finished product.

Various types of tests have been devised, each for a specific purpose. These tests may be broadly classified as visual, destructive, and nondestructive.

VISUAL EXAMINATION

Visual examination consists of looking at a weld, or examination through a magnifying glass. A thorough examination of the weldment may disclose such surface defects as cracks, shrinkage cavities, undercuts, inadequate penetration, lack of fusion, overlaps, and crater deficiencies. Very often weld gages are used to check for proper weld bead size and contour.

The limitation of any visual examination is that there is no way of knowing if *internal defects* exist in the welded area. The outer appearance of a weld may be satisfactory, yet cracks, porosity, slag inclusions, or excessive grain growth can be present which are not visible.

DESTRUCTIVE TESTING

Destructive testing involves the use of sample portions of a welded structure and subjecting them to loads until they fail. The most common types of destructive tests are known as *tensile, shear, weld uniformity, etching,* and *impact.*

Tensile Testing

Tensile testing involves the placement of a weld specimen in a tensile testing machine, as shown in Fig. 37-1 and pulling the piece until it breaks. The specimen is cut either from an all-weld metal area or from a welded butt joint for plate and pipe.

The specimen for an all-weld metal area should conform to dimensions shown in Fig. 37-2. It should be cut from the welded section so its reduced area contains only weld metal.

For a plate and pipe welded butt joint, the specimen should be similar to the ones illustrated in Figs. 37-3 and 37-4.

Before the specimen is placed in the tensile machine an accurate measurement should be taken of the gage length so the percent of elongation can be determined.

The actual tensile strength is found by dividing the maximum load needed to break the piece by the cross-sectional area of the specimen. The cross-sectional area is determined by multiplying the width of the bar by its thickness. For example, assume that the specimen is $1\frac{1}{2}''$ wide and $\frac{1}{4}''$ thick. The computation is carried out as follows:

Cross-sectional area = $1\frac{1}{2} \times \frac{1}{4} = \frac{3}{8}$ sq in
Pull to break the bar = 24,500 lb
Tensile strength = 24,500 ÷ $\frac{3}{8}$ = 65,333 psi

The percent of elongation is found by fitting the broken ends of the two pieces and measuring the new gage length. The percent of elongation is a good indicator of the plasticity of the weld and is calculated with this formula:

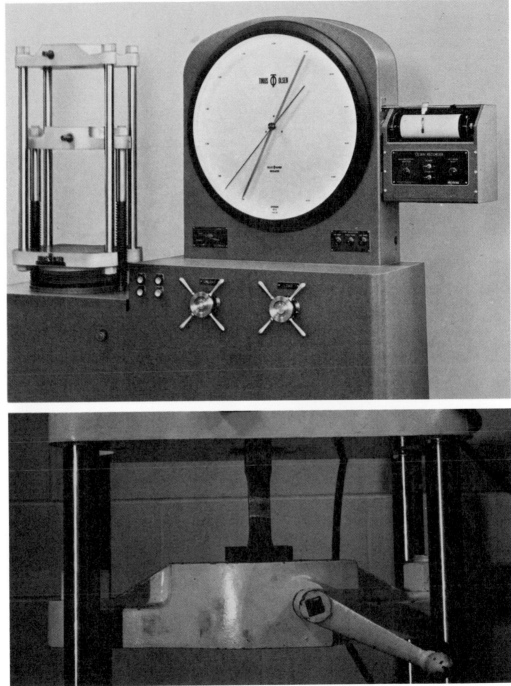

Fig. 37-1. A specimen being tested in a universal testing machine. (Tinius Olson)

$$\frac{FGL - OGL}{OGL} \times 100$$

where: FGL = Final gage length
OGL = Original gage length

Shearing Strength

The shearing strength of a weld can apply either to a transverse or a longitudinal weld.

Transverse shearing strength. To check the shearing strength of a transverse weld, a speci-

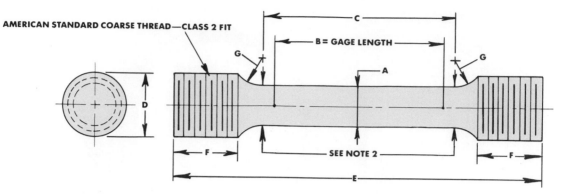

AMERICAN STANDARD COARSE THREAD—CLASS 2 FIT

SPECIMEN   DIMENSIONS OF SPECIMEN (INCHES MINIMUM)

	A	B	C	D	E	F	G
C-1	0.500 ± 0.01	2	2 1/4	3/4	4 1/4	3/4	3/8
C-2	0.437 ± 0.01	1 3/4	2	5/8	4	3/4	3/8
C-3	0.357 ± 0.007	1.4	1 3/4	1/2	3 1/2	5/8	3/8
C-4	0.252 ± 0.005	1.0	1 1/4	3/8	2 1/2	1/2	1/4
C-5	0.126 ± 0.003	0.5	3/4	1/4	1 3/4	3/8	1/8

Note 1: Dimension A, B and C shall be as shown, but alternate shapes of ends may be used as allowed by ASTM specification E-8
Note 2: It is desirable to have the diameter of the specimen within the gage length slightly smaller at the center than at the ends. The difference shall not exceed 1 percent of the diameter.

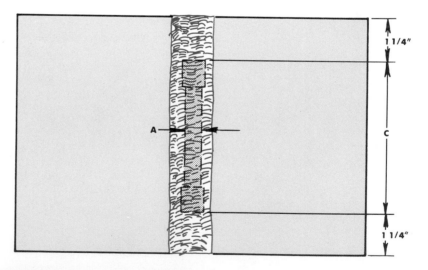

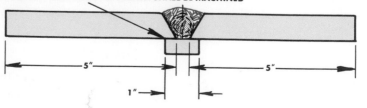

Fig. 37-2. Tensile specimen for an all-weld metal.

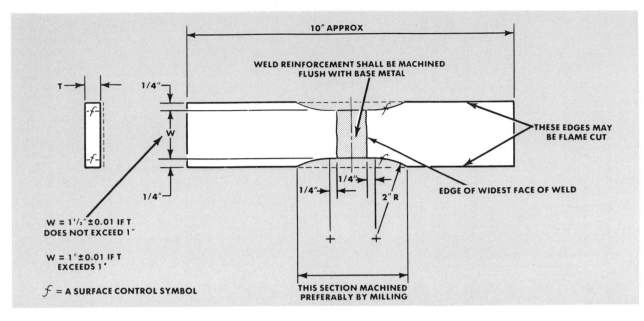

Fig. 37-3. Tensile specimen for flat plate butt weld.

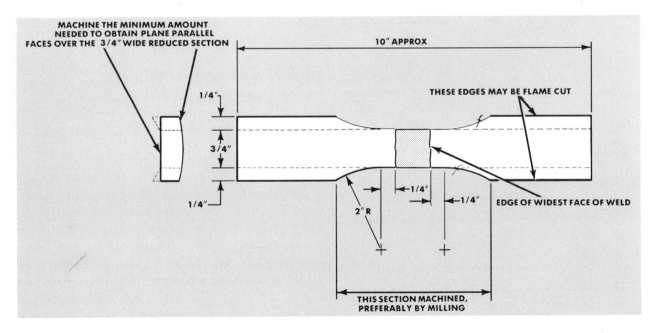

Fig. 37-4. Tensile specimen for pipe butt weld.

men is prepared similar to the one shown in Fig. 37-5. The specimen is then placed in a tensile testing machine and pulled until it breaks. Dividing the maximum load in pounds by twice the width of the specimen will indicate the shearing strength in pounds per linear inch. If the shearing strength in pounds per square inch (psi) is desired, the shearing strength in pounds per

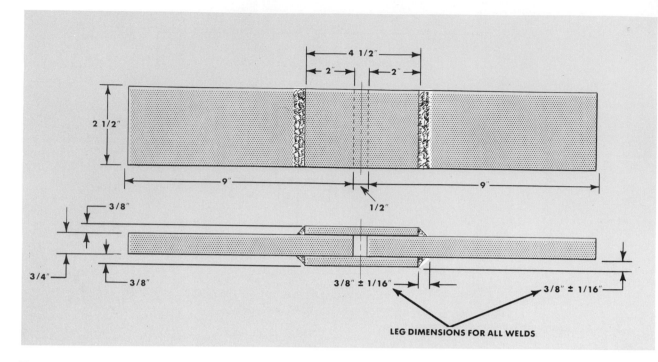

Fig. 37-5. Specimen for determining the shear strength of a transverse weld.

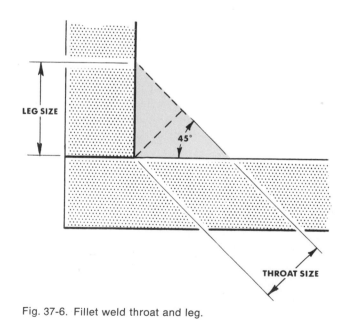

Fig. 37-6. Fillet weld throat and leg.

linear inch is divided by the throat dimension of the weld. See Fig. 37-6. Expressed as formulas, these relationships are shown as:

Shearing Strength (lb/in) =
$$\frac{\text{Maximum load}}{2 \times \text{width of specimen}}$$

or

Shearing strength (psi) =
$$\frac{\text{Shearing strength (lb/in)}}{\text{Throat dimension of weld}}$$

Longitudinal shearing strength. To determine the shearing strength of a longitudinal weld, a specimen as shown in Fig. 37-7 is prepared. The length of each weld is then measured and the piece fractured in a tensile testing machine. The shearing strength in pounds per linear inch is found by dividing the maximum load by the length of the ruptured weld. Expressed as a formula:

Shearing strength (lb/linear inch) =
$$\frac{\text{Maximum Load}}{\text{Length of ruptured weld}}$$

Weld Uniformity Test

The soundness or uniformity and ductility of a weld can be ascertained by the nick-break test, the free-bend test, and the guided-bend test.

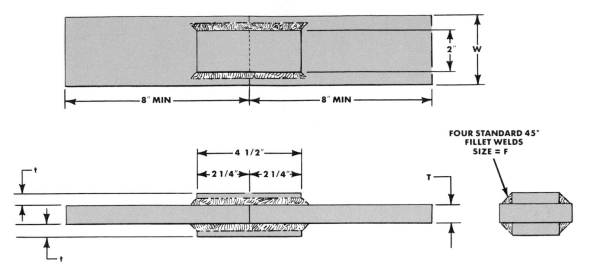

DIMENSIONS			
SIZE OF WELD F, INCHES	1/8	1/2	1/4
THICKNESS T, INCHES MIN	3/8	1/2	1
THICKNESS T, INCHES MIN	3/8	3/4	1 1/4
WIDTH W, INCHES	3	3	3 1/2

LONGITUDINAL FILLET-WELD SHEARING SPECIMEN AFTER WELDING

Fig. 37-7. Specimen for determining the shear strength of a longitudinal weld.

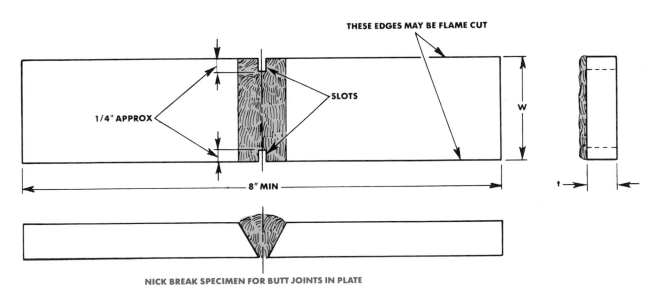

NICK BREAK SPECIMEN FOR BUTT JOINTS IN PLATE

Fig. 37-8. Specimen for a nick-break test.

Nick-break test. A test specimen as illustrated in Fig. 37-8 is prepared and placed on supporting members as shown in Fig. 37-9. A load is applied on this specimen until it breaks. The surface of the fracture is then examined for porosity, gas pockets, slag inclusions, overlaps,

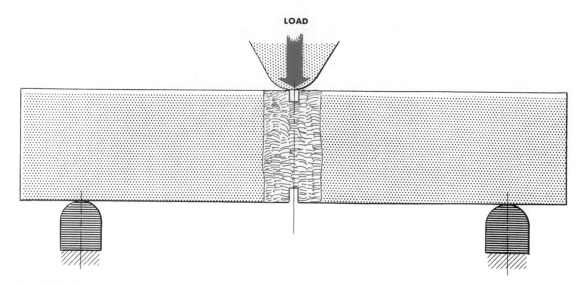

Fig. 37-9. Method of rupturing a nick-break specimen.

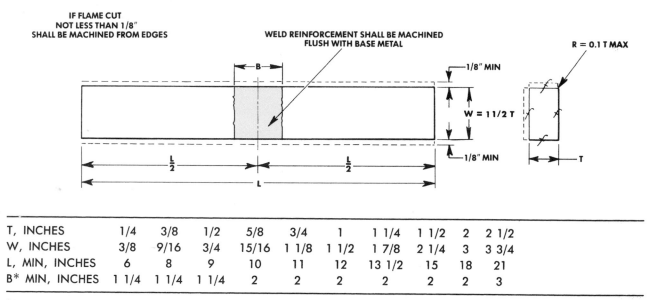

T, INCHES	1/4	3/8	1/2	5/8	3/4	1	1 1/4	1 1/2	2	2 1/2
W, INCHES	3/8	9/16	3/4	15/16	1 1/8	1 1/2	1 7/8	2 1/4	3	3 3/4
L, MIN, INCHES	6	8	9	10	11	12	13 1/2	15	18	21
B* MIN, INCHES	1 1/4	1 1/4	1 1/4	2	2	2	2	2	2	3

*See Fig 13–13

Note: The length L is suggestive only, not mandatory

Fig. 37-10. Specimen for a free-bend test. (AWS)

penetration, and grain size. For a more accurate check of the weld, the fractured pieces should be subjected to an etch test as described in the paragraph on etching testing.

Free-bend test. The free-bend test is used to determine the ductility of the welded specimen. For this test, cut the test piece from the plate so as to include the weld as shown in Fig. 37-10.

Grind or machine the top of the weld so it is flush with the base metal surface. The scratches produced by grinding should run across the weld in the direction of the bend as illustrated in Fig. 37-11. If the scratches extend along the weld they might cause premature failure and give incorrect results. Now measure the distance across the weld and mark it with prickpunch

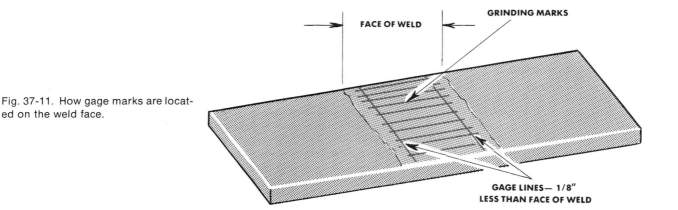

Fig. 37-11. How gage marks are located on the weld face.

FACE OF WELD

GRINDING MARKS

GAGE LINES— 1/8″ LESS THAN FACE OF WELD

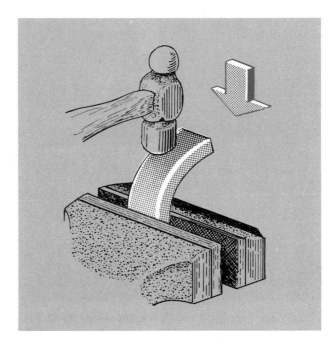

Fig. 37-12. This is one way the initial bend can be made.

marks. The measured length (between gage lines) must be about ¹/₈″ less than the width of the face of the weld as shown by the gage marks in Fig. 37-11.

Bend the specimen by a steady force, with the face containing the gage lines on the outside of the bend. Start the initial bend by placing the piece in a vise as shown in Fig. 37-12, or by using a device as illustrated in Fig. 37-13. After the specimen is given a permanent set, make the final bend in a vise as shown in Fig. 37-14.

Continue the bend until a crack or depression appears, and then immediately remove the load. Measure the distance between the prick-punch marks or gage lines with a flexible rule graduated in hundredths of an inch. This elongation is measured between the gage lines along the convex surface of the weld to the nearest 0.01″. The percent of elongation is obtained by dividing the elongation by the initial gage length and multiplying by 100. For example, suppose the initial gage lines measure 2″, and after the bend

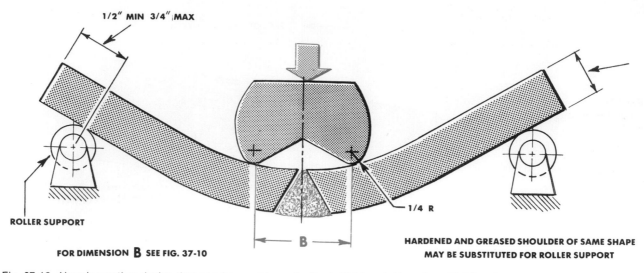

1/2" MIN 3/4" MAX

ROLLER SUPPORT

FOR DIMENSION **B** SEE FIG. 37-10

1/4 R

B

HARDENED AND GREASED SHOULDER OF SAME SHAPE
MAY BE SUBSTITUTED FOR ROLLER SUPPORT

Fig. 37-13. Here is another device that can be used to make the initial bend. (American Welding Society—AWS)

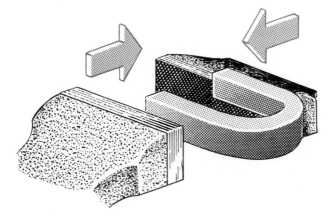

Fig. 37-14. For the final bend, the specimen must be bent in a vise.

the elongation measured 2.5". The calculation then would be:

Increase in elongation = 2.5" − 2" = 0.5"
Percent elongation = 0.5 ÷ 2 × 100 = 25

The ductility of this specimen is therefore considered to be 25 percent in 2".

Guided-bend test. For this test, two specimens, as shown in Fig. 37-15, are required. One piece referred to as the *face-bend specimen* is used to check the quality of fusion; that is, whether the weld is free of defects such as porosity, inclusions, etc. The second piece, re-

ferred to as the *root-bend specimen,* is used to check the degree of weld penetration.

To perform the face-bend test, place the specimen in the guided-bend jig face down and depress the plunger until the piece becomes U-shaped in the die. See Fig. 37-16. If upon examination, cracks greater than 1/8" appear in any direction, the weld is considered to have failed.

In the root-bend test, place the specimen in the jig with the root down or in just the reverse position of the face-bend piece. The results must also show no cracks to be acceptable.

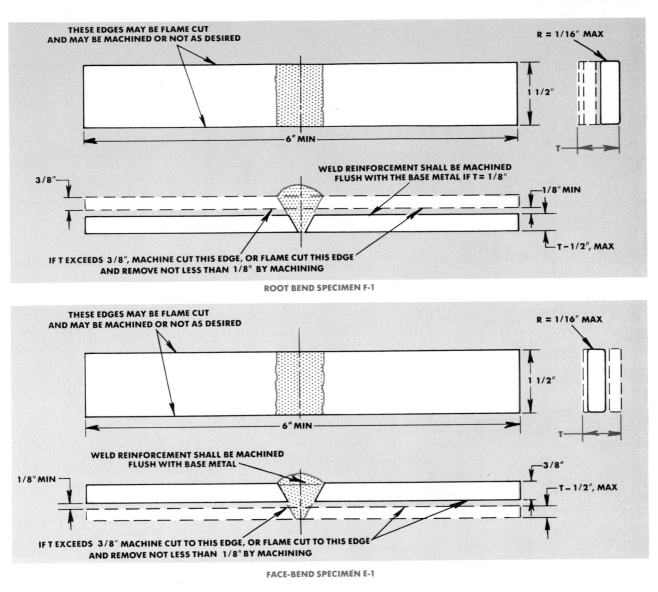

ROOT BEND SPECIMEN F-1

FACE-BEND SPECIMEN E-1

NOTE: THE FACE BEND AND ROOT BEND TEST ARE NOT APPLICABLE IF THE PLATE THICKNESS T IS LESS THAN 3/8"

Fig. 37-15. Specimen for a guided-bend test. (AWS)

Fillet Welded Joint Test

This test is used to ascertain the soundness of fillet welds. *Soundness* refers to the degree of freedom a weld has from defects discernable by visual inspection of any exposed surface of weld metal. These defects include penetrations, inclusions, and gas pockets. For such a test, prepare a specimen as in Fig. 37-17. Then apply force on point *A,* as shown in Fig. 37-18, until a break in the specimen occurs. The force may be applied by a press, a testing machine, or hammer blows.

In addition to checking the fractured weld for soundness, the weld specimen should be etched so that the weld can be examined for cracks.

Etch Test

The etch test is used to determine the soundness of a weld and to make visible the boundary between the weld metal and the base metal.

To make such a test, cut a specimen from the welded joint so it displays a complete transverse section of the weld. The piece may be cut either

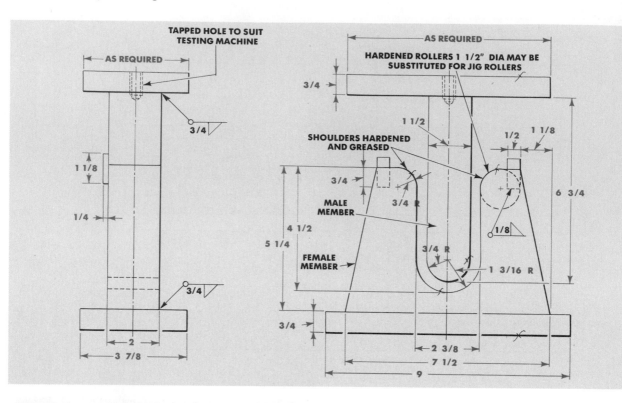

Fig. 37-16. The detail working drawing of a jig (top) for a guided-bend test; jig being used (bottom) to perform a test.

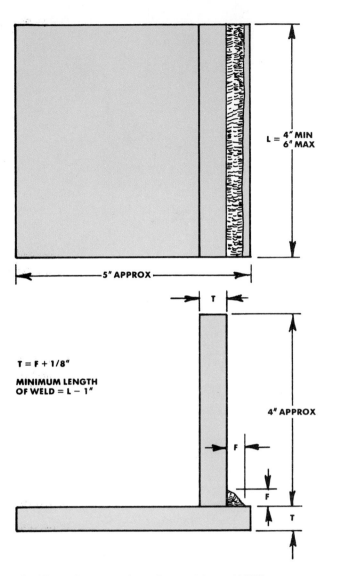

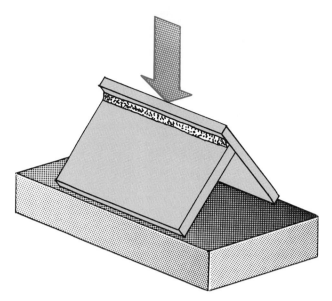

L = 4" MIN 6" MAX

5" APPROX

T

T = F + 1/8"

MINIMUM LENGTH OF WELD = L − 1"

4" APPROX

F

F

T

Fig. 37-17. Specimen for a fillet weld test. (AWS)

Fig. 37-18. Method of rupturing fillet weld specimen. (AWS)

by sawing or flame cutting. File the face of the cut and polish it with grade 00 abrasive cloth. Then place the specimen in one of the following etching solutions:

Hydrochloric acid. This solution should contain equal parts by volume of concentrated hydrochloric (muriatic) acid and water. Immerse the weld in the boiling reagent. Hydrochloric acid will etch unpolished surfaces. It will usually enlarge gas pockets and dissolve slag inclusions, enlarging the resulting cavities.

Ammonium persulfate. Mix one part of ammonium persulfate (solid) to nine parts of water by weight. Vigorously rub the surface of the weld with cotton saturated with this reagent at room temperature.

Iodine and potassium iodide. This solution is obtained by mixing one part of powdered iodine (solid) to twelve parts of a solution of potassium iodide by weight. The latter solution should consist of one part potassium iodide to five parts water by weight. Brush the surface of the weld with this reagent at room temperature.

Nitric acid. Mix one part of concentrated nitric acid to three parts of water by volume. *CAUTION: Always add acid to water when diluting. Nitric acid causes bad stains and severe burns. Wash instantly with water if on skin.*

Either apply this reagent to the surface of the weld with a glass stirring rod at room temperature, or immerse the weld in a boiling reagent provided the room is well ventilated. Nitric acid etches rapidly. It should be used on polished surfaces only, and will show the refined zone as well as the metal zone.

After etching, wash the weld immediately in clear water, preferably hot water; remove the excess water; dip the etched surface in ethyl alcohol; and then remove and dry it in a steady blast warm air.

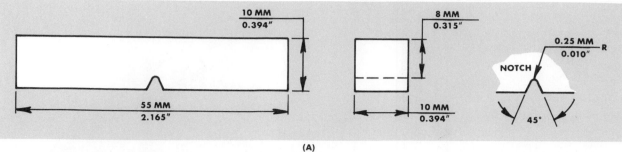

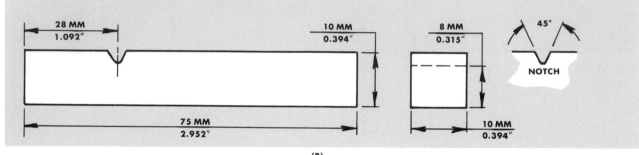

Fig. 37-19. Specimens for impact testing.

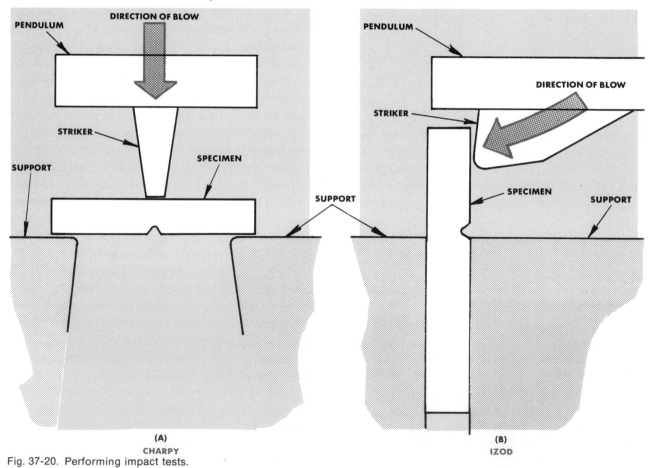

Fig. 37-20. Performing impact tests.

Impact Testing

Impact testing is concerned with the ability of a weld to absorb energy under impact without fracturing. This is a dynamic test in which a specimen is broken by a single blow and the energy absorbed in breaking the piece is measured in foot pounds. The purpose of the test is to compare the toughness of the weld metal with the base metal. It is especially significant in finding if any of the mechanical properties of the base metal have been destroyed due to welding.

The two types of specimens used for impact testing are known as Charpy and Izod. See Fig. 37-19. Both specimens are broken in an impact testing machine. The difference is simply in the manner in which the specimens are anchored. The *Charpy* piece is supported horizontally between two anvils and the pendulum allowed to strike opposite the notch as shown in Fig. 37-20A. The *Izod* specimen is supported as a vertical cantilever beam and struck on the free end projecting above the holding vise. See Fig. 37-20B for example.

NONDESTRUCTIVE TESTING

Nondestructive testing is used to evaluate a structure without destroying it or impairing its actual usefulness. Tests of this nature will disclose all of the common surface and internal defects that normally occur with improper welding procedures or practices.

Currently a variety of testing devices are available which do provide effective data concerning the reliability of a weldment. These devices are often more convenient to use than other regular destructive testing techniques particularly on large and costly welded units.

Magnetic Particle Inspection

The magnetic particle inspection method uses a strong magnetizing current and a finely divided powder suspended in a liquid to detect lack of fusion, very fine cracks, and inclusions or internal flaws which are slightly below the surface in weldments.

Fig. 37-21. Magnetic particle inspection.

In this test the piece to be examined is subjected to a very strong magnetizing current and the areas of inspection are covered with suspended powder. Any impurities or discontinuities in the magnetized material will interrupt the lines of magnetic force causing the particles of suspended powder to concentrate at the defect showing its size, shape, and location. Surface cracks of all kinds are detected by this method. It is one of the most reliable techniques for nondestructive testing. See Fig. 37-21.

Dye Penetrant Inspection

In dye penetrant inspection, surface defects are found by means of proprietary dyes suspended in liquids having high fluidity. These liquids are readily drawn into all surface defects by capillary action. Application of a suitable developer brings out the dye and outlines the defect.

In this test, the surface of the weldment, which must be clean and dry, is coated with a thin film of a penetrant. After allowing a small amount of time for the penetrant to flow into the defects, the part is wiped clean. Only the penetrant in the defects remains. An absorbent material, called a developer, is put on the weldment and allowed to remain until the liquid from the imperfection flows into the developer. The dye now clearly outlines the defects.

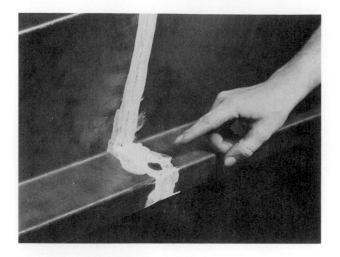

Fig. 37-22. Dye penetrant inspection.

Fig. 37-23. Eddy current testing.

Some of the penetrants used contain a fluorescent dye. The method of applying and developing are the same as for the previously mentioned dye penetrants, however the fluorescent penetrant must be viewed under ultraviolet light, commonly referred to as *black light.* This light causes the penetrants to fluoresce (glow) to a yellow-green color which is more clearly defined than regular dye penetrants. The dye penetrant methods are particularly useful for bringing out defects in nonferrous materials such as aluminum. These materials are nonmagnetic so magnetic particle tests can not be used on them. See Fig. 37-22.

Eddy Current Testing

Eddy current testing uses electromagnetic energy to detect discontinuities in weld deposits and is effective in testing both ferrous and nonferrous materials for porosity, slag inclusions, internal cracks, external cracks, and lack of fusion.

Whenever a coil carrying a high-frequency alternating current is brought close to a metal, it produces a current in the metal by induction. The induced current is called an *eddy current.*

The part to be tested is subjected to electromagnetic energy by being placed in or near high frequency alternating current coils. Differences in metal in the weld deposit change the impedance of the coil. The change in impedance is indicated on electronic measuring instruments, and the size of the defect is shown by the amount of this change. See Fig. 37-23.

Radiographic Inspection

Radiographic inspection is a method of determining the soundness of a weldment by means of rays which are capable of penetrating through the entire weldment. X-Rays and Gamma Rays are two types of electro-magnetic waves used to penetrate opaque materials. A permanent record of the internal structure is obtained by placing a sensitized film in direct contact with the back of the weldment. When these rays pass through a weldment of uniform thickness and structure, they fall upon the sensitized film and produce a negative of uniform density. If the weldment contains gas pockets, slag inclusions, cracks, or lacks penetration, more rays will pass through the less dense areas and will register on the film as dark areas, clearly outlining the defects and showing their size, shape and location.

X-rays are produced by electrons travelling at high speed which are suddenly stopped by impact with a tungsten anode. Gamma rays are given off by radium or by other radioactive substances. Gamma rays are of shorter wave length than X-rays. See Fig. 37-24.

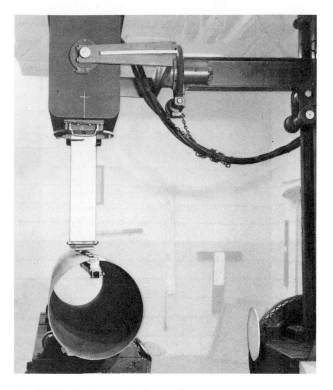

Fig. 37-24. Radiographic inspection.

Fig. 37-25. Ultrasonic inspection.

Hardness Testing

Hardness testing is often used in preference to the more expensive tensile testing methods since comparable results are obtained.

Hardness tests are effective in determining the relative hardness of the weld area as compared with the base metal. This hardness is indicated by values obtained from various hardness testing machines. Hardness numbers represent the resistance offered by the metal to the penetration of an indenter. The standard hardness machines are known as *Brinell* and *Rockwell*. See Fig. 37-26.

In the Brinell test, a 10 mm diameter ball is forced into the surface of a metal by a load of 3000 kg. The load must remain on the specimen 15 seconds for ferrous materials, and 30 seconds for nonferrous materials. Sufficient time is required for adequate flow of the material being tested otherwise the readings will be in error. Brinell hardness numbers are calculated by dividing the applied load by the area of the surface indentation. The diameter of the indentation is read from a calibrated microscope and this number is then converted to a Brinell hardness number from a chart.

The Rockwell hardness tester employs a variety of loads and indenters, consequently different

Ultrasonic Testing

In ultrasonic testing, high frequency vibrations or waves are used to locate and measure defects in both ferrous and nonferrous materials. This method is very sensitive, and is capable of locating very fine surface and subsurface cracks, as well as other internal defects. All types of joints can be evaluated and the exact size and location of defects measured.

Ultrasonic testing utilizes high frequency vibratory impulses to ascertain the soundness of a weld. If a high frequency vibration is sent through a sound piece of metal, a signal will travel through it to the other side of the metal and be reflected back and shown on a calibrated screen of an oscilloscope. Discontinuities of structure interrupt the signal and reflect it back sooner than the signal of the sound material. This is shown on the oscilloscope screen and indicates the depth of the defect. Only one side of the weldment needs to be exposed for testing purposes. See Fig. 37-25.

Fig. 37-26. Hardness testing.

scales can be used. These scales are designated by letters. For example, R_c 60 represents a Rockwell scale with a diamond penetrater and a 150 kg load. Since hardness numbers give relative or comparative hardness values of materials, the scale must always be specified.

Points to Remember

1. The soundness of a weld can be determined by visual inspection or destructive testing.

2. Visual inspection will not show if there are any internal defects in a weld.

3. The true strength of a weld can be determined by tensile, shear, and by impact testing procedures.

4. Tensile testing involves placing a weld specimen in a tensile testing machine and pulling the piece until it breaks.

5. Weld uniformity can be checked by applying a load on a weld specimen until it breaks.

6. A free-bend test is used to determine the ductility of a weld.

7. A guided-bend test will disclose the quality of fusion and degree of penetration.

8. An etch test will show the boundary lines between the weld metal and the base metal.

9. When preparing the mixture for an etch test, always add the acid to the water.

10. Impact testing is used to determine the toughness of the weld metal.

11. In nondestructive testing, special ultrasonic, radiographic, eddy current or magnetic particle, equipment is used to evaluate a weld.

12. Nondestructive testing is used when it is impractical to check weld quality by other kinds of test methods.

QUESTIONS FOR STUDY AND DISCUSSION

1. What is the limitation of any visual inspection method for testing a weld?

2. What is the difference between destructive and nondestructive testing methods?

3. What will a tensile test show?

4. In conducting a tensile test, what is the value of finding the percent of elongation?

5. What is the function of a shearing strength test?

6. What is meant by a longitudinal and transverse weld?

7. What is the function of a nick-break test?

8. When is a free-bent test used?

9. How is a guided-bent test conducted?

10. How can the soundness of fillet welds be determined?

11. What will etching tests disclose?

12. What precautions must be taken when using nitric acid?

13. Of what value is impact testing?

14. Why are nondestructive tests often more valuable than destructive tests?

15. What is the basic principle of magnetic particle testing?

16. How are dye penetrants used for inspection purposes?

17. What is the basic principle of eddy-current testing?

18. What is radiographic inspection?

19. What is ultrasonic testing?

20. Of what value are hardness tests?

CHAPTER 38 reading weld symbols

In the fabrication of metal products, the welder usually has to work from a print which shows in detail exactly how the structure is to be made. He or she will find on the print not only where the welds are to be located, but also the type of joint to be used, as well as the correct size and amount of weld to be deposited at the designated seams. This information is indicated by a set of symbols which have been standardized by the American Welding Society (AWS).

Some of the more common symbols for weldments are included in this chapter.

Building a Weld Symbol

A weld symbol is made up of a base, to which are attached instructions as to the type of weld required, the location of the weld, whether it is a field weld or a shop weld, and other reference data which is necessary to do a complete weld job. While these symbols may be very complex and carry a large amount of data, they may also be quite simple. The welding student must study the various examples and learn to read the symbols before he or she can qualify for apprenticeship or ask for a full time job as a welder.

Base of symbol. The main foundation of the weld symbol is a reference line with an arrow at one end, as shown in Fig. 38-1. Notice that above or below the *base line* the type of weld is indicated, whether it is a fillet, groove, flange, plug, spot or seam weld. Also included is such information as surface contour of a weld, size of a weld bead, length of a weld, how beads are to be finished, and often what type welding process is to be

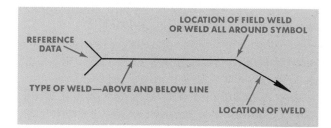

Fig. 38-1. Base of weld symbol.

used. All of this data is indicated by geometric figures, numerical values, and abbreviations.

Designating types of welds. The most important factor in a welding symbol is the type of weld. The type of weld is used in relationship to five basic types of joints: butt, corner, lap, T, and edge. These are represented in Fig. 38-2.

Welds are classed as fillet, plug, spot, seam, or groove. Groove welds are further divided and classified according to the particular shape of the grooved joint.

Each weld has its own specific symbol. For example, a fillet weld is designated by a right triangle, and a plug weld by a rectangle, as in Fig. 38-3. (All of the weld symbols are included in Fig. 38-28, in the Summary and Application section of this chapter.)[1]

---

1. A more complete treatment of symbols as they apply to all forms of manual and automatic machine welding will be found in the pamphlet A2,0-76 *Standard Welding Symbols,* published by the American Welding Society.

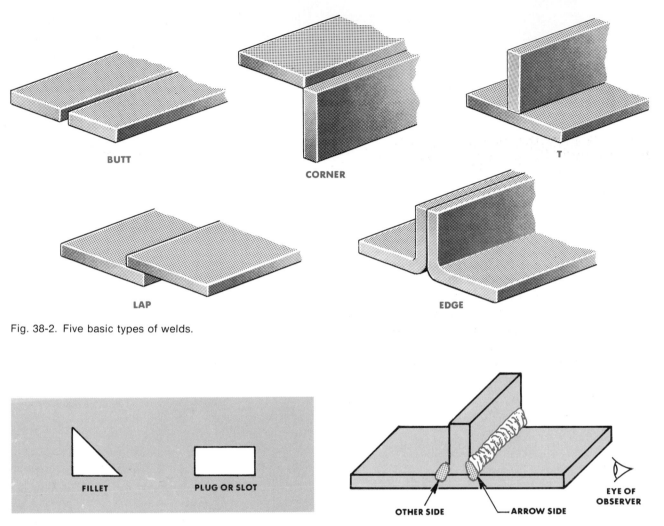

BUTT

CORNER

T

LAP

EDGE

Fig. 38-2. Five basic types of welds.

FILLET

PLUG OR SLOT

Fig. 38-3. Fillet and plug or slot weld symbols.

OTHER SIDE

ARROW SIDE

EYE OF OBSERVER

Fig. 38-4. How symbols are placed.

Location of symbols. Another requirement in understanding weld symbols is the method which is used to specify on what side of a joint a weld is to be made. A weld is said to be either on the *arrow* or *other* side of a joint. The arrow side is the surface that is in direct line of vision, while the other side is the opposite surface of the joint. See Fig. 38-4.

Weld location is designated by running the arrowhead of the reference line to the joint. The direction of the arrow is not important, that is, it can run on either side of a joint and extend upward or downward. See Fig. 38-5. If the weld is to be made on the *arrow side*, the appropriate weld symbol is placed *below* the reference line. If the weld is to be located on the *other side* of the joint, the weld symbol is placed *above* the reference line. When both sides of the joint are to be welded, the same weld symbol appears above and below the reference line. See Fig. 38-6 for examples of these.

The only exception to this practice of indicating weld location is in seam and spot welding. With seam or spot welds, the arrowhead is simply run to the centerline of the weld seam and the appropriate weld symbol centered above or below the reference line. See Fig. 38-6. If no arrow side or other side is important, the symbol is placed astride the reference line to indicate this condition.

Information on weld symbols is placed to read from left to right along the reference line in accordance with the conventions of drafting.

Fillet, bevel and J-groove, flare-bevel groove, and corner-flange weld symbols are shown with the *perpendicular leg* always to the left of the weld symbol. See Fig. 38-7 for example.

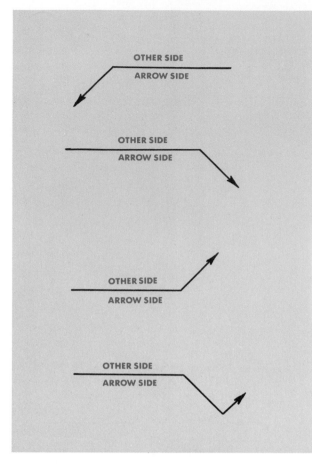

Fig. 38-5. The arrow may run in any direction.

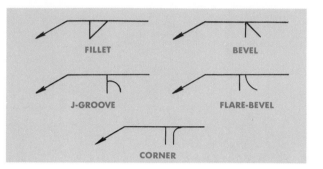

Fig. 38-7. The perpendicular leg of these weld symbols will be found on the left.

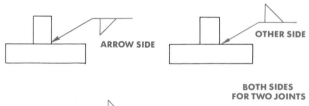

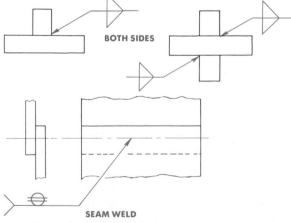

Fig. 38-6. How weld locations are designated.

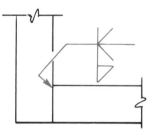

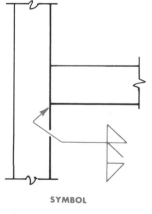

Fig. 38-8. Combined weld symbols.

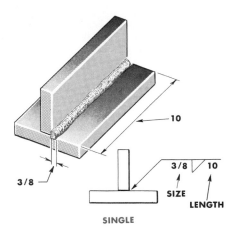

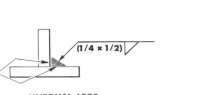

Fig. 38-9. How size and length of fillet welds are indicated.

Combining Weld Symbols

In the fabrication of a product, there are occasions when more than one type of weld is to be made on a joint. Thus a joint may require both a fillet and double-bevel groove weld. When this happens, a symbol is shown for each weld. See Fig. 38-8.

Size of fillet welds. The width of a fillet weld is shown to the left of the weld symbol and is expressed in fractions with or without the inch mark. See Fig. 38-9. When both sides of a fillet are to be welded and both welds have the same dimensions, one or both may be dimensioned. If the welds differ in dimensions, both are dimensioned. Where a note appears on a drawing that governs the size of a fillet weld, no dimensions are usually shown on the symbol.

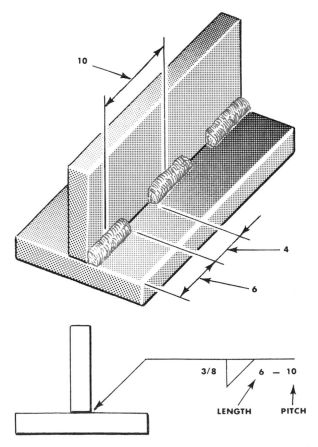

Fig. 38-10. How length and pitch of intermittent fillet welds are indicated.

The length of the weld is shown to the right of the weld symbol by numerical values representing the actual required length.

When a fillet weld with unequal legs is required, the size of the legs is placed in parentheses as shown in Fig. 38-9.

Intermittent fillet welds. The length and pitch increments of intermittent welds are shown to the right of the weld symbol. The first figure represents the length of the weld section and the second figure the pitch (center-to-center spacing) between welds. See Fig. 38-10.

Size of groove welds. There are several types of groove welds. Their sizes are shown as follows:

1. For single-groove and symmetrical double-groove welds which extend completely through the members being joined, no size is included on the weld symbol. See Fig. 38-11.

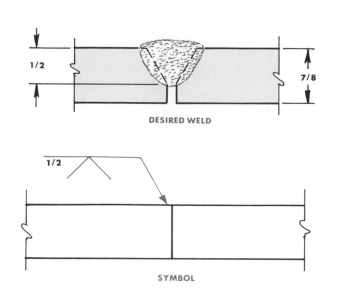

Fig. 38-12. How size is shown on grooved welds with partial penetration. (AWS)

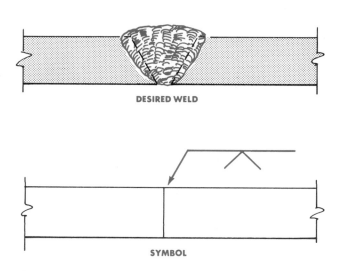

Fig. 38-11. Size is not shown for single and symmetrical double-groove welds with complete penetration.

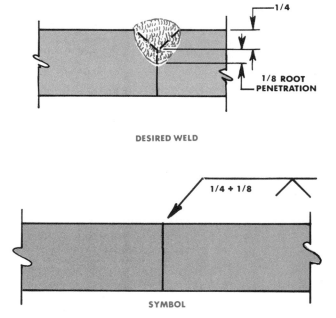

Fig. 38-13. How dimensions are used to show size and root penetration of grooved welds. (AWS)

2. For groove welds which extend only partly through the members being joined, the weld size is included on the left of the weld symbol. See Fig. 38-12.

3. If grooved welds require certain root penetration, this size is indicated by showing the depth of chamfering and the root penetration, separated by a plus mark, and placed to the left of the weld symbol. See Fig. 38-13.

4. Root opening and included angle of groove welds are shown inside the weld symbol. See Fig. 38-14.

5. The size of flare-groove welds is considered as extending only to the tangent points as indicated by dimensional lines. See Fig. 38-15.

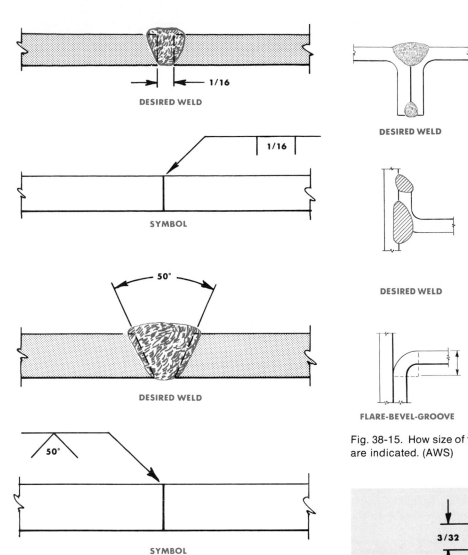

Fig. 38-14. How root opening and included angle are shown for groove welds. (AWS)

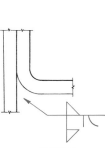

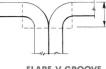

Fig. 38-15. How size of flare-bevel and flare-V grooved welds are indicated. (AWS)

Size of flange welds. The radius and height of the flange is separated by a plus mark and placed to the left of the weld symbol. The size of the weld is shown by a dimension located outward of the flange dimensions. See Fig. 38-16.

Size of plug welds. The size of plug welds is shown to the left of the weld symbol, the depth, when less than full, on the inside of the weld symbol, the center-to-center spacing (pitch) to the right of the symbol, and the included angle of countersink below the symbol. See Fig. 38-17.

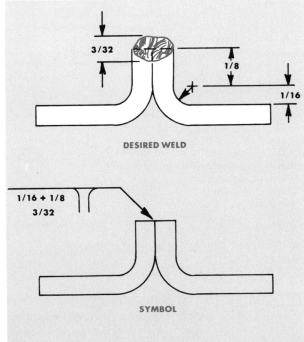

Fig. 38-16. How flange welds are indicated.

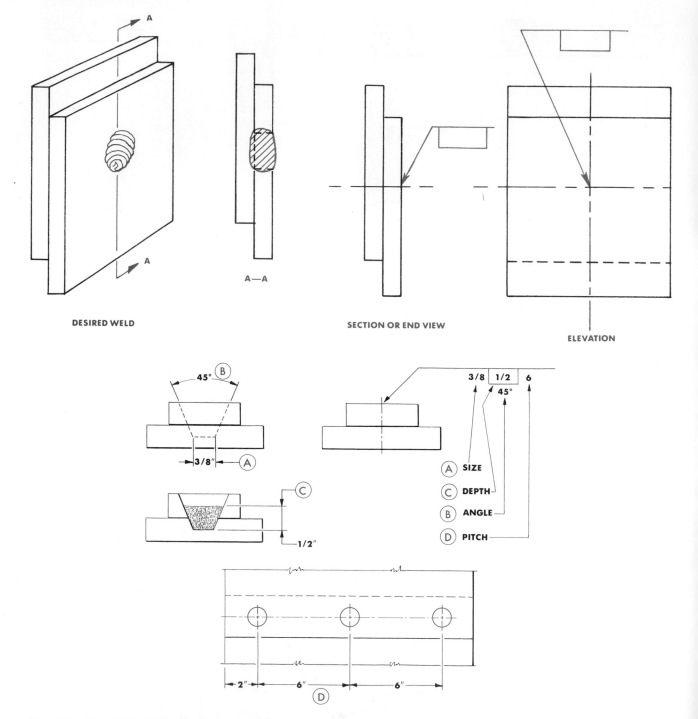

Fig. 38-17. How dimensions apply to plug welds.

Size of slot welds. Length, width, spacing, angle of countersink, and location of slot welds are not shown on the symbol. This data is included by showing a special detail on the print. If the slots are partly filled, the depth of filling is shown inside. Fig. 38-18.

Size of spot welds. Spot welds are dimensioned either by size or strength. Size is desig-

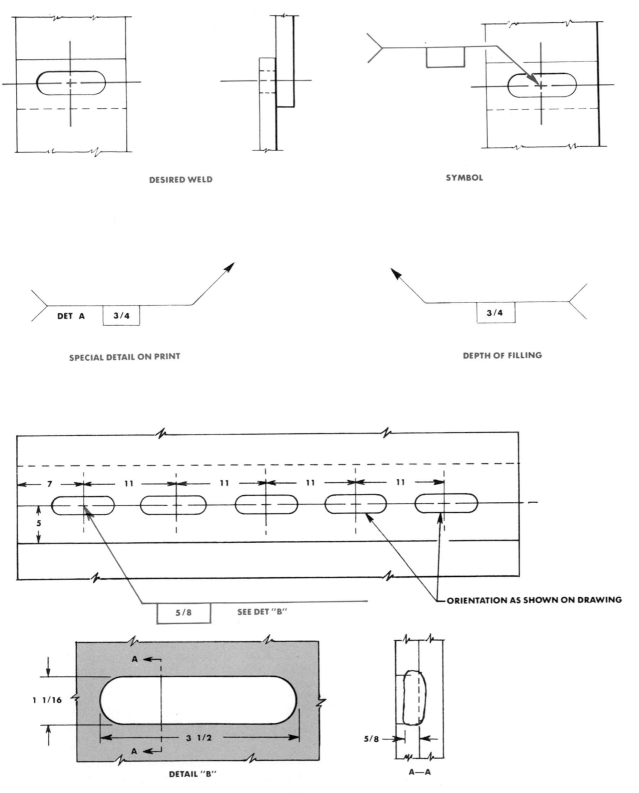

DESIRED WELD

SYMBOL

DET A | 3/4

SPECIAL DETAIL ON PRINT

3/4

DEPTH OF FILLING

ORIENTATION AS SHOWN ON DRAWING

5/8   SEE DET "B"

DETAIL "B"

A—A

PARTIALLY FILLED SLOT WELDS

Fig. 38-18. How slot welds are indicated.

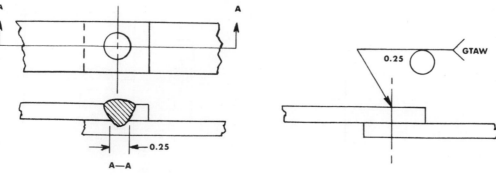

DESIRED WELD

SYMBOL

DIAMETER OF SPOT WELDS
(GAS TUNGSTEN-ARC SPOT)

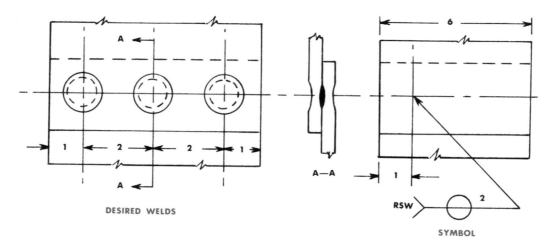

DESIRED WELDS

A—A

SYMBOL

PITCH OF SPOT WELDS
(RESISTANCE SPOT)

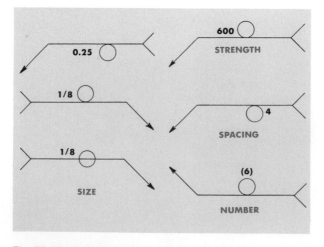

Fig. 38-19. Method of designating spot welds.

nated as the diameter of the weld expressed in fractions, or decimally in hundredths of an inch, and placed to the left of the symbol. The strength is also placed to the left of the symbol and expresses the required minimum shear strength in pounds per spot. The spacing of spot welds is shown to the right of the symbol. When a definite number of spot welds are needed in a joint, this number is indicated in parentheses either above or below the weld symbol. See Fig. 38-19.

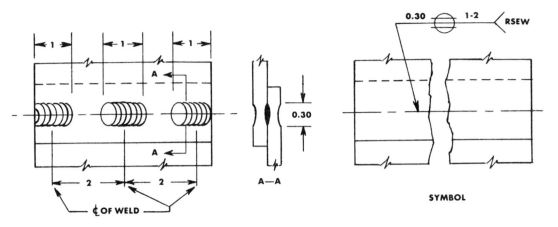

SIZE, LENGTH AND PITCH OF INTERMITTENT SEAM WELDS
(RESISTANCE SEAM)

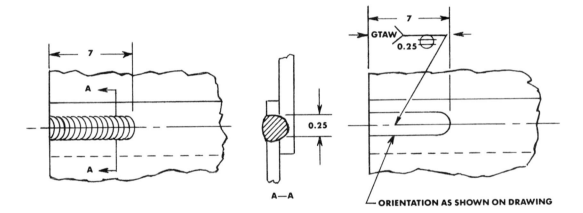

EXTENT OF SEAM WELDS
(GAS TUNGSTEN-ARC SEAM)

Size of seam welds. Seam welds are dimensioned either by size or strength. Size is designated as the width of the weld in fractions, or decimally in hundreths of an inch, and shown to the left of the weld symbol. The length of the weld seam is placed to the right of the weld symbol. The pitch of intermittent seam welds is shown to the right of the length dimension. See Fig. 38-20.

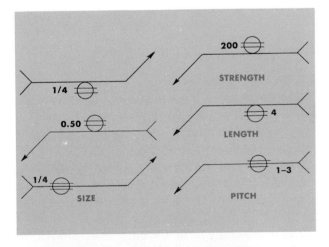

Fig. 38-20. Method of designating seam welds.

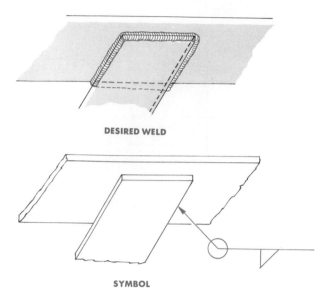

DESIRED WELD

SYMBOL

Fig. 38-21. Weld-all-around symbol. (AWS)

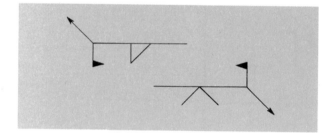

Fig. 38-22. Field weld symbol. (AWS)

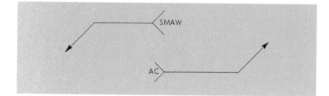

Fig. 38-23. The tail is used to indicate some specific detail or weld process. (AWS)

The strength of the weld, when used, is located to the left of the symbol, and is expressed as the minimum acceptable shear strength in pounds per linear inch.

Weld-all-around symbol. When a weld is to extend completely around a joint, a small circle is placed where the arrow connects the reference line. See Fig. 38-21.

Field weld symbol. Welds that are to be made in the field (welds not made in a shop or at the place of initial construction), are indicated by a vertical flag. See Fig. 38-22.

Reference tail. The tail is included only when some definite welding specification, procedure, reference, weld or cutting process needs to be called out, otherwise it is omitted. This data is often in the form of symbols. See Fig. 38-23 and Table 38-1. Abbreviations in the tail may also call out some specifications which are included on some other part of the print.

TABLE 38-1. MASTER CHART OF WELDING AND ALLIED PROCESSES

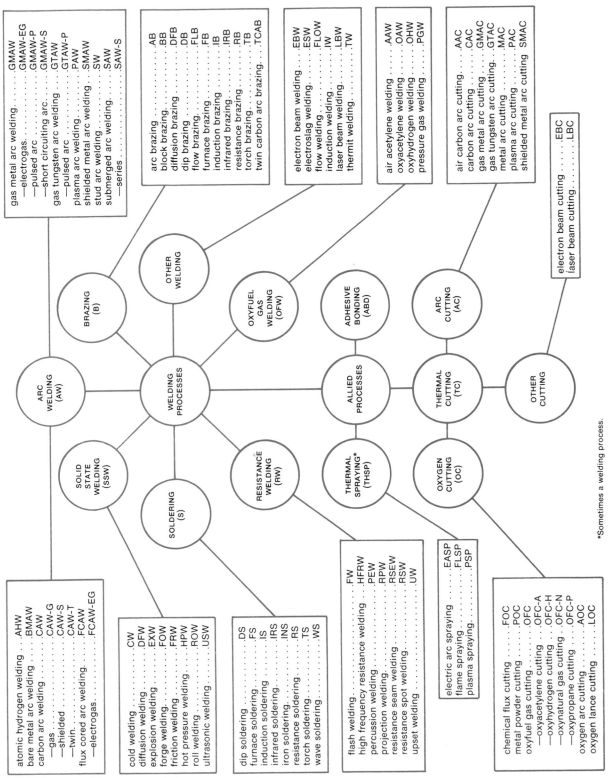

gas metal arc welding . . . . . . GMAW
  —electrogas. . . . . GMAW-EG
  —pulsed arc . . . . GMAW-P
  —short circuiting arc . . . . . GMAW-S
gas tungsten arc welding . . . GTAW
  —pulsed arc . . . . GTAW-P
plasma arc welding . . . . PAW
shielded metal arc welding . . SMAW
stud arc welding . . . . SW
submerged arc welding . . . . . SAW
  —series . . . . SAW-S

arc brazing . . . . AB
block brazing . . . . BB
diffusion brazing . . . . DFB
dip brazing . . . . DB
flow brazing . . . . FLB
furnace brazing . . . . FB
induction brazing . . . . IB
infrared brazing . . . . IRB
resistance brazing . . . . RB
torch brazing . . . . TB
twin carbon arc brazing . . . . TCAB

electron beam welding . . . . EBW
electroslag welding . . . . ESW
flow welding . . . . FLOW
induction welding . . . . IW
laser beam welding . . . . LBW
thermit welding . . . . TW

air acetylene welding . . . . . . AAW
oxyacetylene welding . . . . . OAW
oxyhydrogen welding . . . . . OHW
pressure gas welding . . . . . PGW

air carbon arc cutting . . . . . AAC
carbon arc cutting . . . . CAC
gas metal arc cutting . . . . . GMAC
gas tungsten arc cutting . . . GTAC
metal arc cutting . . . . MAC
plasma arc cutting . . . . PAC
shielded metal arc cutting . SMAC

electron beam cutting . . . . . EBC
laser beam cutting . . . . LBC

OTHER WELDING

BRAZING (B)

OXYFUEL GAS WELDING (OFW)

ADHESIVE BONDING (ABD)

ARC CUTTING (AC)

ARC WELDING (AW)

WELDING PROCESSES

ALLIED PROCESSES

THERMAL CUTTING (TC)

OTHER CUTTING

SOLID STATE WELDING (SSW)

SOLDERING (S)

RESISTANCE WELDING (RW)

THERMAL SPRAYING* (THSP)

OXYGEN CUTTING (OC)

atomic hydrogen welding . . . AHW
bare metal arc welding . . . . BMAW
carbon arc welding . . . . CAW
  —gas . . . . CAW-G
  —shielded . . . . CAW-S
  —twin. . . . . CAW-T
flux cored arc welding . . . . . FCAW
  —electrogas . . . . FCAW-EG

cold welding . . . . CW
diffusion welding . . . . DFW
explosion welding . . . . EXW
forge welding . . . . FOW
friction welding . . . . FRW
hot pressure welding . . . . . . HPW
roll welding . . . . ROW
ultrasonic welding . . . . USW

dip soldering . . . . DS
furnace soldering . . . . FS
induction soldering . . . . IS
infrared soldering . . . . IRS
iron soldering . . . . INS
resistance soldering . . . . . . RS
torch soldering . . . . TS
wave soldering . . . . WS

flash welding . . . . FW
high frequency resistance welding . . HFRW
percussion welding . . . . PEW
projection welding . . . . RPW
resistance seam welding . . . RSEW
resistance spot welding . . . . RSW
upset welding . . . . UW

electric arc spraying . . . . . . EASP
flame spraying . . . . FLSP
plasma spraying . . . . PSP

chemical flux cutting . . . . . . FOC
metal powder cutting . . . . . . POC
oxyfuel gas cutting . . . . OFC
  —oxyacetylene cutting . . . . OFC-A
  —oxyhydrogen cutting . . . . . OFC-H
  —oxynatural gas cutting . . . OFC-N
  —oxypropane cutting . . . . . OFC-P
oxygen arc cutting . . . . AOC
oxygen lance cutting . . . . . . LOC

*Sometimes a welding process.

AWS

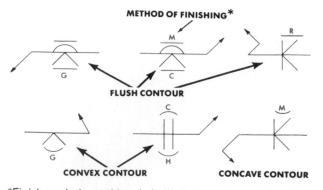

*Finish symbols used herein indicate the method of finishing ("C" = chipping; "G" = grinding; "M" = machining; "R" = rolling; "H" = hammering) and not the degree of finish.

Fig. 38-24. Method of showing surface contour of welds. (AWS)

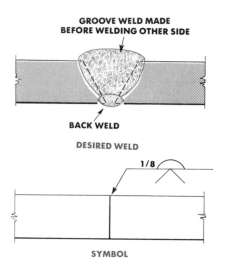

Fig. 38-25. Use of back weld symbol to indicate backweld. (AWS)

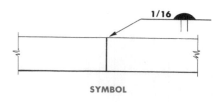

Fig. 38-26. Application of melt-thru symbol.

Surface contour of welds. When bead contour is important, a special flush, concave or convex contour symbol is added to the weld symbol. Welds that are to be mechanically finished also carry a finish symbol along with the contour symbols. See Fig. 38-24.

Back or backing welds. Back or backing welds refer to the weld made on the opposite side of the regular weld. Back welds are occasionally specified to insure adequate penetration and provide additional strength to a joint. This particular symbol is included opposite the weld symbol. No dimensions of back or backing welds except height of reinforcement is shown on the weld symbol. See Fig. 38-25.

Melt-thru welds. When complete joint penetration of the weld through the material is required in welds made from one side only, a special melt-thru weld symbol is placed opposite the regular weld symbol. No dimension of melt-

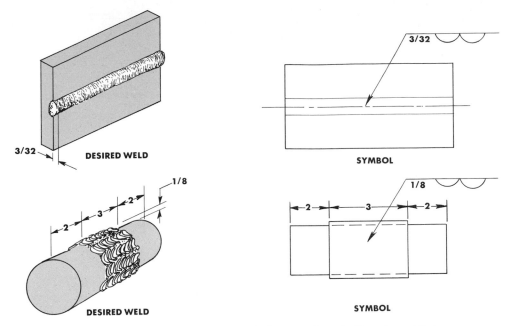

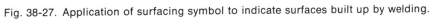

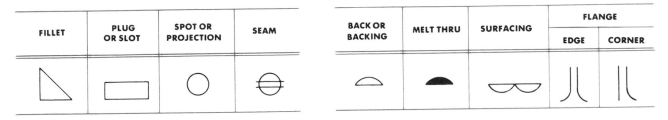

Fig. 38-27. Application of surfacing symbol to indicate surfaces built up by welding.

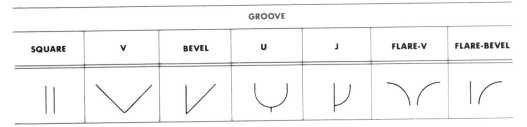

FILLET	PLUG OR SLOT	SPOT OR PROJECTION	SEAM		BACK OR BACKING	MELT THRU	SURFACING	FLANGE	
								EDGE	CORNER

GROOVE

SQUARE	V	BEVEL	U	J	FLARE-V	FLARE-BEVEL

BASIC ARC AND GAS WELD SYMBOLS

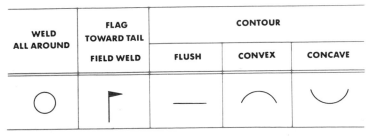

WELD ALL AROUND	FLAG TOWARD TAIL — FIELD WELD	CONTOUR		
		FLUSH	CONVEX	CONCAVE

SUPPLEMENTARY SYMBOLS

Fig. 38-28. Weld symbols with new field weld symbol, used with flag toward tail of symbol. (AWS)

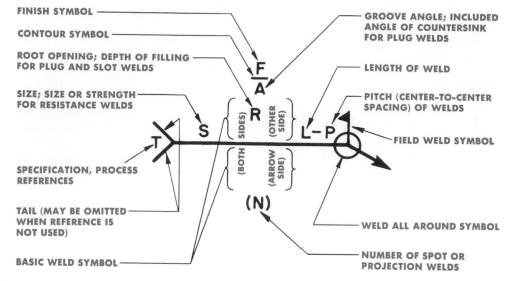

Fig. 38-29. Master symbol. (AWS)

thru, except height of reinforcement, is shown on the weld symbol, as you will note in Fig. 38-26.

Surfacing welds. Welds whose surfaces must be built up by single or multiple pass welding are provided with a surfacing weld symbol. The height of the built-up surface is indicated by a dimension placed to the left of the surfacing symbol. See Fig. 38-27. The extent, location, and orientation of the area to be built up is normally indicated on the drawing.

Summary

A more complete listing of basic arc and gas welding symbols and some supplementary symbols are listed in Fig. 38-28. Fig. 38-29 is what a combined welding symbol might be, based on all that appeared previously in the chapter.

QUESTIONS FOR STUDY AND DISCUSSION

1. Indicate the meaning of the following weld symbols.

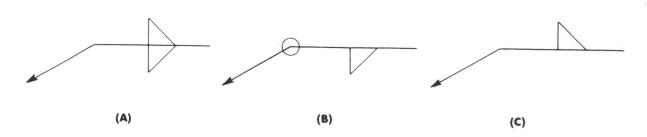

(A)   (B)   (C)

2. What type of weld do these symbols indicate?

(A)   (B)   (C)   (D)

3. How would you interpret these symbols?

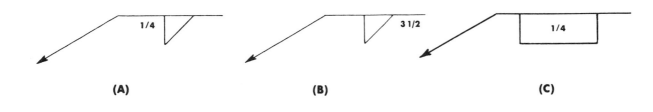

(A)   (B)   (C)

4. These symbols represent what weld specifications?

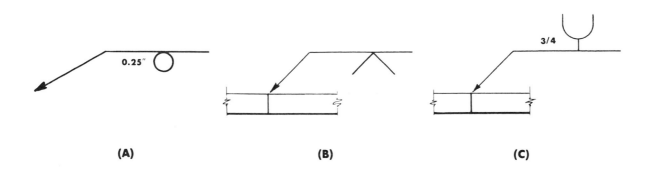

(A)   (B)   (C)

5. What do welds designated with the following symbols represent?

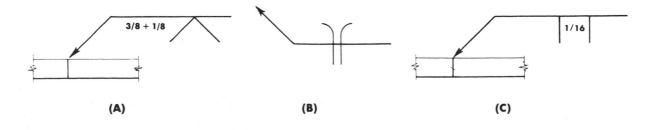

(A)   (B)   (C)

6. What is the meaning of each of these symbols?

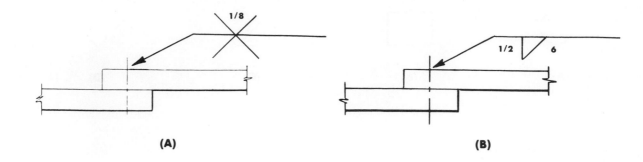

(A)   (B)

7. What do these symbols represent?

(A)   (B)

8. What do these symbols mean?

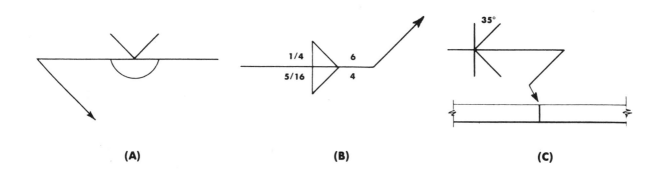

(A)   (B)   (C)

CHAPTER 39 the metric system

The United States is in the process of converting its system of weights and measurements to the standard metric system now in use all over the world. Some industries, such as the medical, chemical, photographic, and pharmaceutical went metric years ago. Currently a large number of manufacturing plants, particularly in the automotive, and tool and die fields, are rapidly going metric. Welding activities also will be affected by a change-over to metric. Gas flow, current selection, welding specifications, and other welding data will be indicated in metric values. Consequently, a welder will need to understand the metric system.

Complete metrification in all industries probably will take many years because people will have to develop a new kind of thinking. Our inch, foot, yard, pound, quart, and mile concepts have been with us so long that only through a gradual process of education can we throw off the old system and start "thinking metric." Surprisingly, the metric system is much simpler.

The United States is about the last major country using the inch-foot-pound, or English system. The reason of course is that for a long time the United States was the largest producing industrial country, and consequently we did not especially care what or how other producing countries produced their goods. Now the manufacturing capabilities of other major countries are becoming much more competitive. There-fore, if we wish to participate in world trade we must become involved in the worldwide coordinated measuring system. Countries on the metric system produce goods having interchangeable parts. We cannot compete if our products are not interchangeable with theirs.

STRUCTURE OF THE METRIC SYSTEM

The metric system simply coordinates into one decimalized system the unrelated measurements which we now use. Its three principal segments are *meters, kilograms,* and *liters. Meters* represent units of linear measurements, *kilograms* are units of mass or weight; and *liters* are designations for volume—liquid or dry. In each case the units are based on decimals, or powers of ten, and identified by prefixes. Thus, instead of dividing by three to determine the number of feet in a yard, or twelve to get inches from feet, or sixteen to find pounds, or four to find quarts in a gallon, we need only the number 10 for all multiplication and division. Using the powers of ten for all calculations, we can assign measured quantities whether the object of measurement involves length, weight or volume.

The following prefixes are used for all metric units used with SI base units.

For small quantities:
   deci (d) = tenths
   centi (c) = hundredths
   milli (m) = thousandths
For large quantities:
   deka (da) = tens
   hecto (h) = hundreds
   kilo (k) = thousands

Linear Metric Values

The meter is the basic measurement of length. Its principal sub-units are millimeter, centimeter, and kilometer. Notice how these values are expressed with their appropriate powers of ten and the following prefixes:

One thousand meters (m) (10^3 meters) =
                                    1 kilometer (km)
One hundred meters (m) (10^2 meters) =
                                    1 hectometer (hm)
Ten meters (m) (10^1 meters) =
                                    1 dekameter (dam)
A meter (m) (10^0 meters) =
                                    1 meter (m)

One tenth of a meter (m) (10^{-1} meters) =
                                    1 decimeter (dm)
One hundredth of a meter (m)
                    (10^{-2} meters) = 1 centimeter (cm)
One thousandth of a meter (m)
                    (10^{-3} meters) = 1 millimeter (mm)

Thus kilometer means 1000 meters, a centimeter is $^1/_{100}$ of a meter, and a millimeter is $^1/_{1000}$ of a meter.

Suppose you want to change 5.75 meters into centimeters or millimeters. Dividing by the powers of 10 we get:

5.75 meters ÷ by 100 (10^{-2}) = 0.0575 cm

5.75 meters ÷ by 1000 (10^{-3}) = 0.00575 mm

or

0.0575 cm × 100 = 5.75 meters

0.00575 mm × 1000 = 5.75 meters

Since we are so accustomed to using inches, feet, and miles, it might be a good idea to

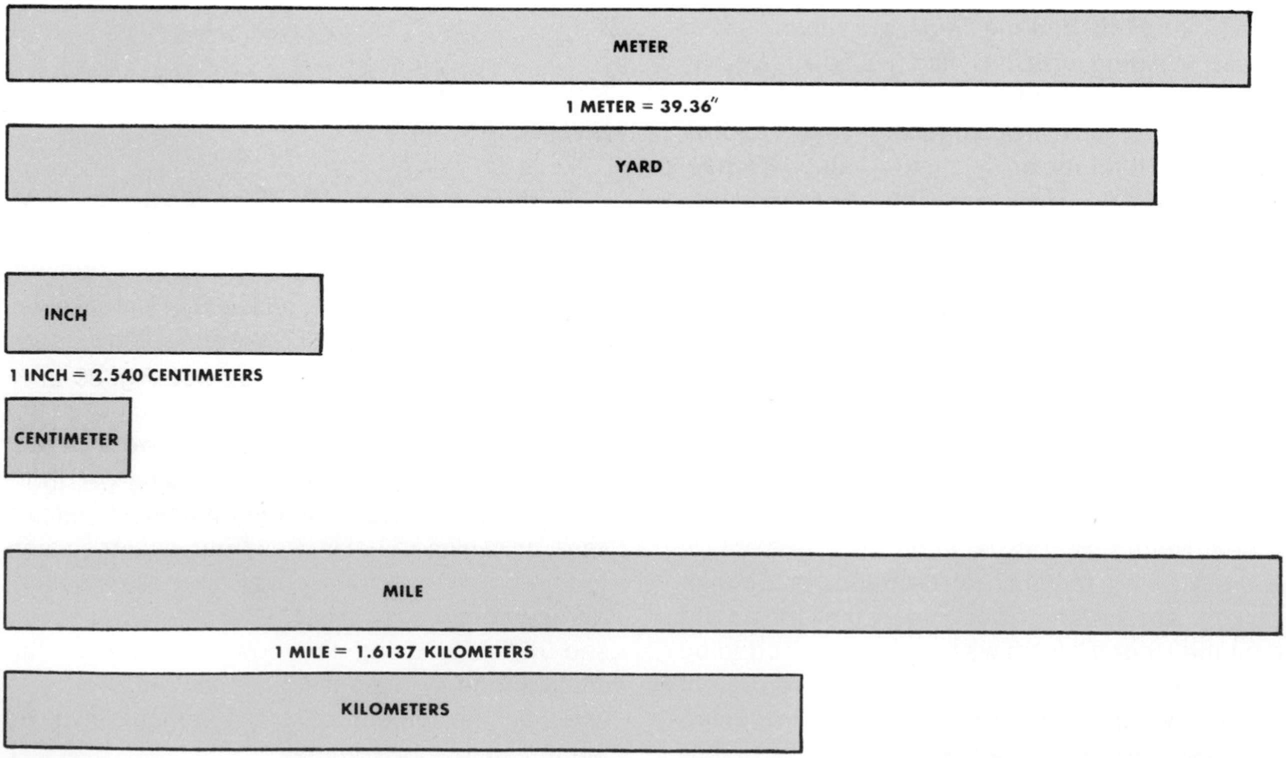

METER

1 METER = 39.36"

YARD

INCH

1 INCH = 2.540 CENTIMETERS

CENTIMETER

MILE

1 MILE = 1.6137 KILOMETERS

KILOMETERS

Fig. 39-1. Comparison of linear values.

compare these units and their corresponding metric replacements. Notice in Fig. 39-1 that a meter is slightly longer than a yard (39.37 inches) the centimeter is smaller than the inch (0.4 inches) and the kilometer is ⁵/₈ths of a mile.

Meter conversion. Until such time when metric is used exclusively there may be occasions when a given size may have to be converted from one value to another. The following factors will provide the approximate equivalents:

Converting metric measures to our customary system:

millimeters × 0.039	=	inches
centimeters × 0.39	=	inches
meters × 39.4	=	inches
centimeters × 0.33	=	feet
meters × 3.28	=	feet
meters × 1.09	=	yards
kilometers × 0.62	=	miles

Converting our customary system into metric values:

inch × 25.4	= mm
inch × 2.5	= cm
inch × 0.025	= m
feet × 30.5	= cm
feet × 0.305	= m
yard × 0.91	= m
mile × 1.6	= km

Assume that a given dimension is 3¹/₂″ and is to be converted into mm. Thus:

$$3.50'' \times 25.4 = 88.9 \text{ mm}$$

or

$$88.9 \text{ mm} \times 0.039 = 3.5''$$

Let's take another example. Suppose you are driving 50 miles per hour. The speed in kilometers would be:

$$50 \times 1.6 = 80 \text{ km/hr}$$

Mass (Weight) Metric Values

The kilogram is the basic metric unit of weight. The numerical values and prefixes for the kilogram are the same as those for linear measurements. Thus:

One thousand grams (10^3 grams) =
1 kilogram (kg)

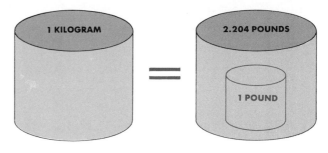

Fig. 39-2. Comparison of kilogram and pounds.

One hundred grams (10^2 grams) =
1 hectogram (hg)

Ten grams (10^1 grams)   =
1 dekagram (dag)

A gram (10^0 grams)   =
1 gram (g)

One tenth of a gram (10^{-1} grams) =
1 decigram (dg)

One hundredth of a gram (10^{-2} grams) =
1 centigram (cg)

One thousandth of a gram (10^{-3} grams) =
1 milligram (mg)

Gram conversion factors. The following values are used to change pounds and ounces into metric equivalents or vice versa:

1 lb	=	0.453 kg
1 kg	=	2.20 lb
1 oz	=	28.34 g
1 g	=	0.035 oz
1 lb	=	453.59 g
1 g	=	0.002 lb

Notice that a kilogram, which is probably used most frequently for common purposes, is a little more than two pounds (2.2 lb). See Fig. 39-2.

Volume (Liquid) Values

The basic metric unit of volume is the liter. Again the same values and prefixes apply, thus:

One thousand liters (10^3 liters)   =
1 kiloliter (kl)

One hundred liters (10^2 liters)   =
1 hectoliter (hl)

Ten liters (10^1 liters)   =
1 dekaliter (dal)

A liter (10^0 liters)   =
1 liter (l)

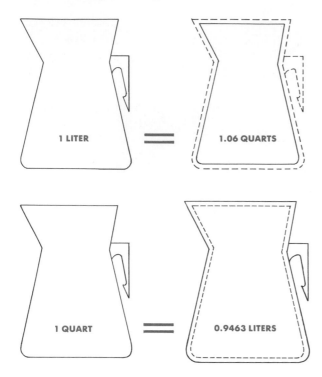

Fig. 39-3. Comparison of liters and quarts.

One tenth of a liter (10^{-1} liters) =
                                 1 deciliter (dl)
One hundredth of a liter (10^{-2} liters) =
                                 1 centiliter (cl)
One thousandth of a liter (10^{-3} liters) =
                                 1 milliliter (ml)

Liter conversion factors. The following factors are used to change liters to pints, quarts, or gallon from one system to another:

1 pint   = 0.47 liters
1 liter   = 2.1 pints
1 quart = 0.95 liters
1 liter   = 1.06 quarts
1 gallon = 3.8 liters
1 liter   = 0.26 gallons

The liter is the most standard for common usage. Notice in Fig. 39-3 that an ordinary quart is comparable to one liter.

Supplementary Metric Values

Temperature. Temperature in our English system is stated at so many degrees *Fahrenheit* (°F). In the metric system temperature is expressed in Kelvin (°K) or *Celsius* (°C) (formerly called centigrade). Degrees *Kelvin* is used by scientists. Kelvin temp. = C + 273.15. Degrees Kelvin is degrees Celsius −273.15°K.

To change Fahrenheit to Celsius the conversion formula is:

$$°C = \frac{5}{9} \times (°F - 32°)$$

Example: Change 72° to °C

$$°C = \frac{5}{9} \times (72 - 32)$$

$$\frac{5}{9} \times (40)$$

$$\frac{5 \times 40}{9} = 22.2°$$

To change Celsius to Fahrenheit the conversion formula is:

$$°F = \frac{9 \times °C}{5} + 32$$

Example: Change 22.2 °C to °F

$$°F = \frac{9 \times 22.2}{5} + 32$$

$$\frac{199.8}{5} + 32$$

$$39.9 + 32 = 72°$$

Metric area. The basic method of determining area in the metric system is the same as finding area in the English system, that is, one side is multiplied by the other side.

Example: Find the area of a surface 5″ by 9″
                          5″ × 9″ = 45 sq in
or
      12.5 cm × 22.5 cm = 281 cm²

Approximate area conversions from English system to metric:

in² × 6.5 = cm²
ft² × 0.09 = m²
yd² × 0.8 = m²
mi² × 2.6 = km²

Approximate area conversion from metric to English system:

cm² × 0.16 = in²
m² × 1.2 = yd²
km² × 0.4 = mi²

Metric Cubic. Metric cubic values are determined in the same manner as in figuring English cubic calculations. Thus the volume of a container 4″ wide 6″ deep and 10″ long would be:

$$4'' \times 6'' \times 10'' = 240 \text{ cu in}$$

or

$$10 \text{ cm} \times 15 \text{ cm} \times 25 \text{ cm} = 3750 \text{ cm}^3$$

Approximate volume conversion factors:

ft³ × 0.03 = cubic meters (m³)
yd³ × 0.76 = cubic meters (m³)
m³ × 35   = cubic feet (ft³)
m³ × 1.3   = cubic yard (yd³)

Points to Remember

1. The principal base of a metric system consists of meters, kilograms and liters.

2. Meters represent units of linear measurements, kilograms are units of mass or weight, and liters are units of volume.

3. Quantities of any metric base are designated by prefixes.

4. Prefixes are always used with a unit; thus millimeter, dekagram, centiliter.

5. Dividing or multiplying by the powers of 10, we get the quantities that represent the prefixes; thus one thousand meters (10^3) is a kilometer, one thousandth of a meter (10^{-3}) is a millimeter.

6. For small quantities the basic prefixes are *deci* (tenths), *centi* (hundredths), and *milli* (thousandths).

7. For large quantities the basic prefixes are *deka* (tenths), *hecto* (hundredths), and *kilo* (thousandths).

8. A meter is slightly longer than a yard, the centimeter is smaller than the inch, and the kilometer is approximately 0.6 miles.

9. To change Fahrenheit to Celsius, use the formula

$$°C = \frac{5}{9} \times (°F - 32°)$$

10. Metric square area and metric cubic volume is determined in the same way as the English system.

QUESTIONS FOR STUDY AND DISCUSSION

1. Why is it important that the United States convert to the metric system of weights and measurements?

2. In the metric system, what is the basic unit of length?

3. How many centimeters are there in one meter?

4. A kilometer is approximately what part of a mile?

5. What prefix is used to designate 1000 in the metric system?

6. How is one hundredth of a gram properly designated?

7. A kilogram is the equivalent of how many pounds?

8. What metric unit is used to designate volume?

9. If a weld is designated to be 5³/4″ long, how many centimeters would it be? (This measurement is often expressed in millimeters.)

10. The width of a weld is to be ¹/4″. How much would this be in millimeters?

11. If gas flow is to be set at 21 cubic feet per hour (cfh) how much would this be in liters per minute (lpm)?

12. What is the cm² of an area 12′ × 15′?

13. A 42 gallon tank contains how many liters?

14. A box 8 cm wide, 15 cm high and 20 cm long would contain how many cubic inches of sand?

15. Pure aluminum melts at 1220°F. What is this temperature in Celsius?

CHAPTER 40 certification of welders

To protect human lives and to make certain that products or structures fabricated by welding will function safely and effectively, certain safeguards are generally established to cover the quality of welding that must be done. The safeguards are usually stipulated in some document which clearly defines the nature and conditions of the required work.

The listing of specifications governing quality is not always as simple as it might appear because of the diversity of products and structures involving a great variety of welding processes and a wide gradation of welding skills. State and local laws sometimes specify in great detail what these requirements must be. At other times there may be only limited regulations, as established by the manufacturer, especially if high performance standards are not required. In all instances safeguards relate directly to the performance skills of the welder. Degrees of welding competences may be expressed in the form of codes, standards or specifications.

Code

A code consists of a set of regulations covering permissible materials, service limitations, fabrication, inspection, testing procedures and qualifications of welding operators. These codes have been established by a number of nationally recognized agencies such as:

American Welding Society (AWS)
American Society of Mechanical Engineers (ASME)

American Petroleum Institute (API)
American National Standards Institute (ANSI)

Codes usually deal with a specific field of work such as ship building, piping, boiler work, building construction, tanks, aircraft and many others. A typical example of what a code may include is the ASME code for boiler construction. This code lists specifications covering:

Section I   —Power Boilers
Section II   —Materials
Section III   —Boilers of Locomotives
Section IV   —Low-Pressure and Heating Boilers
Section V   —Miniature Boilers
Section VI   —Rules for Inspection
Section VII   —Suggested Rules for Care of Power Boilers
Section VIII—Unfired Pressure Vessels
Section IX   —Welding Qualifications

A code is sometimes enacted into law and consequently is often the most enforceable of any safety regulation. A properly worded code is written in mandatory language, using imperative words such as *shall* or *must.* Inclusion of other words like *should* or *it is desirable* would raise the question of their enforceability and therefore are excluded.

Standards

Standards are specific regulations which cover the quality of a particular product to be fabricated by welding. By and large standards deal with work quality rather than work procedure. Thus standards may cover type of material to be used, test strength of required welds, characteristics of filler metals, preheating and postheating temperatures and other essentials which have a direct bearing on the quality of the finished product.

Standards are usually developed by the manufacturer and apply only to its own welding personnel and work or product to be produced. The stringency of the standards depends on the nature of the work or product and the demands of the consumer. Thus for some jobs the required competency of the welders may not be unduly high, whereas for other tasks the performance requirements could be extremely critical. On many occasions the established standards are based on a nationally recognized code and may be supplemented by other demands which are dictated by the parties for whom the work is to be done.

Specifications

Specifications are specific descriptions of fabricating procedure. Among other manufacturing instruction they include such welding data as location of welds, welding process to be used and method of testing the soundness of welds. These specifications are usually formulated by the design engineer and are included on production prints or on separate specification sheets. The nature of the specifications are governed by established standards and quality of work required. Welders must then produce the type of welds indicated by these specifications.

Certification Requirements

There is no one set of certification requirements dealing with all segments of the welding trade. Each area whether it involves welding pipe, aircraft parts, building structures, boilers or ships will have its own certification requirements.

Although certification requirements may vary somewhat for one classification of welders to another, in general, they all specify comparable tests which welders have to take. Most tests involve one or more of the following:

1. Tension test to establish the strength of a weld.

2. Guided bend tests to determine the ductility of the weld bead.

3. A fillet weld test to check the lack of fusion or cracks and proper weld contour.

4. A radiographic test or other testing techniques to detect porosity, cracks, inclusions and penetration.

These tests may have to be carried out in one or several positions, such as flat, horizontal, vertical, or overhead.

Qualifying tests are performed on specimens sometimes called *coupons,* of the same material to be used in the product involved. Each test weld is done on certain size pieces and then subjected to some destructive or nondestructive tests. See Chapter 37. A welder is considered qualified only if his test specimens meet the required standards of quality.

Certifying Agency

Generally manufacturers have their own testing programs for qualifying welders. In such a program someone in the plant is designated as the certifying agent. A welder then reports to the agent and performs certain welding tasks that will demonstrate his or her ability to meet the requirements of the established standards.

The results of the test are analyzed and if found satisfactory are so stated and recorded by the company. The welder, Fig. 40-1, is then said to be qualified to work on the contracted job.

Certification Procedure

The question often arises among welders, "How can I become certified?" First of all one must remember that there are no standardized certification requirements which will lead to a general permanent certificate. Secondly, certification of welders is assumed by the manufacturer, or supplier who has contracted to provide certain types of products or perform special kinds of welding jobs. Consequently, as a welder, you cannot go to just a single agency and apply for permanent certification. Each time you apply for a welding job, or if already employed and are designated to work on a different weld-

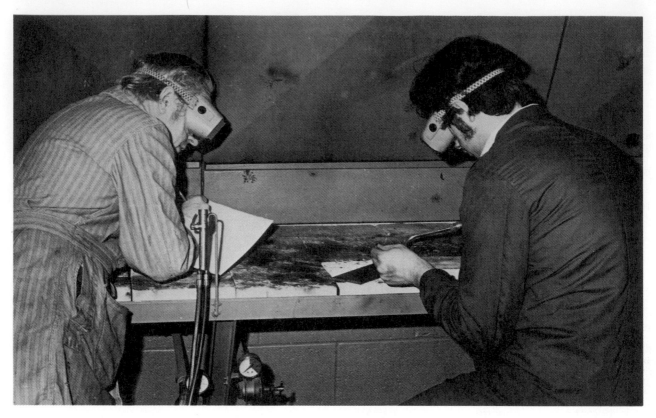

Fig. 40-1. Welders usually are required to pass specified tests to be certified.

ing assignment, you have to subject yourself to some qualifying examination. If you pass the designated tests you are then certified for that particular job. If you move to some other assignment then you have to be certified again. For example, just because you have qualified to weld pipe does not automatically certify you to weld boilers. Each time you have to be re-examined even though some of the actual tests may be similar to those you have previously taken.

QUESTIONS FOR STUDY AND DISCUSSION

1. Why are strict regulations often established for welders?

2. What is the difference between a *code, standards* and *specifications?*

3. What are some of the agencies that prescribe welding codes?

4. Why are codes designated for only one category of work?

5. Who is generally responsible for certifying welders?

6. What are some of the tests which welders must take to be certified?

7. Why isn't it possible to apply for permanent certification status?

8. How often do you have to be certified?